CAMBRIDGE LIBRARY COLLECTION

Books of enduring scholarly value

Mathematics

From its pre-historic roots in simple counting to the algorithms powering modern desktop computers, from the genius of Archimedes to the genius of Einstein, advances in mathematical understanding and numerical techniques have been directly responsible for creating the modern world as we know it. This series will provide a library of the most influential publications and writers on mathematics in its broadest sense. As such, it will show not only the deep roots from which modern science and technology have grown, but also the astonishing breadth of application of mathematical techniques in the humanities and social sciences, and in everyday life.

A History of the Mathematical Theories of Attraction and the Figure of the Earth

Newton's *Principia* paints a picture of the earth as a spinning, gravitating ball. However, the earth is not completely rigid and the interplay of forces will modify the earth's shape in subtle ways. Newton predicted a flattening at the poles, yet others disagreed. Plenty of books have described the expeditions which sought to measure the shape of the earth, but very little has appeared on the mathematics of a problem, which remains of enduring interest even in an age of satellites. Published in 1874, this two-volume work by Isaac Todhunter (1820–84), perhaps the greatest of Victorian historians of mathematics, takes the mathematical story from Newton, through the expeditions which settled the matter in Newton's favour, to the investigations of Laplace which opened a new era in mathematical physics. Volume 1 traces developments from Newton up to 1780, including coverage of the work of Maupertuis, Clairaut and d'Alembert.

Cambridge University Press has long been a pioneer in the reissuing of out-of-print titles from its own backlist, producing digital reprints of books that are still sought after by scholars and students but could not be reprinted economically using traditional technology. The Cambridge Library Collection extends this activity to a wider range of books which are still of importance to researchers and professionals, either for the source material they contain, or as landmarks in the history of their academic discipline.

Drawing from the world-renowned collections in the Cambridge University Library and other partner libraries, and guided by the advice of experts in each subject area, Cambridge University Press is using state-of-the-art scanning machines in its own Printing House to capture the content of each book selected for inclusion. The files are processed to give a consistently clear, crisp image, and the books finished to the high quality standard for which the Press is recognised around the world. The latest print-on-demand technology ensures that the books will remain available indefinitely, and that orders for single or multiple copies can quickly be supplied.

The Cambridge Library Collection brings back to life books of enduring scholarly value (including out-of-copyright works originally issued by other publishers) across a wide range of disciplines in the humanities and social sciences and in science and technology.

A History of the Mathematical Theories of Attraction and the Figure of the Earth

From the Time of Newton to that of Laplace

VOLUME 1

ISAAC TODHUNTER

CAMBRIDGE
UNIVERSITY PRESS

University Printing House, Cambridge, CB2 8BS, United Kingdom

Cambridge University Press is part of the University of Cambridge.
It furthers the University's mission by disseminating knowledge in the pursuit of education, learning and research at the highest international levels of excellence.

www.cambridge.org
Information on this title: www.cambridge.org/9781108084574

This edition first published 1873
This digitally printed version 2015

ISBN 978-1-108-08457-4 Paperback

HISTORY OF

THE THEORIES OF ATTRACTION

AND

THE FIGURE OF THE EARTH.

VOLUME I.

Cet admirable Ouvrage [Newton's *Principia*] contient les germes de toutes les grandes découvertes qui ont été faites depuis sur le système du monde: l'histoire de leur développement par les successeurs de ce grand géomètre serait à la fois le plus utile commentaire de son Ouvrage, et le meilleur guide pour arriver à de nouvelles découvertes.

LAPLACE. *Connaissance des Tems pour l'an* 1823.

A HISTORY

OF THE

MATHEMATICAL THEORIES OF ATTRACTION

AND

THE FIGURE OF THE EARTH,

FROM THE TIME OF NEWTON TO THAT OF LAPLACE.

BY

I. TODHUNTER, M.A., F.R.S.

IN TWO VOLUMES.

VOLUME I.

London:
MACMILLAN AND CO.
1873.

Cambridge:
PRINTED BY C. J. CLAY, M.A.
AT THE UNIVERSITY PRESS.

PREFACE.

IN the volumes now offered to students I have written the history of an important branch of science in the manner in which I formerly treated the Calculus of Variations and the mathematical theory of Probability; and in the present work, as in those, I undertake a task hitherto unattempted. For although much has been published on the History of Astronomy, yet the progress of the mathematical development of the principle of Attraction has been left almost untouched. The last of the six volumes which constitute the great work of Delambre is devoted to the Astronomy of the eighteenth century; but the Astronomy discussed is almost entirely that of observation, and the investigations of the eminent mathematicians who contributed to fill up the outline traced by Newton are scarcely noticed. There are indeed interesting and valuable works in which the results obtained by theory are stated in popular language for the benefit of general readers; such is the well-known history by Bailly in French, with its continuation by Voiron; and in English we have various excellent productions of the same kind, especially Narrien's *Historical Account of the Origin and Progress of Astronomy*, and Grant's *History of Physical Astronomy*. But the object of these works is quite distinct from that which I have kept in view in my contributions to scientific history. I desire not merely to record the results which may have been obtained but to trace the analysis which led to those results, to estimate its value, and to discriminate between its failure and its success, its error and its truth. So far as I know the only example of a mathematical treatise bearing on the history of Physical Astronomy is Gautier's *Essai Historique sur le problême des trois corps:* but as this treats of the Lunar and Planetary Theories, omitting the Figure of the Bodies, it has nothing in common with the present work.

In the fifth volume of the *Mécanique Céleste* Laplace arranges the whole subject of Physical Astronomy in six divisions, and gives brief sketches of the progress of the theory of all: in every case sound knowledge practically begins with Newton. Laplace's first division is devoted to the Figure and Rotation of the Earth; and this has suggested to me the subject of the present work. I undertake accordingly to trace the history of the Theories of Attraction and of the Figure of the Earth from Newton to Laplace. The two subjects are necessarily associated in origin, and have been historically always united; they are discussed together by Laplace in the second volume of his great work. I have confined myself to a single division of the wide subject of Physical Astronomy, for the extent and difficulty of the whole might deter even a professional cultivator of the science; and the numerous unfinished fragments of works intended to bear on the *Mécanique Céleste* furnish an impressive warning against the rashness of any extravagant design.

I will now give an outline of the plan of my work. The first Chapter is necessarily occupied with Newton, the founder of Physical Astronomy. The power revealed in all his efforts is nowhere more conspicuous than in his treatment of our two subjects.

In the theory of attraction, among other important results, he shewed that the attraction of a spherical shell on an external particle is the same as if the shell were collected at its centre, and that the attraction on an internal particle is zero. These two propositions constitute a complete theory of the attraction of a sphere in which the density varies as the distance from the centre. Moreover the result with respect to an internal particle was extended by Newton to the case in which the bounding surfaces of the shell are similar, similarly situated, and concentric ellipsoids of revolution.

Newton originated the idea of investigating the Figure of the Earth on the supposition that it might be treated as a homogeneous fluid rotating with uniform angular velocity. He assumed as a postulate that there could be relative equilibrium in such a case if the form were that of an oblate ellipsoid of revolution; and he determined the ratio of the axes and the law of variation of gravity at the surface. The investigation, though not free from imperfection, is a rare example of success in the first discussion of a most difficult problem, and constitutes an enduring monument to the surpassing ability of its author.

The second Chapter is devoted to Huygens. To him we owe the important condition of fluid equilibrium, that the resultant force at any point of the free surface must be normal to the surface at that point; and this has indirectly promoted the knowledge of our subject. But Huygens never accepted the great principle of the mutual attraction of particles of matter; and thus he contributed explicitly only the solution of a theoretical problem, namely the investigation of the form of the surface of rotating fluid under the action of a force always directed to a fixed point.

The third Chapter treats of various miscellaneous investigations connected with the subject in the course of one generation after the publication of the *Principia*. No real addition was made to Newton's theoretical results, while the measurements of arcs of the meridian in France led the Cassinis to adopt the hypothesis that the form of the Earth was not oblate but oblong.

The fourth Chapter relates to Maupertuis. He wrote various memoirs, among which were two in the form of commentaries on Newton's theories of Attraction and the Figure of the Earth. These theories were rendered more accessible by the translation from their original geometrical expression into the familiar analytical language of the epoch. By adhering to Newton's conclusions Maupertuis must have contributed much to maintain the truth among his countrymen, in opposition to the errors recommended by the authority of Des Cartes and the Cassinis.

The important postulate assumed by Newton was first considered by Stirling, a mathematician of great power: the fifth Chapter shews that he obtained, at least implicitly, an approximate demonstration of the required result.

In the sixth Chapter an account is given of various memoirs by Clairaut which preceded the publication of his important work on the Figure of the Earth. Clairaut explicitly demonstrated the truth of Newton's postulate approximately. He also gave the theorem, called Clairaut's theorem, which establishes a connection between the ellipticity of the earth and the coefficient of the term expressing the increase of gravity in passing from the equator to the pole.

The seventh Chapter narrates briefly the circumstances of the measurement of an arc of the meridian in Lapland. I have undertaken to develop the progress of the Mathematical Theories of Attraction and of the Figure of the Earth; but I do not profess to include the practical operations conducive to our knowledge of the exact dimensions of the Earth. These consist mainly of observations of pendulums, and measurements of arcs; and an account of them drawn from the original sources would form an interesting and instructive work. But the more difficult matters to which I have devoted the present volumes have furnished ample employment without any serious divergence into the department of practical application. I have therefore limited myself to short notices of the earlier pendulum experiments, and of the two great measurements in Lapland and Peru; these measurements deserve some attention on account of their historical interest and their decisive testimony to the oblate form of the Earth.

The eighth Chapter treats of various miscellaneous investigations between 1721 and 1740. Desaguliers maintained, with a zeal not uniformly discreet, the oblate form against the Cassinian hypothesis; on the other hand, the measurements in France were still held to be in favour of that hypothesis. Towards the end of the period the Academy of Paris proposed the Tides as the subject of a Prize Essay; and this led to the important researches of Maclaurin.

The ninth Chapter is devoted to Maclaurin. He completely solved the problem of the attraction of an ellipsoid of revolution on an internal or superficial particle; and his method and results admitted of obvious extension to the case of an ellipsoid not of revolution. The extent to which he proceeded for the case of an external particle requires to be stated with accuracy, in order to correct errors of opposite kinds which are current. The most general result yet attained may be stated thus: the potentials of two confocal ellipsoids at a given point external to both are as their masses. This theorem was first established by Laplace, but Maclaurin demonstrated it for the particular case in which the external point is on the prolongation of an axis of the ellipsoids. In the theory of the Figure of the Earth, Maclaurin's main achievement was an exact demonstration of Newton's postulate, of which hitherto only approximate investigations had been given.

In the tenth Chapter the contributions of Thomas Simpson are noticed. This eminent mathematician explicitly shewed that if the angular velocity of rotation exceeds a certain value, the oblatum is not a possible form of relative equilibrium for a fluid mass; and it followed implicitly from his results that for any value of the angular velocity less than the limit, more than one figure for relative equilibrium would exist. Simpson also gave a remarkable investigation of the attraction at the surface of a very extensive class of nearly spherical bodies.

The eleventh Chapter consists of an analysis of the celebrated work by Clairaut. The first part of the work treats on the principles of fluid equilibrium; here Clairaut far surpassed his predecessors in extent and accuracy, and left the theory in the form which it still retains, with the single exception of the improvement effected by Euler, who introduced the notion of the pressure at any point of the fluid, together with the appropriate symbol by which it is denoted. The second part of the work treats on the Figure of the Earth. For the case of a homogeneous fluid Clairaut closely followed Maclaurin. The case of a heterogeneous fluid had been hitherto practically untouched, and Clairaut invented for it a beautiful process which has remained substantially unchanged to the present time; the chief result is a certain equation connecting the ellipticity of the strata with their density, which appears in two forms: these I have called respectively Clairaut's primary equation, and Clairaut's derived equation.

The twelfth Chapter narrates briefly the circumstances of the measurement of an arc of the meridian in Peru. I have carefully examined the extensive literature, much of which is controversial, arising from this memorable expedition; and by means of exact references I have afforded assistance to any student who wishes to render himself familiar with all the circumstances.

The thirteenth Chapter is devoted to the earlier half of the writings of D'Alembert which bear on our subjects. They are extensive in amount, and may have served indirectly to diffuse the interest in such investigations which the writer must have felt himself; but on account of errors in principle and inaccuracy of detail their direct value is small. In various attempts which D'Alembert made to criticise the work of Clairaut he was I believe almost uniformly wrong, so far as regards the Figure of the Earth, and barely right on some unimportant points of Hydrostatics. It is stated in the life of D'Alembert published in the *Biographical Dictionary of the Society for the Diffusion of Useful Knowledge* that "He and Clairaut were rivals, and no work of either appeared without finding a severe critic in the other; but D'Alembert, the more cautious and profound of the two, was generally on the right side of the question:..." The judgment is pronounced by a most eminent authority to which I usually bow with reverence; but so far as the subjects of the present work extend, I should venture to reverse it.

The fourteenth Chapter is devoted principally to Boscovich, whose writings furnish elementary accounts of the most important results which had been obtained up to their date. I have also given a brief notice of the poem by Stay, for which Boscovich supplied notes and supplementary dissertations.

The fifteenth Chapter treats of various miscellaneous investigations between the years 1741 and 1760. It includes a brief notice of a Prize Essay on the Figure of the Earth, published by Clairaut, some years after his treatise.

The sixteenth Chapter is occupied with the later half of the writings of D'Alembert. The general character is the same as of the earlier half; the investigations themselves are disfigured by serious errors, but they serve to suggest interesting and important matter.

The works of Frisi are noticed in the seventeenth Chapter: they resemble those of Boscovich in the fact that they served to teach the subject rather than to promote its progress.

The eighteenth Chapter treats of various miscellaneous investigations between the years 1761 and 1780. The first three of Laplace's memoirs belong to this period, but for convenience the consideration of them is postponed. The Chapter includes an account of a memoir by Lagrange in which he proceeded by analysis to the point Maclaurin had reached by geometry. The operations carried on at Schehallien for ascertaining the density of the Earth are noticed, and references are supplied to the subsequent labours on the same subject. Here the first volume ends, which contains the history of our subjects during the century which followed the publication of Newton's *Principia*.

The nineteenth Chapter takes the first three memoirs of Laplace. The principal object of these memoirs may be said to be the solution of a problem which is an extension of Newton's postulate. Newton assumed that an oblatum was a possible form of relative equilibrium for rotating fluid; the present problem is to shew that an oblatum is the *only* possible form, at least under certain restrictions. I call the problem Legendre's, because he was the first who solved it with tolerable success. D'Alembert attempted the investigation, but failed. Laplace did not solve the problem completely; but he shewed that for a very large class of nearly spherical figures, the relative equilibrium was impossible. He also obtained the expression for the law of gravity which would hold universally.

The twentieth Chapter is devoted to a memoir which is conspicuous in the history of the Theory of Attraction, namely the earliest of Legendre's. The limit reached by Maclaurin is now for the first time left behind; Legendre shews that the theorem with respect to confocal ellipsoids is true for *any position* of the external point when the ellipsoids are solids of revolution. Legendre introduces here the memorable expressions, hitherto unknown, which are now usually called *Laplace's coefficients*; and also, at the suggestion of Laplace, the function now called the Potential function takes its place in the subject.

The twenty-first Chapter brings before us a scarce treatise by Laplace, and gives an analysis of that half of it which relates to Attraction and the Figure of the Earth. Here was published for the first time, the demonstration of the theorem relating to the action of confocal ellipsoids at an external point which I call by Laplace's name. The subjects of the Attraction of Ellipsoids and of the homogeneous Figure of the Earth appear in this treatise in nearly the same form as in the *Mécanique Céleste*.

The twenty-second Chapter relates to Legendre's second memoir. Here Legendre solves the problem which I call by his name. He assumes that the fluid is in the form of a figure of revolution, and that it does not deviate widely from the spherical form.

The twenty-third Chapter notices Laplace's fourth, fifth, and sixth memoirs. The fourth and fifth memoirs contain the theory of the attraction of spheroids, and the theory of Laplace's functions, in the form they assume in the *Mécanique Céleste*. The sixth memoir relates to Saturn's ring.

The twenty-fourth Chapter is devoted to Legendre's third memoir. The object of this memoir is to demonstrate Laplace's theorem respecting confocal ellipsoids by a more direct process than Laplace himself had employed. Legendre does demonstrate the theorem, without expanding his expressions in series, but the process is excessively long and complicated.

The twenty-fifth Chapter analyses Legendre's fourth memoir. Here we have a great development of Clairaut's process for the case of heterogeneous fluid. A general equation is obtained analogous to Clairaut's primary equation; and from this it is shewn that the strata must be ellipsoidal.

The twenty-sixth Chapter is devoted to Laplace's seventh memoir. This contains some numerical discussion of the lengths of degrees, and of the lengths of the seconds pendulum; there is also a theory of the heterogeneous figure of the Earth, which substantially agrees with that in Legendre's fourth memoir.

The twenty-seventh Chapter treats of miscellaneous investigations between the years 1781 and 1800. Among other matters we have here to notice Cousin's Introduction to the study of Physical Astronomy, a memoir by Lagrange, and a memoir by Trembley; the last is of the same unsatisfactory character as various memoirs by the same writer which I have examined in my *History of the Mathematical Theory of Probability*.

The twenty-eighth Chapter gives an account of the first two volumes of the *Mécanique Céleste*, so far as they relate to our subjects. Laplace in effect reproduced with small change the last four of his seven memoirs; and the result is a treatise not yet superseded.

The twenty-ninth Chapter traces the history of investigation with respect to Laplace's Theorem. Ivory, Legendre, Gauss and Rodrigues all gave complete discussions of the attraction of ellipsoids; while Biot and Plana also commented on parts of the theory. The method of Ivory is the simplest of all, and has obtained a permanent position in our elementary works; insomuch that it is usual to speak of *Ivory's theorem*, although the more correct phrase would be *Ivory's demonstration of Laplace's theorem*.

The thirtieth Chapter treats on an equation which Laplace seems to have regarded with peculiar favour, and which occurs often in his writings. The equation however did not satisfy Ivory, and he criticised it with severity. The result of the discussion may be said to have established the accuracy of Laplace's equation when used, as he himself used it, with due caution. But at the same time the objects which Laplace sought by the aid of his equation are now generally obtained without it; so that practically the equation is at present rarely employed.

The thirty-first Chapter elucidates the partial differential equation for the symbol which denotes the potential function. Laplace had originally assumed that a certain equation held both for an external particle, and for a component particle of the body considered; but Poisson shewed that the two cases required different forms of the equation.

The thirty-second Chapter discusses a method which Laplace gave for solving Legendre's problem, with the objection brought against it by Liouville, and the treatment which Poisson substituted in place of Laplace's.

The thirty-third Chapter passes in review various memoirs which Laplace published during the first quarter of the present century.

The thirty-fourth Chapter is devoted to that part of the fifth volume of the *Mécanique Céleste* which relates to our subjects; it consists chiefly of a republication of the memoirs noticed in the thirty-third Chapter.

Strictly speaking the period of history which I proposed to describe closes here; but it seemed convenient to include within my range all the writings of three mathematicians who had already been prominent in my work, and who may be naturally associated with their predecessors, especially with Laplace. These writers are Poisson, Ivory and Plana.

The thirty-fifth Chapter contains an account of all Poisson's contributions which had not been previously examined. The most important of these are an elaborate memoir on the Attraction of Spheroids, and a memoir giving a new investigation of Laplace's theorem respecting confocal ellipsoids.

The thirty-sixth Chapter gives a brief sketch of the numerous articles and memoirs which Ivory produced, mainly in support of opinions of his own which were both peculiar and erroneous. The great promise which his early success held out was not followed by any corresponding merit in the essays of his later years.

The thirty-seventh Chapter is devoted to Plana, who wrote several papers chiefly in the form of comments on Lagrange, Legendre and Laplace.

The last Chapter treats of various miscellaneous investigations during the first quarter of the present century. It is by accident the history finishes with a paragraph relating to Bowditch; but on account of his moral and intellectual eminence, and of his unselfish devotion to science, the name of one of the most distinguished mathematicians beyond the Atlantic may justly close a roll which commences with that of Newton.

The period of time which I have traversed will be found to correspond with some accuracy to a distinct boundary line in the subject. The labours of more recent date present to us many indications of what may be more appropriately called new methods rather than mere developments of those already discussed. Among them we may mention the investigations respecting the Potential by Green and Gauss, and the numerous researches on the attraction of Ellipsoids by Chasles; all these writers will occupy conspicuous places in any future record of the subjects. Sir John Herschel spoke of my *History of Probability* as embracing the series of the *Pleiocene analysts* in distinction from the Post-Pleiocene; and the illustration might be similarly applied in the present case.

Such then is the outline of the history which the present volumes contain. The principles on which I have executed my task are the same as those adopted in my former works; and I may refer especially to the preface to my *History of Probability* for an account of them. I will only state here that I have not thought it necessary to preserve the exact notation of the original authors; that notation frequently varies much in various places, and it is really advantageous for the sake of brevity and clearness to use the same symbols throughout. For example the ratio of the centrifugal force at the equator to the gravity there is denoted in some English books by the letter m; Clairaut uses ϕ; D'Alembert in the sixth volume of his *Opuscules Mathématiques* uses ω; Laplace in the *Mécanique Céleste,* Vol. v. page 7, uses ϕ, and in Vol. v. page 23 he uses $\alpha\phi$. For the ratio of the centrifugal force at the equator to the attraction there, which is very approximately the same thing as the preceding ratio, the letter j is used throughout the present work.

I have been very sparing in the introduction of new terms, for this practice seems carried to an embarrassing extent in some modern mathematical works. I have however found it necessary to have short designations for two things which occur perpetually in these investigations. The body formed by the revolution of an ellipse round its minor axis I call an *oblatum*, and the body formed by the revolution of an ellipse round its major axis I call an *oblongum.* In English books the former has usually been called an *oblate spheroid;* and the latter a *prolate spheroid.* Something is gained in conciseness by using one word instead of two for a name which is frequently required; but the chief reason of the change arises from the fact that the word *spheroid* has been much used in a different sense, namely to denote a body which differs but little from a sphere. It would be very convenient if this sense of the word *spheroid* could be so established as to render superfluous the formal enunciation of the condition of resemblance to a sphere. Perhaps the use of a word to express a form only approximately determined is felt to be somewhat unlike the ordinary precision of mathematical language; and this may account for the frequent repetition of the condition even after it has been explicitly adopted. Moreover the great French writers have often employed the word *spheroid* in a sense so wide as to render it practically equivalent to *body;* an example will be found in the title of a memoir by Poisson on page 388 of the second volume.

I have found it convenient to give a name to a certain ratio which is of importance in our subject, namely the ratio of the difference of gravity at the equator and at the pole to gravity at the equator. This ratio is one of the elements connected by Clairaut's theorem, and I have accordingly called it *Clairaut's fraction.*

There is one term, perhaps the most objectionable of all that have become permanent in mixed mathematics, which is used throughout the work, namely *centrifugal force.* It is with great reluctance that I have felt myself constrained to yield to universal authority and to employ language which experience shews to be most perplexing and misleading. The well-trained student will however have learned that the so-called centrifugal force is a fiction; the simple fact is that a dynamical problem relating to a body which is rotating uniformly, can be reduced to a statical problem by supposing the rotation to cease and a certain force to be introduced.

This History assumes on the part of the reader some elementary acquaintance with the subjects on which it treats. For the Theory of Attractions the Chapter in my work on *Statics,* to which I have occasionally referred, will be sufficient. For the Figure of the Earth the student may consult three well-known English treatises, namely one in Airy's *Mathematical Tracts,* one in O'Brien's *Mathematical Tracts,* and Pratt's Chapter on the subject in his *Mechanical Philosophy,* afterwards enlarged and published separately in a Treatise on *Attractions, Laplace's Functions and the Figure of the Earth:* Pratt's Treatise is the most comprehensive of these English treatises, and the easiest to procure. An interesting work was published at Paris in 1865, entitled *Traité Elémentaire de Mécanique Céleste. Par H. Resal.* About a third of this volume is devoted to our subjects; and it gives a very instructive account of them: but the extreme inaccuracy of the printing is a serious diminution of the value of the work.

The mathematical expressions which are called *Laplace's coefficients* and *Laplace's functions* play a very important part in the higher investigations of our subjects. The treatises of O'Brien, Pratt, and Resal, which have just been cited contain a sufficient account of these expressions for elementary purposes. The student who wishes to become intimately acquainted with them will have recourse to the work by Heine which is named on page 24 of the second volume; this is an admirable volume enriched with numerous references to the original authorities.

It may be naturally expected that a person who has devoted much time to the study of the history of science will feel disposed to attribute considerable value to the pursuit. The interest which attaches to the struggle of the human mind with serious difficulties, to its gradual progress and final triumph, may be at least as great as that which is excited by an account of the vicissitudes of civil history. An acquaintance with the origin and the course of any science will often give great assistance in the comprehension of its present state, and may even point out the most promising direction for future efforts. Moreover a familiarity with what has been already accomplished or attempted in any subject is conducive to a wise economy of labour; for it may often prevent a writer from investigating afresh what has been already settled, or it may warn him by the failure of his predecessors, that he should not too lightly undertake a labour of well-recognised difficulty. The opinions of Laplace and Arago, which are quoted in my title-pages, are justly entitled to great weight on these points.

That the subjects here treated historically are of no common importance and influence may be easily seen. A knowledge of the figure and dimensions of the Earth is the basis of all the numerical results of Astronomy, and therefore of the greatest practical value. Moreover the researches into the theories of Attraction and of the Figure of the Earth have been fertile in yielding new resources for mathematicians; it will be sufficient to point to the Transformation of Multiple Integrals, the theory of the Potential, and the elaborate doctrine of Laplace's functions, which have all sprung up in the cultivation of this field of Physical Astronomy. Humboldt has drawn attention to this circumstance in his *Cosmos;* the following passage occurs on pages 156 and 157 of the fifth edition of Sabine's translation of the first Volume: "Except the investigations concerning the parallax of the fixed stars, which led to the discovery of aberration and nutation, the history of science presents no problem in which the object obtained,—the knowledge of the mean compression of the Earth, and the certainty that its figure is not a regular one,—is so far surpassed in importance by the incidental gain which, in the course of its long and arduous pursuit, has accrued in the general cultivation and advancement of mathematical and astronomical knowledge."

It may appear that some apology is due for the extent to which the work has grown; this must be found in the extent and intricacy of the materials which had to be analysed. Indeed Ivory, who devoted much attention to the subject of the Figure of the Earth, asserts that it has been attended with greater difficulty and has occasioned a greater number of memoirs than any other branch of the system of the world. I have had some trouble in keeping within the limits of two volumes, and have been compelled to omit many developments which I should gladly have printed. I have also published separately various papers which have grown out of my historical studies; to these I refer in the appropriate places, but it may be convenient to give a list of them here. They are the following:

On Jacobi's Theorem respecting the relative equilibrium of a revolving ellipsoid of fluid, and on Ivory's discussion of the Theorem. *Proceedings of the Royal Society,* Vol. XIX.

Note relating to the Attraction of Spheroids. *Proceedings of the Royal Society,* Vol. XX.

Note on an erroneous extension of Jacobi's Theorem. *Proceedings of the Royal Society,* Vol. XXI.

On the Arc of the Meridian measured in Lapland. *Transactions of the Cambridge Philosophical Society,* Vol. XII.

On the equation which determines the form of the strata in Legendre's and Laplace's theory of the Figure of the Earth. *Transactions of the Cambridge Philosophical Society,* Vol. XII.

On the Proposition 38 of the Third Book of Newton's *Principia. Monthly Notices of the Royal Astronomical Society,* Vol. XXXII.

On the Arc of the Meridian measured in South Africa. *Monthly Notices of the Royal Astronomical Society,* Vol. XXXIII.

The account which is given of the memoirs and treatises will be found ample enough in most cases to supply all that a student will ever want to read of them; but this does not apply to the *Mécanique Celeste,* which I desire to illustrate not to supersede. In other words all that I say relative to that great work is intended as a commentary for the use of those who are consulting the original. I have usually cited it by *sections,* but in some cases, which occur almost exclusively in the fifth volume, I have for greater distinctness cited it by *pages.* The pages meant are those of Laplace's own edition; but the student who uses the national edition will be able to adjust the references by observing that in the fifth volume the 85 pages with which we are concerned correspond to 103 pages in the national edition.

It is well known that Laplace does not give any specific

references to the labours of his predecessors and contemporaries; in his great treatises on Physical Astronomy and Probability he embodied with his own results much that he derived from others, and as these treatises have become the standards of authority for the subjects to which they relate, it has followed that with uncritical readers Laplace has not unfrequently obtained credit for what was not distinctively his own production. A student of the course of science will often discover that important investigations which first came under his notice in the works of Laplace, are really due to other mathematicians; and by a natural reaction the conjecture will arise that further research will lead to the restitution of much more to the rightful owners; and thus there may be a recoil from an undue admiration to a suspicious depreciation. But a complete evolution of the history will restore the reputation of Laplace to its just eminence. The advance of mathematical science is on the whole remarkably gradual, for with the single exception of Newton there is very little exhibition of great and sudden developments; but the possessions of one generation are received, augmented, and transmitted by the next. It may be confidently maintained that no single person has contributed more to the general stock than Laplace.

In the life of Laplace in the *English Cyclopædia*, which we may safely attribute to the late Professor De Morgan, there are some valuable remarks suggested by the want of specific information in the writings of Laplace as to what was done by himself and what was done by others; and it is stated that no one has yet supplied the deficiency. With respect to Laplace it is said: "Had he consulted his own glory, he would have taken care always to note exactly that part of his own work in which he had a forerunner; and it is not until this shall have been well and precisely done, that his labours will receive their proper appreciation." In the present history and in that of Probability I have gone over a third part of the collected mathematical works of Laplace; and to that extent the evidence of his great power and achievements is I hope fully and fairly manifested.

I have not hesitated to criticise all that has come before me; and there is scarcely any memoir or treatise of importance left without the suggestion of corrections or additions. I cannot venture to hope that I have uniformly escaped without any obscurity or error. My readers will I trust excuse such blemishes, arising partly from the nature of the task and partly from the circumstance that only such leisure could be found for it as remained amidst continuous occupation in elementary teaching and writing. The work has thus furnished ample employment for seven years of labour, with the exception of a necessary digression in order to explain and illustrate some peculiarities in the Calculus of Variations. It was perhaps rash for a mere volunteer to undertake so extensive a task; but in spite of the imperfections with which it may have been accomplished, I am willing to hope that the result will be a permanent addition to the literature of Physical Astronomy.

It is not from any desire to challenge comparisons with illustrious men, but merely to justify my estimate of the labour involved, that I venture to quote the following opinion expressed by the late Professor James Forbes in his *Review of the Progress of Mathematical and Physical Science*, and to extend its application from pure to mixed mathematics: "Specimens of what a history of pure mathematics would be, and must be, are to be found in the able 'Reports' of Dr Peacock and Mr Leslie Ellis, in the Transactions of the British Association for 1833, and 1846. A glance at these profound and very technical essays will shew the impossibility of a popular mode of treatment, while the difficulty and labour of producing such summaries may be argued from their exceeding rarity in this or any other language."

I have to record my great obligations to the Rev. J. Sephton, Head Master of the Liverpool Institute, formerly Fellow of St John's College, for his most valuable assistance in conducting the work through the Press. To the Syndics of the University Press I am indebted for their liberality in defraying the expenses of the printing.

I. TODHUNTER.

ST JOHN'S COLLEGE, CAMBRIDGE,
July, 1873.

CONTENTS.

VOLUME I.

VOLUME II.

CHRONOLOGICAL LIST OF AUTHORS.

The figures refer to the Articles of the Volumes.

THE following Table gives the DATES OF BIRTH AND DEATH of the principal writers on ATTRACTION and the FIGURE OF THE EARTH:

Name	Birth	Death
BIOT	1777	1862
BOSCOVICH	1711	1787
BOUGUER	1698	1758
BOWDITCH	1773	1838
CASSINI, J. D.	1625	1712
CASSINI, J.	1677	1756
CASSINI DE THURY	1714	1784
CASSINI IV.	1748	1845
CAVENDISH	1731	1810
CLAIRAUT	1713	1765
COULOMB	1736	1806
COUSIN	1739	1800
D'ALEMBERT	1717	1783
DELAMBRE	1749	1822
EULER	1707	1783
FRISI	1728	1784
GAUSS	1777	1855
HUYGENS	1629	1695
IVORY	1765	1842
LA CAILLE	1713	1762
LA CONDAMINE	1701	1774
LAGRANGE	1736	1813
LA LANDE	1732	1807
LAPLACE	1749	1827
LEGENDRE	1752	1833
MACLAURIN	1698	1746
MAIRAN	1678	1771
MASKELYNE	1732	1811
MAUPERTUIS	1698	1759
NEWTON	1642	1727
PLANA	1781	1864
PLAYFAIR	1748	1819
POISSON	1781	1840
SIMPSON	1710	1761

THE following Table gives references to the principal numerical discussions of the Figure and Dimensions of the Earth in chronological order.

1755 BOSCOVICH. *De Litteraria Expeditione.*
1760 BOSCOVICH. Stay's *Philosophiæ Recentioris.*
1768 FRISI. *De Gravitate.*
1770 French translation of Boscovich's work.
1783 LAPLACE's fifth Memoir.
1789 LAPLACE's seventh Memoir.
1799 LAPLACE. *Mécanique Céleste,* Vol. II.
1826 AIRY. *Philosophical Transactions.*
1828 IVORY. *Philosophical Magazine.*
1830 AIRY. *Encyclopœdia Metropolitana.*
1832 BOWDITCH. Translation of the second volume of the *Mécanique Céleste,* pages 450 to 455.
1837 BESSEL. *Astronomische Nachrichten,* Vol. XIV.
1842 BESSEL. *Astronomische Nachrichten,* Vol. XIX.
1844 BIOT. *Traité Elémentaire d'Astronomie Physique,* Vol. III.
1859 SCHUBERT. *Petersburg Mémoires,* seventh series, Vol. I.
1861 CLARKE. *Memoirs of the Royal Astronomical Society,* Vol. XXIX.
1866 HERSCHEL. *Familiar Lectures on Scientific Subjects.*

CHAPTER I.

NEWTON.

1. NEARLY two centuries have passed away since the publication of the greatest work known in the history of science. Newton's *Philosophiæ Naturalis Principia Mathematica* appeared in 1687. The volume is in quarto; it contains a title-leaf, a dedication to the Royal Society on another leaf, a preface on two pages, some Latin verses by Halley on two pages, then the text consisting apparently of 510 pages, followed by errata on one leaf. I say the text consists apparently of 510 pages; there are, however, no pages numbered from 384 to 399 inclusive: the third Book begins on page 401, and so perhaps some of this was struck off before the second Book was finished, and a gap was left in the number of pages which proved too large.

The second edition of the *Principia* appeared in 1713, edited by Cotes; the third in 1726, edited by Pemberton. Newton was born in 1642, and died in 1727.

2. Newton's researches on Attractions form Sections XII. and XIII. of the first Book of the *Principia*. Section XII. contains Propositions 70...84; it relates to the attraction of spherical bodies. Section XIII. contains Propositions 85...93; it relates to the attraction of bodies which are not spherical. These Sections remain unchanged in the other two editions of the *Principia*.

3. In his Proposition 70, Newton shews that a particle will be in equilibrium if placed at any point of the hollow part of an indefinitely thin spherical shell, which attracts according to

the law of the inverse square of the distance. Newton's demonstration is remarkable for its simplicity. Let any indefinitely small double cone be described with the position of the attracted particle as vertex; the areas of the indefinitely small surfaces which the cone intercepts on the shell are ultimately as the squares of the distances of the elements from the vertex: thus the elements exert equal attractions in opposite directions. Therefore the entire shell exerts no action in any direction.

We assume here and in the other propositions that the attracting body is homogeneous unless the contrary is stated.

4. In his Proposition 71, Newton shews that an indefinitely thin spherical shell attracts an external particle towards the centre of the shell, with a force which varies inversely as the square of the distance of the particle from the centre of the shell. Newton's demonstration is geometrical; it can, however, be easily translated into an analytical form.

Let a be the radius of the shell, c the distance of the particle from the centre of the shell, ds an element of the length of the circle which by revolution round the straight line joining the particle to the centre generates the surface of the shell, r the distance of this element from the particle, y its distance from the axis of revolution. Then the element of surface generated by the revolution of ds is $2\pi y ds$; and the attraction of this element along the axis is $\frac{2\pi k\rho y ds}{r^2}\cos\theta$; where k is the thickness of the shell, ρ is the density, and θ is the angle between the direction of r and the axis. Let p denote the perpendicular from the centre of the shell on the direction of r. We have

$$p = c\sin\theta, \qquad r^2 - 2rc\cos\theta + c^2 = a^2;$$

hence $$\frac{dr}{d\theta} = -\frac{rc\sin\theta}{r - c\cos\theta}, \qquad \frac{ds}{d\theta} = \frac{ar}{r - c\cos\theta}.$$

Thus $$\frac{2\pi k\rho y ds}{r^2}\cos\theta = \frac{2\pi k\rho y\cos\theta}{r^2}\cdot\frac{ard\theta}{r - c\cos\theta}$$

$$= \frac{2\pi k\rho y a dp}{cr(r - c\cos\theta)} = \frac{2\pi k\rho a p dp}{c^2\sqrt{(a^2 - p^2)}}.$$

Hence the resultant attraction of the shell will be found by integrating this expression between appropriate limits. If we take 0 and a as the limits of p, we obtain the attraction of either of the two parts into which the shell is divided by the curve of contact of straight lines drawn from the particle to *touch* the shell; hence these two parts exert equal attractions, and the attraction of the whole shell is

$$2 \times \frac{2\pi k\rho a}{c^2} \int_0^a \frac{pdp}{\sqrt{(a^2 - p^2)}},$$

which varies inversely as c^2.

The value of the definite integral is a; and thus the attraction of the whole shell is $\frac{4\pi k\rho a^2}{c^2}$

We see from this investigation that if any right cone be taken having its vertex at the position of the particle, and its axis coincident with the straight line drawn from the particle to the centre of the shell, we can determine the attraction which is exerted by the portion of the shell cut off by the cone: we have only to give an appropriate value to the upper limit of p in the integration. We may observe too that if any indefinitely small cone be taken having its vertex at the position of the particle, the two distinct portions of the shell which it intercepts exert equal attractions.

We may observe that Proposition 71 has been very well treated by Professor Thomson: see *Cambridge and Dublin Mathematical Journal,* Vol. III. page 146.

5. Propositions 72...76 extend the conclusions obtained respecting indefinitely thin spherical shells to spheres.

It appears that Newton arrived at his theorems respecting the attraction of spheres in 1685. See the *Mécanique Céleste,* Vol. v., page 87; Rigaud's *Historical Essay on the first publication of the Principia,* page 27 of the Appendix.

6. Newton's Propositions 77 and 78 relate to the case in which the law of attraction is that of the direct distance.

Between Propositions 78 and 79 a Lemma occurs.

Let x and y be the co-ordinates of a point on a circle; r the distance of the point from any fixed origin. We have

$$r^2 = x^2 + y^2;$$

therefore $\quad rdr = xdx + ydy = (x - c)\,dx + ydy + cdx.$

Let c be the distance of the centre of the circle from the origin, the centre being on the axis of x. Then $(x-c)\,dx+ydy=0$; therefore $rdr = cdx$. This result constitutes the Lemma; it is of course demonstrated geometrically by Newton. Throughout this Chapter we shall translate Newton's geometrical processes into modern mathematical language.

7. In his Proposition 79, Newton finds the attraction of a zone of an indefinitely thin spherical shell on a particle at the centre of the shell.

Take the axis of the zone for that of x, and a line at right angles to this through the centre of the shell for the axis of y; let a be the radius of the sphere. Then $2\pi a dx$ represents an element of the zone; and the attraction of this element will be $kf \,.\, 2\pi a \cdot \frac{x}{a}\,dx$, where k denotes the thickness of the shell, and f is a constant which denotes the attraction of a unit of matter, condensed at a point, on a particle at the distance a. Hence the attraction of the zone $= kf \,.\, 2\pi \int x dx$, the integral being taken between proper limits. If the zone be the segment cut off by the plane $x = x_1$, we have to integrate between the limits x_1 and a. Thus we obtain $kf\pi\,(a^2 - x_1^2)$, that is $kf\pi y_1^2$, where y_1 is the radius of the base of the segment.

8. Newton's Proposition 80 investigates the attraction of a sphere on an external particle, the law of attraction being expressed by any function of the distance.

Divide the sphere into elements by describing spherical surfaces indefinitely close to each other from the external particle as centre. Let r be the radius of one of the surfaces of one of the segments of shells thus obtained, and y the radius of the base of the segment; let $\phi\,(r)$ denote the law of attraction. Then by Art. 7

we have $\pi dr\phi(r)y^2$ for the attraction of the segment. Let c be the distance of the external particle from the centre of the sphere; then by Art. 6 we have $dr = \frac{cdx}{r}$; thus the attraction becomes $\frac{c\pi}{r}\phi(r)\,y^2dx$. Hence the resultant attraction of the sphere is $c\pi\int_{c-a}^{c+a}\frac{\phi(r)}{r}y^2dx$, where a is the radius of the sphere.

9. Newton's Proposition 81 amounts to a transformation of the integral obtained in Art. 8.

We have $y^2 = a^2 - (c-x)^2$, and also $y^2 = r^2 - x^2$;

therefore $$r^2 = a^2 - c^2 + 2cx.$$

Put $\frac{c^2-a^2}{2c} = b$, and $x - b = x'$; thus

$$r^2 = 2c(x-b) = 2cx',\ y^2 = -2bc + 2cx - x^2 = 2cx' - (x'+b)^2.$$

Hence the resultant attraction

$$= c\pi\int\frac{2(c-b)x' - x'^2 - b^2}{r}\phi(r)\,dx',$$

the limits of x' being $c-a-b$ and $c+a-b$.

As soon as $\phi(r)$ is known we can substitute for r in terms of x', and effect the integration. Newton gives three examples:

$$(1)\ \phi(r) = \frac{\mu}{r}, \qquad (2)\ \phi(r) = \frac{\mu}{r^3}, \qquad (3)\ \phi(r) = \frac{\mu}{r^4},$$

where μ in each case is a constant.

10. Newton's Proposition 82 shews that the calculation of the attraction of a sphere on an internal particle may be made to depend on the calculation of the attraction on an external particle.

We have found in Art. 8 for the attraction of an element of the sphere $\pi dr\,\phi(r)\,y^2$, where r is the distance of the particle from every point of the element. In the same manner $\pi dr'\,\phi(r')\,y^2$ will express the attraction of the corresponding element on another particle which is at the distance r' from every point of the element. The two particles and the centre of the sphere are of course on the same straight line. Suppose the second particle within the sphere; let c be the distance of the first particle from the centre of the sphere, c' that of the second, a the radius of the sphere. Let c and c' be taken so that $cc' = a^2$.

In the diagram let

$$SP = c, \quad SI = c', \quad EP = r, \quad EI = r'.$$

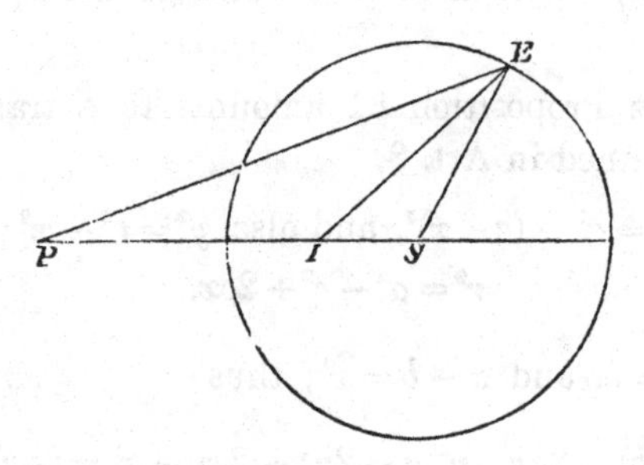

As $cc' = a^2$ the triangles PSE and ESI are similar; thus we have

$$\frac{r'}{r} = \frac{c'}{a}.$$

In finding the attraction on the internal particle, we may if we please suppose the matter to be removed which forms a sphere having its centre at the internal particle and radius equal to $a - c'$: thus the limits of integration become $r' = a - c'$ and $r' = a + c'$.

Suppose $\phi(r) = \frac{\mu}{r^n}$; the attraction on the internal particle

$$= \pi \int y^2 \phi(r')\, dr' = \pi\mu \int \frac{y^2}{r'^n} dr';$$

the limits being $a - c'$ and $a + c'$. Now put $\frac{rc'}{a}$ for r'; thus we get $\pi\mu \left(\frac{a}{c'}\right)^{n-1} \int \frac{y^2}{r^n} dr$, and the limits of r are $\frac{a^2}{c'} - a$ and $\frac{a^2}{c'} + a$, that is, $c - a$ and $c + a$.

Hence the attraction on the internal particle at the distance c' from the centre is equal to the product of $\left(\frac{a}{c'}\right)^{n-1}$ into the attraction on the external particle at the distance c from the centre.

And $$\left(\frac{a}{c'}\right)^{n-1} = \frac{(cc')^{\frac{n-1}{2}}}{c'^{n-1}} = \left(\frac{c'}{c}\right)^{\frac{1}{2}} \left(\frac{c}{c'}\right)^{\frac{n}{2}}$$

This is the result which Newton intended to give. He says that the attraction on the particle at I is to the attraction on the particle at P, in ratione composita ex subduplicatâ ratione distantiarum a centro IS et PS, et subduplicatâ ratione virium centripetarum, in locis illis P et I, ad centrum tendentium. It seems to me that instead of P et I we ought to read I et P.

11. Newton's Propositions 83 and 84 shew briefly that there would be no difficulty in calculating the attraction of a homogeneous segment of a sphere on a particle situated on the axis of the segment.

12. Newton's Propositions 85, 86, and 87 involve simple general statements, which need not be repeated here.

Propositions 88 and 89 shew that if the law of attraction is that of the direct distance, the resultant attraction exerted by a body or a system of bodies is the same as if the body or system were collected at its centre of gravity.

13. Proposition 90 finds the attraction of a circular lamina on a particle which is situated on the straight line drawn through the centre of the lamina at right angles to its plane. Then Proposition 91 shews how from this we can deduce the attraction of a solid of revolution on a particle situated at any point of the axis. Newton makes this depend on the problem of finding the area of a certain curve; that is, in modern language, he leaves only a single integration to be effected. He takes the case of a right cylinder for an example; and he also states the result for the case of an ellipsoid of revolution, which he calls a spheroid. He shews by a special investigation that a shell bounded by two concentric similar and similarly situated ellipsoidal surfaces of revolution exerts no attraction on a particle placed at any point within the hollow part; the demonstration is very striking and well known: see *Statics*, Chapter XIII. Of course this result includes Newton's Proposition 70 as a particular case; but the demonstrations differ and should be carefully compared.

Hence follows the important result that along the same radius vector from the centre the attraction of an ellipsoid of revolution on an internal particle varies as the distance from the centre.

Newton contented himself with considering ellipsoids of revolution; but the processes and results of Proposition 91, as we now know, may be easily extended to ellipsoids which are not solids of revolution.

14. Proposition 92 shews how we may find experimentally the law of attraction of given matter. Form the given matter into such a shape that the resultant attraction can be obtained when the law of attraction is assumed; for example, the shape of a sphere. Then ascertain by experiment what the resultant attraction really is at various distances; and thus we shall be guided in assuming a law of attraction and verifying the assumption.

15. Proposition 93 treats of the attraction of an infinite plane lamina, deducing it from Proposition 90. A scholium to this Proposition gives some interesting remarks relating to the motion of a particle acted on by a force the direction of which is always parallel to a fixed straight line.

16. Newton's Propositions on Attractions are illustrated by a good commentary in the edition of the *Principia* which is known as the Jesuits' edition. They had been previously discussed by Maupertuis, as we shall see in another Chapter. Notes by Plana on some of the Propositions will be found in the *Memorie della Reale Accademia...di Torino*, second series, Vol. XI., 1851.

17. We pass now to the investigations made by Newton with respect to the Figure of the Earth; they are contained in Propositions XVIII., XIX., and XX. of the third Book of the *Principia*: these Propositions remain substantially the same in the second and third editions as in the first, but modifications occur arising from additional information as to the facts involved.

Before we consider these Propositions we ought to advert to Newton's remarkable conjecture which is contained in Proposition X. Newton here suggests that the mean density of the Earth may be five or six times that of water:...verisimile est quod copia materiæ totius in Terrâ quasi quintuplo vel sextuplo major sit quàm si tota ex aqua constaret. We may now consider it certain that the mean density is between five and six times that of water. Laplace draws attention to Newton's remarkable conjecture in the *Connaissance des Tems* for 1823, page 328.

It will be convenient to give the enunciations of Newton's Propositions XVIII., XIX. and XX.

XVIII. Axes Planetarum quæ ad eosdem axes normaliter ducuntur minores esse.

XIX. Invenire proportionem axis Planetæ ad diametros eidem perpendiculares.

XX. Invenire et inter se comparare pondera corporum in Terræ hujus regionibus diversis.

18. Proposition XVIII. contains a general statement that the planets are not accurately spherical. In the first edition Cassini and Flamsteed are quoted as authorities for this statement with respect to Jupiter; in the second edition instead of these names we are referred to astronomers in general.

19. Proposition XIX. undertakes to determine the ratio of the axes of a planet. This important process consists of various steps. In the first edition Newton begins by saying briefly he finds from calculation that the centrifugal force at the equator is to the force of attraction there as 1 to $290\frac{4}{5}$. In the second edition the details of the calculation are supplied, and the ratio obtained is that of 1 to 289: this ratio is that which is now usually given in our elementary books, and it will be convenient to adopt it as we proceed with an account of Newton's investigation.

Suppose two slender canals of homogeneous fluid, one along the polar radius of the earth, and the other along an equatorial radius. The resultant attraction on the equatorial canal must be greater than that on the polar canal in the ratio of 289 to 288 in order that there may be relative equilibrium. For in proceeding along any given radius inside the earth the attraction varies as the distance, and the centrifugal force varies as the distance; hence the ratio of the latter to the former is constant along the equatorial radius; so that the effect of the centrifugal force may be considered equivalent to removing $\frac{1}{289}$ of the force of attraction.

20. Newton's next step is to compare the attraction of an oblate ellipsoid of revolution on a particle at its pole with the attraction of the same body on a particle at its equator, the ellipticity being supposed very small. He states his results without giving his process at full. It will be remembered that he had found an expression for the attraction of an ellipsoid of revolution at any point of its axis: see Art. 13.

I. Suppose an oblate ellipsoid of revolution formed from an ellipse, such that the major semi-axis CA is to the minor semi-axis CQ as 101 is to 100. The reader can easily draw the diagram for himself. Newton says that the attraction at Q would be to the attraction of a sphere having C for centre and CQ for radius, as 126 is to 125. If ϵ denote the ellipticity we know from our modern works that this ratio is that of $1 + \frac{4\epsilon}{5}$ to 1; see *Statics*, Chapter XIII.; this agrees closely with Newton's numerical example.

II. Suppose a prolate ellipsoid of revolution formed from the same ellipse. Newton says that the attraction at A would be to the attraction of a sphere having C for centre and CA for radius, as 125 is to 126. If ϵ denote the ellipticity we know from our modern works that this ratio is that of $1 - \frac{4\epsilon}{5}$ to 1; see *Statics*, Chapter XIII.: this agrees with Newton's numerical example.

In the first edition Newton put the fraction $\frac{2}{15}$ after 126 and 125 in I. and II. The fraction was removed by Cotes: see the *Correspondence of Newton and Cotes*, page 69.

III. Now return to the oblate ellipsoid of revolution. Suppose a particle at A: Newton says that the attraction on it will be a mean proportional between the attractions of the sphere and of the prolate ellipsoid of revolution in II. We will develop his argument. Begin with the sphere having CA for radius; if we change the radius which lies along CQ into CQ we deduce the oblate ellipsoid of revolution; if in this we change the radius which is at right angles to CA and CQ into a radius equal to CQ, we deduce the prolate ellipsoid of revolution. Now each of these changes may be assumed to have affected the attraction to the same amount; and so the attraction of the oblate ellipsoid of revolution is approximately an arithmetical mean between the attractions of the sphere and of the prolate ellipsoid of revolution. Moreover the arithmetical mean between two nearly equal quantities is practically equivalent to the geometrical mean. Hence, finally, the attraction of the sphere with centre C and radius CA is to the attraction of the oblate ellipsoid of revolution on the particle at A as 126 is to $125\frac{1}{2}$.

IV. Thus we have

Attraction of oblate ellipsoid of revolution at the pole

$$=\frac{126}{125}\times \text{attraction of sphere of radius 100 at its surface;}$$

attraction of sphere of radius 101 at its surface

$$=\frac{126}{125\frac{1}{2}}\times \text{attraction of oblate ellipsoid of revolution at its equator;}$$

attraction of sphere of radius 100 at its surface

$$=\frac{100}{101}\times \text{attraction of sphere of radius 101 at its surface.}$$

Hence we find by multiplication that the ratio of the attraction of the oblate spheroid of revolution at the pole to the attraction at the equator is expressed by $\frac{126}{125}\times\frac{126}{125\frac{1}{2}}\times\frac{100}{101}$, that is, by $\frac{501}{500}$ nearly.

21. In future I shall use the single word *oblatum* instead of oblate ellipsoid of revolution, and the single word *oblongum* instead of prolate ellipsoid of revolution.

22. I shall now make some remarks on the statement by Newton which forms the paragraph III. of Art. 20.

If we cut the three solids by two adjacent planes at right angles to AC we obtain slices, two of which are circular and the other elliptical. It is not difficult to believe that in passing from the larger circular slice to the elliptical slice we diminish the attraction by the same amount as we do in passing from the elliptical slice to the smaller circular slice. In fact, the decrement of mass is about the same in the two cases, and the mass lost is at about the same situation with respect to the attracted particle in the two cases.

It is easy to test the statement by the aid of the modern formula; see *Statics*, Chapter XIII.

First take an oblatum of density unity; let r be its greatest radius and $r\sqrt{(1-e^2)}$ its least radius. The attraction at the equator

$$= 2\pi r(1-e^2)\int_0^{\frac{\pi}{2}} \sin^3\theta\,(1-e^2\sin^2\theta)^{-1}\,d\theta$$

$$= \frac{4\pi r(1-e^2)}{3}\left\{1+\frac{4}{5}e^2+\frac{4.6}{5.7}e^4+\ldots\right\}.$$

Next take an oblongum; let r be its greatest radius and $r\sqrt{(1-e^2)}$ its least radius. The attraction at the pole

$$= 4\pi r(1-e^2)\int_0^{\frac{\pi}{2}} \sin\theta\cos^2\theta\,(1-e^2\cos^2\theta)^{-1}\,d\theta$$

$$= \frac{4\pi r(1-e^2)}{3}\left\{1+\frac{3e^2}{5}+\frac{3e^4}{7}+\ldots\right\}.$$

The attraction of a sphere of radius r at its surface $= \dfrac{4\pi r}{3}$.

Suppose e so small that we may reject e^4 and higher powers of e; then the first of these attractions reduces to $\dfrac{4\pi r}{3}\left(1-\dfrac{e^2}{5}\right)$, and the second to $\dfrac{4\pi r}{3}\left(1-\dfrac{2e^2}{5}\right)$, so that the first is an arithmetical mean between the second and the third. But this statement does not hold if we carry our approximations as far as e^4 inclusive; it will be found then that the first of the attractions is rather better represented by the mean proportional between the second and the third than by their arithmetical mean.

It will be convenient to quote Newton's own words, premising that in his diagram PCQ is the polar diameter and ACB an equatorial diameter.

Est autem gravitas in loco A in Terram, media proportionalis inter gravitates in dictam Sphæroidem et Sphæram, propterea quod Sphæra, diminuendo diametrum PQ in ratione 101 ad 100, vertitur in figuram Terræ; et hæc figura diminuendo in eadem ratione diametrum tertiam, quæ diametris duabus AB, PQ perpendicularis est, vertitur in dictam Sphæroidem, et gravitas in A, in casu utroque, diminuitur in eadem ratione quam proximè.

The words *in eadem ratione,* which occur at the end of this extract, seem to have been formerly misunderstood; it was supposed Newton intended to affirm that the attractions of the three bodies were in the same ratio as their volumes. But this is not the case. The volume of the oblatum is $\frac{4\pi r^3}{3}\sqrt{(1-e^2)}$; the volume of the oblongum is $\frac{4\pi r^3}{3}(1-e^2)$; and the volume of the sphere is $\frac{4\pi r^3}{3}$: these volumes are not in the ratio of the attractions exactly nor approximately to the order of e^2. Hence the following passage, which occurs in a note in the Jesuits' edition of the *Principia,* is erroneous: " ...attractiones sphæræ, sphæroidis compressæ, et sphæroidis oblongatæ, sunt respectivè ut quantitates materiæ in illis corporibus contentæ quam proximè."

The words *in eadem ratione,* which occur at the end of the extract from Newton, must be understood to mean only *to the same amount;* and must not be taken in exactly the same sense as in the middle of the passage.

It is obvious, however, that it would have been simpler and more natural to say, that the attraction of the oblatum is an *arithmetical mean* between those of the sphere and the oblongum, than to say that it is a *mean proportional* between them, to the order of accuracy which Newton adopts.

23. Newton now compares the resultant attraction on the fluid in a slender canal having AC for its axis, with that on the fluid in a slender canal having QC for its axis. He finds that

these resultants are in a ratio compounded of the ratios of the lengths and of the ratios of the attractions at the extremities A and Q: see Art. 33. Thus the resultant attraction on the fluid in the canal AC is to that on the fluid in the canal QC, as 101×500 is to 100×501, that is, as 505 is to 501. Hence Newton infers, that if the centrifugal force at any point of AC is to the attraction at that point as 4 is to 505, the weights of the fluids in the two canals will be equal, and the canals in relative equilibrium.

24. The last step in the preceding Article is more obvious to us, who have the modern theory of the equilibrium of fluids, than it would have been before that theory was constructed. We see that in the state of relative equilibrium the pressure at C must be the same in every direction round C: the pressure on a given area at right angles to AC will be measured by the weight of a column of fluid having that area for its section; and similarly for the pressure on an area at right angles to QC.

The canals of fluid which Newton considered were rectilinear, meeting at the centre. Other writers, especially Clairaut, considered canals of various forms, curvilinear as well as rectilinear, meeting at any point of the body. The more simple case to which Newton restricted himself may be conveniently described as that of *central columns;* so that the word *canal* may be in future used in Clairaut's more general sense.

25. It is now necessary to explain carefully the sense in which the words *attraction, gravity* and *weight* will be used in this history.

By the *attraction* of the Earth at any point, I understand that force which the Earth would exert, supposing it did not rotate on its axis. By *gravity* I denote the force which arises from the combination of the attraction and the so-called *centrifugal force;* and *weight* may be considered as an effect produced by gravity as the cause. As we may measure the cause by the effect, it will be found that it is often indifferent whether we use the word *gravity* or the word *weight:* but it is convenient to have both words at our service. The word *weight* is thus left in its ordinary sense as denoting an effect which is actually produced in the existing constitution of things, and actually observed.

It would be convenient if we had a word for the effect which corresponds to *attraction* as a cause; but such a word is not very often required, because practically we are not concerned with an Earth at rest, but with one which rotates. For want of such a word I employ the phrase *resultant attraction* in Arts. 19 and 23.

The distinctions which we have here drawn actually exist, and various modes have been adopted for preventing confusion. Some writers, indeed, leave us to gather from the context the sense in which they use their terms. This is the case with Newton himself. Thus, for instance, in the passage quoted in Art. 22, he uses *gravitas* for what I call *attraction*. In his Proposition XX. he uses *gravitas* for what I call *gravity*. In his Proposition XIX. he uses *pondus* sometimes to express the effect produced by what I call *attraction*, and sometimes in the sense I give to *weight*. To secure accuracy I have used not his words but my own.

Maupertuis used *gravité* for my *attraction*, and *pesanteur* for my *gravity;* and Clairaut followed Maupertuis. See Maupertuis's *Figure de la Terre*, page 158, and Clairaut's *Figure de la Terre*, page xiii. In the *Mécanique Céleste*, Vol. v. page 2, we have the same use of *gravité*.

Maupertuis had previously used *pesanteur réduite* for my *gravity:* see the Paris *Mémoires* for 1734, page 97.

Bouguer used *pesanteur primitive* for my *attraction*, and *pesanteur actuelle* for my *gravity:* see the Paris *Mémoires* for 1734, page 22, and Bouguer's *Figure de la Terre*, page 169.

Maclaurin used *gravitation* as Maupertuis used *pesanteur*, and *gravity* as Maupertuis used *gravité:* so also did Thomas Simpson. See Maclaurin's *Fluxions*, page 551, and Simpson's *Mathematical Dissertations*, page 22. The word *gravitation* has been employed by some eminent modern writers in about the same sense as my *attraction;* as, for example, in Airy's article on *Gravitation* in the *Penny Cyclopædia*, and by Thomson and Tait in their *Natural Philosophy*, Vol. I. page 167.

Boscovich used *gravitas primitiva* for my *attraction*, and *gravitas residua* for my gravity: see his *De Litteraria Expeditione*, page 403. But in the Bologna *Commentarii*, Vol. IV. page 382, he uses *tota gravitas* for the *gravitas residua* of his book.

In the French translation of Boscovich's book, we have *gravité primitive* for his *gravitas primitiva*, and *gravité absolue* for his *gravitas residua*; see the pages 8 and 384 of the translation.

The word *pesanteur* is used by Bailly in his *Histoire de l'Astronomie Moderne*, Vol. III. page 4, as equivalent to my *attraction*; but in general the sense assigned by Maupertuis and Clairaut to *pesanteur* has been adopted by their successors. In French the word *poids* is almost equivalent to my *weight*; see Maupertuis's *Figure de la Terre*, page 155.

26. We now return to Art. 23. The result there obtained is, that if the ellipticity be $\frac{1}{101}$, then for relative equilibrium the centrifugal force at the equator must be $\frac{4}{505}$ of the attraction there. Newton now forms a proportion. He says:

Verum vis centrifuga partis cujusque est ad pondus ejusdem ut 1 et 289, hoc est, vis centrifuga, quæ deberet esse ponderis pars $\frac{4}{505}$, est tantum pars $\frac{1}{289}$. Et propterea dico, secundum Regulam auream, quod si vis centrifuga $\frac{4}{505}$ faciat ut altitudo aquæ in crure $ACca$ superet altitudinem aquæ in crure $QCcq$ parte centesimâ totius altitudinis: vis centrifuga $\frac{1}{289}$ faciet ut excessus altitudinis in crure $ACca$ sit altitudinis in crure altero $QCcq$ pars tantum $\frac{1}{229}$.

These are the numbers of the second and third editions; in the first edition Newton has $\frac{1}{290}$ instead of $\frac{1}{289}$, and $\frac{3}{689}$ instead of $\frac{1}{229}$. In his diagram ca is parallel and adjacent to CA, and cq is parallel and adjacent to CQ.

27. If we put Newton's investigation into a modern form it will stand thus. Let ϵ be the ellipticity, supposed very small; then the attraction of an oblatum at its pole is to the attraction of the oblatum at its equator as $1+\frac{\epsilon}{5}$ is to 1; see *Statics*, Chapter XIII. Hence, as in Art. 23, the ratio of the resultant attraction on the fluid in AC to the resultant attraction on the fluid in QC is expressed by $\frac{1+\epsilon}{1+\frac{\epsilon}{5}}$, that is, by $1+\frac{4\epsilon}{5}$ approximately. Therefore for relative equilibrium we must have the centrifugal force at any point in AC equal to $\frac{4\epsilon}{5}$ of the attraction at that point.

Newton, in fact, sees that the fraction which we have found to be $\frac{4\epsilon}{5}$ must be proportional to ϵ; hence it may be denoted by $k\epsilon$, where k is some constant. Then, when $\epsilon=\frac{1}{101}$, he finds that $k\epsilon=\frac{4}{505}$, so that $k=\frac{404}{505}=\frac{4}{5}$.

28. The result obtained by Newton then is that if the Earth is homogeneous and its shape the same as if it were entirely fluid, the ellipticity must be $\frac{1}{230}$, supposing it to be very small; that is, the ellipticity must be $\frac{5}{4}$ of the ratio of the centrifugal force to the attraction at the equator. The result is very important in the theory of the subject; but we know now that the ellipticity is about $\frac{1}{300}$, and we are confident that the Earth is not homogeneous.

29. Newton proceeds to some remarks on the oblateness of Jupiter.

Let j denote the ratio of the centrifugal force to the attraction at the equator, and ϵ the ellipticity; then we have shewn that $j=\frac{4\epsilon}{5}$.

In his first edition Newton erroneously asserts that j is independent of the density. He says:

Si Planeta vel major sit vel densior, minorve aut rarior quàm Terra, manente tempore periodico revolutionis diurnæ, manebit proportio vis centrifugæ ad gravitatem, et propterea manebit etiam proportio diametri inter polos ad diametrum secundum æquatorem.

Accordingly he considers j to vary inversely as the square of the time of rotation; so that the value of j for Jupiter becomes $\frac{29}{5}$ times its value for the Earth: and hence Jupiter's ellipticity is taken to be $\frac{29}{5}$ times that of the Earth, so that the ratio of the difference of the axes to the minor axis is about $\frac{1}{39\frac{3}{5}}$.

In the second edition Newton corrects his error. He says:

Si Planeta major sit vel minor quàm Terra manente ejus densitate ac tempore periodico revolutionis diurnæ, manebit proportio vis centrifugæ ad gravitatem...

Accordingly he now rightly considers j to vary inversely as the density as well as inversely as the square of the time of rotation; so that, taking the density of Jupiter to be $\frac{1}{5}$ of the density of the Earth, the ratio of the difference of the axes to the minor axis becomes $\frac{29}{5} \times \frac{5}{1} \times \frac{1}{229}$, that is, about $\frac{1}{8}$.

In the third edition, the density of Jupiter is taken as $\frac{94\frac{1}{2}}{400}$ of the density of the Earth; and the ratio of the difference of the axes to the minor axis becomes about $\frac{1}{9\frac{1}{3}}$.

30. In the first and second editions these words occur at the end of Proposition XIX.:

Hæc ita se habent ex Hypothesi quod uniformis sit Planetarum materia. Nam si materia densior sit ad centrum quàm ad circumferentiam, diameter, quæ ab oriente in occidentem ducitur, erit adhuc major.

Thus Newton considered that if the Earth, instead of being of uniform density, were denser towards the centre than towards the surface, the ellipticity would be increased; see also Art. 37. But Newton was wrong. Assuming the original fluidity of the Earth, the ellipticity is diminished by increasing the density of the central part, supposed spherical, and making it solid. This was shewn by Clairaut, who pointed out Newton's error: see Clairaut's *Figure de la Terre*, pages 157, 223, 224, 253...256. Clairaut, however, ought to have remarked that Newton omitted the passage in his third edition.

In his third edition, as I have just said, Newton omitted the above passage. He says instead:

Hoc ita se habet ex hypothesi quod corpus Jovis sit uniformiter densum. At si corpus ejus sit densius versùs planum æquatoris quàm versùs polos, diametri ejus possunt esse ad invicem ut 12 ad 11, vel 13 ad 12, vel forte 14 ad 13.

Then, after stating some observations as to the ratio of the axes of Jupiter, Newton says:

Congruit igitur theoria cum phænomenis. Nam planetæ magis incalescunt ad lucem Solis versùs æquatores suos, et propterea paulo magis ibi decoquuntur quàm versùs polos.

It might then appear that in his third edition Newton had recognised his error; but we shall find, when we discuss Proposition XX., that a distinct trace of the error still remains: see Arts. 38 and 41.

31. I do not feel certain as to the meaning of the sentence "congruit...polos," which I have quoted in the preceding Article. Since heat expands bodies it would appear that the equatorial parts ought by the Sun's action to be rendered *less* dense than the polar parts. The same difficulty presented itself to Boscovich: see page 475 of his *De Litteraria Expeditione*, and pages 89 and 380 of Stay's *Philosophiæ Recentioris*, Vol. II. Clairaut says in his *Figure de la Terre*, pages 223, 224:

De la même manière, on voit combien il était inutile à M. Newton, lorsque sa théorie lui donnait pour Jupiter, une ellipticité moindre que

celle qui résulte des observations, d'aller imaginer que l'équateur de cette planète étant continuellement exposé aux ardeurs du soleil, était plus dense que le reste de la planète. Il n'avait qu'à supposer simplement que le noyau était plus dense que le reste de la planète...

The word *moindre* in this passage is wrong; for Newton's theoretical value of Jupiter's ellipticity in the third edition is *greater* than the value in the observations he quotes.

32. I will now briefly indicate the changes which Newton's Proposition XIX. underwent in the later editions with respect to the facts which it involves.

In the first edition, Newton takes the mean semi-diameter of the Earth to be 19615800 Paris feet, *juxta nuperam Gallorum mensuram:* this alludes to Picard's measurement of the length of an arc of the meridian in France.

In the second edition, Newton refers to the measurements made by Picard, by Norwood, and by Cassini; according to Cassini's measurement, the semi-diameter is 19695539 Paris feet, supposing the Earth spherical. This Cassini is the first of the distinguished family; his name was Jean Dominique Cassini, but he is often called simply Dominique Cassini.

In the third edition, Newton refers also to the measurement made by the son of Dominique Cassini, who is known as Jacques Cassini. Picard had obtained 57060 toises for the length of a degree; the arc measured by D. and J. Cassini gave 57061 toises for the mean length of a degree. We shall see as we proceed with the history that these measurements were subsequently re-examined and corrected.

33. We now come to Newton's Proposition XX., the object of which is to compare the weights of a given body at different places on the Earth's surface.

Newton begins with an important result, which is deduced from his principle of balancing columns: see Art. 23. At any point of the Earth's surface, let f denote the force of gravity, *resolved along the radius;* let r be the distance of this point from the centre; let x be the distance from the centre of any point on the same radius: then the force of gravity at this point *resolved along the radius* will be $\frac{fx}{r}$, because along the same radius within the Earth both the resolved attraction and the resolved centrifugal force vary as the distance. Hence the resolved weight of a column of fluid extending from the surface to the centre, is measured by $f\int_0^r \frac{x}{r}\,dx$, that is, by $\frac{1}{2}fr$. And as the resolved weight of every column must be the same, for relative equilibrium, we must have fr constant, and so f must vary inversely as r.

This is equivalent to an expansion of Newton's brief outline: it becomes more obvious to modern readers by the aid of the theory of the equilibrium of fluids. It is clear that the final result would be true if the resolved force within the Earth varied as *any* direct power of the distance instead of as the *first* power.

Thus we may say that the weight of a given body at any point of the Earth's surface *when resolved along the radius* varies inversely as the radius. Newton, however, omits the words which I have printed in Italics. Since the Earth is very nearly a sphere, the omission will be of no consequence practically, but theoretically it is important to be accurate.

34. Newton proceeds thus:

Unde tale confit Theorema, quod incrementum ponderis, pergendo ab Æquatore ad Polos, sit quam proximè ut Sinus versus latitudinis duplicatæ, vel quod perinde est ut quadratum Sinus recti Latitudinis.

The result here stated may be thus investigated. Let g denote the weight of a given body at any point of the Earth's surface; let r denote the radius at that point, and ϕ the angle which the normal at that point makes with the radius. Then, *assuming that the direction of gravity coincides with the normal*, the resolved part of the weight along the radius will be $g\cos\phi$. Therefore, by Art. 33, we have $g\cos\phi = \frac{\lambda}{r}$, where λ is some constant, so that

$g = \frac{\lambda}{r \cos \phi}$. Let G denote the weight of the given body at the equator, a the radius of the equator; then

$$g - G = \lambda \left(\frac{1}{r \cos \phi} - \frac{1}{a} \right).$$

By neglecting powers of the ellipticity beyond the first, it is found that $\frac{1}{r \cos \phi} - \frac{1}{a}$ varies as the square of the sine of the latitude.

The words printed in Italics in the above investigation involve a principle which is familiar to us in the modern theory of fluids; as we shall see hereafter, this principle was used by Huygens. Newton, however, tacitly assumes that the direction of gravity coincides with the radius. It is true, that to the order of approximation which we adopt $\cos \phi$ may be put equal to unity, and so there is no practical error involved in Newton's assumption: but theoretically his investigation of the very important proposition now before us is thus rendered obscure and imperfect.

35. In Newton's second and third editions, we have after the passage last quoted these words: "Et in eadem circiter ratione augentur arcus graduum Latitudinis in Meridiano." This is a fact in the theory of the Conic Sections with which we are now familiar. Let p denote the perpendicular from the centre of an ellipse on the tangent at any point; then the radius of curvature varies as $\frac{1}{p^3}$. Thus the increase of the radius of curvature in proceeding from the equator towards the pole varies as $\frac{1}{p^3} - \frac{1}{a^3}$; and by neglecting powers of the ellipticity beyond the first it is found that this varies as the square of the sine of the latitude.

36. In the first edition Newton calculated the relative weights of a given body at Paris, Goree, Cayenne, and the Equator. As observations of the length of a seconds pendulum had been made at Paris, Goree, and Cayenne, the relative weights of a body at those places were known, and thus a test of the accuracy of the theory was furnished.

37. The following sentences occur in the first edition; they are repeated substantially in the second edition, but are omitted in the third edition.

Hæc omnia ita se habebunt, ex Hypothesi quod Terra ex uniformi materia constat. Nam si materia ad centrum paulò densior sit quàm ad superficiem, excessus illi erunt paulò majores; propterea quod, si materia ad centrum redundans, qua densitas ibi major redditur, subducatur et seorsim spectetur, gravitas in Terram reliquam uniformiter densam erit reciprocè ut distantia ponderis à centro; in materiam verò redundantem reciprocè ut quadratum distantiæ à materia illa quam proximè. Gravitas igitur sub æquatore minor erit in materiam illam redundantem quàm pro computo superiore, et propterea Terra ibi propter defectum gravitatis paulò altius ascendet quàm in precedentibus definitum est.

The preceding sentences contain a portion of truth. Suppose that a mass of rotatory homogeneous fluid has taken the form which Newton assigns for relative equilibrium. Then gravity at the pole is to gravity at the equator, inversely as the corresponding distances from the centre; and if we suppose the oblatum to become solid, this ratio is not changed. Next suppose that the central part is made denser than the rest, this central part being spherical in shape. Thus the gravity is increased both at the pole and at the equator; but the additional gravity at the pole is to that at the equator inversely as the squares of the corresponding distances. Therefore the whole gravity at the equator bears to the whole gravity at the pole a *less* ratio than for the case of the homogeneous body.

But now Newton in some way returns, as it were, to the supposition of fluidity. It is not obvious whether the whole mass is supposed to be fluid, or whether the central spherical part is still left solid. In either case a new investigation would have to be supplied, in order to determine the figure of the fluid part for relative equilibrium; and no use could be made of a result obtained from the balancing at the centre of *homogeneous* columns. As we have said in Art. 30, the investigations of Clairaut bring out the ellipticity *less* than for the homogeneous case, and not *greater* as Newton stated.

According to Clairaut, Newton's error lay in thinking that gravity at the ends of the columns must be inversely proportional to the lengths of the columns for relative equilibrium, whether the fluid is homogeneous or not. See Clairaut's *Figure de la Terre,* pages 224 and 256, and Stay's *Philosophiæ Recentioris,* Vol. II. page 370.

38. In the first edition, as we have stated, Newton referred to pendulum observations at only three places, Paris, Goree, and Cayenne. These observations indicated a rather greater variation in the length of the seconds pendulum than the theory suggested. So Newton says:

et propterea (si crassis hisce Observationibus satìs confidendum sit) Terra aliquanto altior erit sub æquatore quàm pro superiore calculo, et densior ad centrum quàm in fodinis prope superficiem.

He then points out the advantage which would be derived from a set of experiments for determining the relative weights of a given body at various places on the Earth's surface.

39. In the second edition Newton gave a table of the lengths of a degree of the meridian and of the lengths of the seconds pendulum in different latitudes. This table was computed by the aid of his theory, taking from observation the length of a degree of the meridian in the latitude of Paris, and also the length of the seconds pendulum there. The table is repeated in the third edition; but the lengths of the degrees are not the same as in the second edition. The lengths are expressed in toises; at the equator, at 45°, and at the pole, the lengths are respectively 56909, 57283, and 57657 in the second edition; while in the third edition they are 56637, 57010, and 57382: the difference is, of course, owing to the adoption of a fresh result from the measurement in France. After the table in the second edition we have these words:

Constat autem per hanc Tabulam, quod graduum inæqualitas tam parva sit, ut in rebus Geographicis figura Terræ pro Sphærica haberi possit, quodque inæqualitas diametrorum Terræ facilius et certius per experimenta pendulorum deprehendi possit vel etiam per Eclipses Lunæ, quam per arcus Geographice mensuratos in Meridiano.

I cannot understand how the ratio of the axes could be found by pendulum experiments or by eclipses better than by measured arcs. In the third edition the words which follow "haberi possit" are omitted, and instead of them we have "præsertim si Terra paulò densior sit versùs planum æquatoris quàm versùs polos."

40. In the second and third editions Newton referred to many more pendulum observations than in the first edition. We see from pages 69...89 of the *Correspondence of Newton and Cotes,* that the arrangement of this part of the work for the second edition was a matter of some trouble. The figures in the final draft of Newton were corrected by Cotes: compare pages 85 and 92 of the *Correspondence* with the second edition of the *Principia.* The facts are stated nearly in the same terms in the second and third editions.

41. The more numerous observations to which Newton could now appeal, concurred with the smaller number before used, in giving a greater variation to the length of the seconds pendulum than theory suggested. Accordingly, the sentence which we quoted in Art. 38, appears in the second and third editions, omitting the words *si crassis...sit.* Newton adds, however, "nisi forte calores in Zona torrida longitudinem Pendulorum aliquantulum auxerint."

The supposition that the pendulum observations required a greater ellipticity than $\frac{1}{230}$ was shewn to be untenable by Clairaut: see his *Figure de la Terre,* page 252.

42. In the second edition Newton seems to come to the conclusion that we should correct the theory by observation, and thus take $31\frac{7}{12}$ miles as the excess of the equatorial semi-diameter over the polar semi-diameter. In the third edition, however, he seems to consider that we may hold to the value, 17 miles, furnished by theory.

43. In the second edition Proposition XX. ends with a paragraph in which Newton adverts to the hypothesis, founded on

some measurements by Cassini, that the Earth is an oblongum: Newton deduces results from this hypothesis which are contrary to observations. The paragraph does not appear in the third edition, although the oblong form continued to find advocates for some years after the death of Newton.

44. Newton's investigations in the theories of Attraction and of the Figure of the Earth may justly be considered worthy of his great name. The propositions on Attraction are numerous, exact, and beautiful; they reveal his ample mathematical power. The treatment of the Figure of the Earth is, however, still more striking; inasmuch as the successful solution of a difficult problem in natural philosophy is much rarer than profound researches in abstract mathematics. Newton's solution was not perfect; but it was a bold outline, in the main correct, which succeeding investigators have filled up but have not cancelled. Newton did not demonstrate that an oblatum is a possible form of relative equilibrium; but, assuming it to be such, he calculated the ratio of the axes. This assumption may be called Newton's *postulate* with respect to the Figure of the Earth: the defect thus existing in his process was supplied about fifty years later by Stirling and Clairaut. The difficulty arose from the imperfect state of the theory of fluid equilibrium, which undoubtedly must have produced many obstacles for the earliest investigators in mixed mathematics. Clairaut subsequently gave methods which are sound and satisfactory to a reader who can translate them into modern language; but even these may have appeared obscure to Clairaut's contemporaries. Euler, in the Berlin *Mémoires* for 1755, first rendered Hydrostatics easily intelligible by introducing a symbol p to measure the pressure at any point of a fluid.

45. Besides the defect in Newton's theory which we have pointed out, Laplace finds another, saying in the *Mécanique Céleste*, Vol. v. page 5, "Il suppose encore, sans démonstration, que la pesanteur à la surface, augmente de l'équateur aux pôles, comme le carré du sinus de la latitude." But Laplace is not right. Newton did not absolutely assume the proposition; he gave a demonstration though it was imperfect; see Art. 34.

Laplace's language is inaccurate moreover; it is *not gravity* that increases as the square of the sine of the latitude, but the *variation in gravity.* Laplace proceeds to observe that Newton regarded the Earth as homogeneous, while observations prove incontestably that the densities of the strata increase from the surface to the centre. Laplace's language could scarcely be stronger if borings had actually been executed from the surface to the centre, and had thus rendered the strata open to inspection. He means, of course, that by combining observations made at various places on the surface of the Earth with the suggestions of theory, we are led to infer that the density increases from the surface to the centre. See the *Mécanique Céleste,* Vol. v. page 12.

Laplace truly says that, notwithstanding the imperfections, the first step thus made by Newton in the theory must appear immense.

46. A notice of the *Principia* was given in the *Philosophical Transactions,* Vol. XVI. 1687, I presume by Halley, who was then Secretary of the Society; the simple but expressive words we find on page 297 are still as applicable as they were then:

> and it may be justly said, that so many and so Valuable Philosophical Truths, as are herein discovered and put past dispute, were never yet owing to the Capacity and Industry of any one Man.

CHAPTER II.

HUYGENS.

47. We have now to examine an essay by Huygens entitled *Discours de la Cause de la Pesanteur*.

A small quarto volume was published at Leyden, in 1690, entitled *Traité de la Lumiere...Par C.H.D.Z. Avec un Discours de la Cause de la Pesanteur*. The letters C.H.D.Z. stand for Christian Huygens de Zulichem.

The volume consists of two parts. Pages 1...124 relate to Light; they are preceded by a Preface, and a Table of Contents, on 6 pages, which belong to this part of the volume. After page 124 is a Title-leaf for the part relating to Weight; then a preface on pages 125...128; then a leaf containing a Table of Contents; and then the text on pages 129...180.

48. We of course pass over the part relating to Light, merely remarking that it is memorable as laying the foundation of the Undulatory Theory.

The part relating to Weight is said to appear in a Latin version in the *Opera Reliqua* of Huygens: hence it is sometimes cited by a Latin title, *De causa gravitatis*, or *De vi gravitatis*. My references will all be made to the original edition in French, published during the author's lifetime.

49. The last paragraph of the Preface gives information as to the date of composition:

La plus grande partie de ce Discours a esté écrite du temps que je demeurois à Paris, et elle est dans les Registres de l'Academie Royale des Sciences, jusques à l'endroit où il est parlé de l'alteration des Pendules par le mouvement de la Terre. Le reste a esté adjouté plusieurs années apres : et en suite encore l'Addition, à l'occasion qu'on y trouvera indiquée au commencement.

The former part of the Discourse, which we are here told had been written many years since, is of no value.

50. The theory of Huygens to account for Weight is expounded on pages 129...144 of the work; we may say briefly that this theory is utterly worthless. Huygens assumes the existence of a very rare medium moving about the Earth with great velocity, not always in the same direction. This rare matter is surrounded by other bodies, and so prevented from escaping; and it pushes towards the Earth any bodies which it meets. This vortex has passed away, as well as those similar but more famous delusions with which the name of Des Cartes is connected.

51. Two incidental matters of some interest may be noticed.

On his page 138, Huygens says that there is an invisible ponderable matter present even in the space from which air has been exhausted: so that it would appear he took the partial exhaustion produced by an air-pump for complete exhaustion.

On his page 141, he starts a difficulty which we now know has been removed by experiments:

...De plus, en portant un corps pesant au fond d'un puits, ou dans quelque carriere ou mine profonde, il y devroit perdre beaucoup de sa pesanteur. Mais on n'a pas trouvé, que je sache, par experience qu'il en perde quoy que ce soit.

52. The really valuable part of the Discourse commences on page 145. Huygens says that at Cayenne the seconds pendulum had been found to be shorter than at Paris. As soon as he heard this, he attributed it to the rotation of the Earth. Accordingly he gives a very good explanation; assuming that there is at the surface of the Earth a force of constant magnitude directed towards the centre, and that there is also a centrifugal force. He shews by calculation, that the centrifugal force at the equator

is about $\frac{1}{289}$ of the central force. He calculates that the seconds pendulum at Cayenne should be $\frac{5}{6}$ of a line shorter than at Paris; Richer made it $1\frac{1}{4}$ lines shorter by observation.

Huygens calculates that the plumb-line at Paris deviates nearly 6 minutes from the position it would take if there were no centrifugal force.

53. On his page 152, Huygens states a principle which has since generally been called by his name; he says the surface of the sea is such that at every point the *direction of the plumb-line is perpendicular to the surface.* The principle may be stated more generally thus: the direction of the resultant force at any point of the free surface of a fluid in equilibrium must be normal to the surface at that point.

54. On page 152, we arrive at the *Addition* to which Huygens referred in his preface: see Art. 49. His attention was turned to the subject again by examining an account of some more pendulum experiments, and by reading Newton's *Principia.* Huygens first calculates the ratio of the axes of the Earth. He adopts Newton's principle of the balancing of the polar and equatorial columns; but retains his own hypothesis, that the attractive force is central and constant at all distances. Thus he makes the ratio of the axes to be that of 577 to 578.

55. Huygens next finds the equation to the generating curve of the Earth's surface. He considers it difficult to use his own principle of the plumb-line, stated in Art. 53; and so he uses the principle of balancing columns. He extends this principle beyond the application which Newton made of it: see Art. 24. Huygens contemplates canals of various forms, not necessarily passing through the centre. He says on his page 156: "et mesme, cela doit arriver de quelque maniere qu'on conçoive que le canal soit fait, pourvû qu'il aboutisse de part et d'autre à la surface."

Let the constant force be denoted by λ, the angular velocity by ω, and the equatorial radius by a; take the axis of x to coincide with the polar diameter, and the axis of y with an equatorial diameter. Then, by modern methods, we find for the equation to the curve which by revolution generates the surface of the Earth

$$\lambda \sqrt{(x^2+y^2)} - \frac{\omega^2 y^2}{2} = \lambda a - \frac{\omega^2 a^2}{2} \ldots\ldots\ldots\ldots\ldots\ldots(1).$$

This coincides with Huygens's result.

We may deduce the ratio of the axes from (1); we shall thus get the same value as Huygens obtained before he investigated the equation to the curve.

Put $y=0$ in (1), thus: $x = a\left(1 - \frac{\omega^2 a}{2\lambda}\right)$; therefore the ratio of the axes is that of $1 - \frac{1}{2\times 289}$ to 1, that is, the ratio of 577 to 578.

If ϵ and j have the same meaning as in Art. 29, we see that Huygens's result may be expressed thus: $\epsilon = \frac{j}{2}$.

56. Huygens says on his page 159, that even if we do not suppose the central force to be constant, his result remains almost unchanged. It is important to demonstrate this: and we shall accordingly shew that the result is approximately true, whatever may be the law of the force, which is assumed to be central.

Let $\phi(r)$ denote the force at the distance r from the centre. Then, by modern methods, we find for the equation to the generating curve

$$\int \phi(r)\, dr - \frac{\omega^2 y^2}{2} = \text{constant}.$$

Let a denote the equatorial radius, and b the polar. By putting $y=0$, we determine the value of the constant, and the equation becomes

$$\int_0^r \phi(r)\, dr - \frac{\omega^2 y^2}{2} = \int_0^b \phi(r)\, dr.$$

Now put $y=a$; thus

$$\int_0^a \phi(r)\, dr - \frac{\omega^2 a^2}{2} = \int_0^b \phi(r)\, dr:$$

this is an analytical expression of Newton's principle of the balancing of central columns. We may put the expression in the form

$$\int_b^a \phi(r)\, dr = \frac{\omega^2 a^2}{2}.$$

If $a - b$ is very small this gives approximately

$$(a-b)\,\phi(a) = \frac{\omega^2 a^2}{2},$$

thus
$$\frac{a-b}{a} = \frac{1}{2}\frac{a\omega^2}{\phi(a)},$$

that is
$$\epsilon = \frac{j}{2}.$$

Moreover, we can shew that the diminution of the radius in passing from the equator to the pole will vary approximately as the square of the sine of the latitude. For we have

$$\int_0^r \phi(r)\, dr - \frac{\omega^2 y^2}{2} = \int_0^b \phi(r)\, dr.$$

that is
$$\int_0^r \phi(r)\, dr - \frac{\omega^2 y^2}{2} = \int_0^a \phi(r)\, dr - \frac{\omega^2 a^2}{2},$$

therefore
$$\int_r^a \phi(r)\, dr = \frac{\omega^2}{2}(a^2 - y^2).$$

Hence if $a - r$ be small, we have approximately

$$(a-r)\,\phi(a) = \frac{\omega^2}{2}(a^2 - y^2).$$

Thus $a - r$ varies as $1 - \frac{y^2}{a^2}$, that is, approximately as the square of the sine of the latitude.

57. The particular case in which the central force varies inversely as the square of the distance deserves to be noticed

specially. In this case instead of equation (1) of Art. 55 we obtain

$$\frac{\mu}{\sqrt{(x^2+y^2)}}+\frac{\omega^2 y^2}{2}=\frac{\mu}{a}+\frac{\omega^2 a^2}{2} \text{ .... (2)},$$

where $\frac{\mu}{r^2}$ represents the central force.

Put $y=0$, then $x=\dfrac{a}{1+\dfrac{\omega^2 a^3}{2\mu}}$.

Thus the ratio of the axes is that of 1 to $1+\dfrac{\omega^2 a^3}{2\mu}$, and, taking $\dfrac{1}{289}$ for $a\omega^2 \div \dfrac{\mu}{a^2}$, this ratio becomes that of 578 to 579, which is almost identical with that obtained in Art. 54.

58. If $\omega^2 a=\lambda$, the equation (1) of Art. 55 is equivalent to $y^2-a^2=\pm\, 2ax$, giving two parabolas, as Huygens observes. He seems in consequence to accept without hesitation, for relative equilibrium, a figure of revolution in which the two parts meet so as to produce an abrupt change of direction: see his page 157.

59. In his pages 159...168 Huygens makes some interesting remarks on various points in Newton's *Principia*. Huygens does not admit that all particles of matter attract each other, but he does admit a resultant force exerted by the Sun or by a Planet, and varying inversely as the square of the distance from the centre of the body. He states that he himself had not extended the action of *pesanteur* so far as from the Sun to the Planets, nor had he thought of the law of the inverse square: he fully recognises Newton's merits as to these points.

60. We must notice the value he obtained for the increase of gravity in proceeding from the equator to the pole. He adopts the ratio of the axes which he had found for the case of a constant force, and assumes that it will hold when the central force varies inversely as the square of the distance. Hence since the polar radius is $\dfrac{1}{578}$ part shorter than the equatorial radius, gravity

increases by $\frac{1}{289}$ part in passing from the equator to the pole. And by reason of the absence of centrifugal force at the pole there is another increase of $\frac{1}{289}$ part. Thus on the whole there is an increase of $\frac{2}{289}$. He thinks that observation does not confirm this large increase.

We know now that the increase is not so large as Huygens made it. His error arises from his assuming that the Earth's attraction is a single central force varying inversely as the square of the distance from the centre, instead of calculating the value of it from the form of the Earth.

61. Huygens expresses himself as much pleased with Newton's method of comparing the attraction at the surfaces of the Earth, the Sun, Jupiter and Mars: see his page 167. Huygens alludes to the very different estimates which had been made of the Sun's distance from the Earth: Newton took this to be 5000 times the Earth's diameter, Cassini to be 10000 times; Huygens himself had taken it to be 12000 times.

62. Huygens also refers with pleasure to the researches of Newton respecting the motion of projectiles in a resisting medium. Huygens says he had himself formerly investigated this subject, assuming the resistance to vary as the velocity; after he had finished his investigations he learned from the experiments made by the Academy of Sciences at Paris that the resistance in air and in water varied as the square of the velocity.

He here gives the results of his original investigations without the demonstrations: see his pages 170...172. It will furnish a good exercise for students to verify these results, which must have been obtained with some difficulty in the early days of the Integral Calculus. The results will be found to be all correct, except that on the middle of page 171 we ought to read *terminal velocity* instead of *velocity with which the ground is reached.* The phrase *terminal velocity* is due to Huygens; see his page 170. Huygens makes a few remarks on motion in a medium where the resistance varies as the square of the velocity; but he considers only a particular case of vertical motion, and a particular case of oblique motion. The general problem, he truly says, is very difficult if not impossible.

63. Huygens finishes with a statement of properties of the exponential or logarithmic curve; he does not give demonstrations, but they can be easily supplied.

64. On the whole we may say that the chief contribution of Huygens to our subject is the important principle of fluid equilibrium, which we have noticed in Art. 53. He also first solved a problem in which the form of the surface of a fluid in relative equilibrium under a given force was accurately determined; see Art. 55. The result has become permanently connected with our history for a reason which we will now explain.

The assumption that the attraction of the Earth varies inversely as the square of the distance from a fixed point is equivalent to the hypothesis that the density of the Earth is infinite towards the centre. This remark is in fact due to Clairaut; see the *Philosophical Transactions*, Vol. XL. page 297. It is sometimes ascribed to Huygens himself; as in Barlow's *Mathematical Dictionary*, article *Earth*. But, as we have seen, Huygens preferred to consider the attractive force as *constant;* and this is very different from the notion involved in Clairaut's remark. Laplace is not quite accurate in the *Mécanique Céleste*, Vol. V. page 5, where he omits all notice of the constant force, and says that Huygens supposed the force to vary inversely as the square of the distance from a point.

65. An important error has been sometimes made by representing the researches of Huygens on the Figure of the Earth as preceding those of Newton in the order of time: for example, this is asserted in Barlow's article just cited. Svanberg also has completely misrepresented the relative positions of Newton and Huygens: see his *Exposition des opérations faites en Lapponie*...pages iii...v. The truth is that before the *Addition* to Huygens's *Discourse* the only remark on the subject is the suggestion on page 152, that the Figure of the Earth is that of a sphere flattened at the poles;

and even this occurs in the part which treats on pendulums, written, as Huygens himself states, long after the greater part of the *Discourse.* The researches on the Figure of the Earth are really contained in the *Addition*, which as Huygens himself states was written after reading the *Principia.*

There are two causes which might have led to this error in dates. In the first place, as Huygens was senior to Newton, it was natural in histories of science to give an account of the life and works of Huygens before those of Newton; this, for example, is the course adopted by Bailly in his *Histoire de l'Astronomie Moderne.* Then a hasty glance at his Vol. III. page 9 might mislead an incautious reader. In the second place, it was natural to notice the partial and imperfect attempts of Huygens before proceeding to Newton's nearly complete solution; this, for example, is the course adopted by Clairaut in the *Introduction* to his *Figure de la Terre.*

CHAPTER III.

MISCELLANEOUS INVESTIGATIONS UP TO THE YEAR 1720.

66. THE present Chapter will contain an account of various miscellaneous investigations up to the year 1720.

It is my design to write the history of the *Theories* of Attraction and of the Figure of the Earth; and I have endeavoured to include all the memoirs and works which relate to these subjects. I do not profess to discuss the measurements of arcs and the observations of pendulums; but I shall briefly notice the more important of these operations in their proper places.

67. There are many writers to whom the student may be referred for accounts of the attempts made in ancient times, and in the early days of modern science, to ascertain the figure and dimensions of the Earth. Thus, for example, we may mention Cassini's *De la Grandeur et de la Figure de la Terre*, and Stay's *Philosophiæ Recentioris* Vol. II. with the notes by Boscovich. More recent works are the article by Professor Airy on the *Figure of the Earth* in the *Encyclopædia Metropolitana*, and the article by the late T. Galloway on *Trigonometrical Survey* in the *Penny Cyclopædia*.

68. Some interest attaches to the operations of Richard Norwood, which he has recorded in his *Seaman's Practice*, published in 1637. He says on his page 4:

Upon the 11*th* of *June*, 1635, I made an Observation near the middle of the City of *York*, of the Meridian Altitude of the Sun, by an Arch of a *Sextant* of more than 5 Foot Semidiameter, and found the apparent Altitude of the Sun that Day at Noon to be 59 deg. 33 min.

I had also formerly upon the 11*th* of *June*, *Anno* 1633, observed in the City of *London*, near the Tower, the apparent Meridian Altitude of the Sun, and found the same to be 62 deg. 1 min.

And seeing the Sun's Declination upon the 11*th* day of *June*, 1635, and upon the 11*th* day of *June*, 1633, was one and the same, without any sensible difference; and because these Altitudes differ but little, we shall not need to make any alteration or allowance, in respect of Declination, Refraction, or Parallax: Wherefore subtracting the lesser apparent Altitude, namely 59 deg. 33 min. from the greater 62 deg. 1 min. there remains 2 deg. 28 min. which is the difference of Latitude of these two Cities, namely, of *London* and *York*.

It will be seen that Norwood does not expressly say with what instrument he observed the Sun's altitude at London; he lays stress on the fact that the observations at London and at York were made on the same day of the month. He determined the distance between York and London in the manner which he explains on his page 6:

Yet having made Observation at *York*, as aforesaid, I measured (for the most part) the Way from thence to *London*; and where I measured not, I paced, (wherein through Custom I usually come very near the Truth) observing all the way as I came with a Circumferentor all the principal Angels of Position, or Windings of the Way, (with convenient allowance for other lesser Windings, Ascents and Descents)......; so that I may affirm the Experiment to be near the Truth.

Norwood made the distance between York and London 9149 chains, each of 99 feet. He deduced for the length of a degree of the meridian 367196 feet. This is nearer to the truth than might have been expected from the rough mode of measurement: the modern result would be somewhat less than 365000 feet.

It has been supposed that Norwood's work had been forgotten before Newton's time; but Rigaud is strongly against this supposition: see his *Historical Essay*...page 4. Newton does not refer in his first edition to Norwood's value of a degree; but he does in the second edition. Newton quotes the 367196 feet, which he says is 57300 Paris toises. The number of toises obtained will, of course, depend on the proportion of the English foot to the French foot. Cassini made the English foot to be $\frac{15}{16}$ of the French foot; see the *De la Grandeur et de la Figure de la Terre*, pages 154, 251, and 282: this would give 57374 toises. Bailly in his *Histoire de l'Astronomie Moderne*, Vol. II. page 342, gives 57442 toises, and draws attention in a note to Newton's smaller value. The comparison of English and French standards of length has, of course, been carried to minute accuracy in modern times. See, for example, Airy, *Figure of the Earth*, page 217.

69. Richer made observations of the length of the seconds pendulum at Cayenne in 1672: Varin, Des Hayes, and Du Glos made similar observations at Goree and at Guadaloupe in 1682. These observations are given in the *Recueil d'Observations faites en Plusieurs Voyages*....Folio, Paris, 1693. Newton states the results in the third edition of the *Principia*, omitting the name of Du Glos. It would seem from Newton's words that the same length was obtained at Martinique as at Guadaloupe: but the original account does not mention pendulum observations at Martinique.

These observations had, however, been published before 1693; see Lalande's *Bibliographie Astronomique*, page 327: thus they were accessible to Newton for his first edition, as we have mentioned in Art. 36.

Richer's observations are also given in Vol. VII. of the ancient *Mémoires* of the Paris Academy.

70. In Number 112 of the *Philosophical Transactions*, which is dated March 25, 1675, there is an account of Picard's survey of an arc of the meridian; the Number forms part of Volume X. of the *Transactions;* the account occupies pages 261...272 of the volume; it begins thus:

A Breviate of Monsieur Picarts *Account of the Measure of the Earth.*

This Account hath been printed about two years since, in *French;* but very few Copies of it being come abroad, (for what reasons is hard to divine;) it will be no wonder, that all this while we have been silent of it. Having at length met with an Extract thereof, and been often desired to impart it to the Curious; we shall no longer resist those desires, but faithfully communicate in this Tract what we have received upon this Argument from a good hand.

The account notices an attempt made by the Arabians to measure an arc of the meridian:

a Station being chosen, and thence Troups of Horsemen let out, that went in a straight line, till one of them had raised a degree of *Latitude*, and the other had deprest it; at the end of both their marches, they who raised it, counted $56\frac{2}{3}$ miles, and they who deprest it, reckon'd 56 miles just.

This is not quite faithful to a description given by Picard, from which it may have been derived, which can be seen in Bailly's *Histoire de l'Astronomie Moderne*, Vol. I. page 581. Picard does not mention *Horsemen;* and he does not explicitly say which of the two parties obtained the longer measure.

71. In Number 126 of the *Philosophical Transactions*, which is dated June 20, 1676, and forms part of Volume XI., we have a notice of what Norwood effected. The following is the beginning of the notice:

Advertisement concerning the Quantity of a Degree *of a Great Circle, in* English *measures.*

Some while since an account was given concerning the *Quantity of a Degree of a great Circle*, according to the tenour of a printed *French* Discourse, entituled *De la Mesure de la Terre.* The Publisher not then knowing what had been done of that nature here in *England*, but having been since directed to the perusal of a Book, composed and published by that known Mathematician *Richard Norwood* in the year 1636, entituled *The Seaman's Practice*, wherein, among other particulars, the compass of the *Terraqueous Globe*, and the *Quantity of a Degree* in *English* measures are deliver d, approaching very near to that, which hath been lately observ'd in *France;* he thought, it would much conduce to mutual confirmation, in a summary Narrative to take publick notice here of the method used by the said *English* Mathematician, and of the result of the same; which, in short, is as follows:

The "Publisher" here means H. Oldenburg who was Secretary to the Royal Society.

An English translation of Picard's account of his survey of an arc of the meridian was published in 1687. The bulk of the volume in which it was included seems to have consisted of a translation of Memoirs on the Natural History of Animals. The Natural History was translated by Alexander Pitfield, and Picard's account by Richard Waller. See *Philosophical Transactions*, Number 189, page 371.

72. *A Discourse concerning Gravity, and its Properties, wherein the Descent of Heavy Bodies, and the Motion of Projects is briefly, but fully handled. Together with the Solution of a Problem of great Use in Gunnery. By E. Halley.*

This memoir is published in Number 179 of the *Philosophical Transactions;* the Number is for January and February, 1686, and forms part of Volume XVI.: the memoir occupies pages 3...21 of the number.

I notice this memoir for the sake of a fact to which Newton refers in the second edition of the *Principia*, Book III. Prop. XX. Halley says:

......'Tis true at S. *Helena* in the *Latitude* of 16 Degrees *South*, I found that the *Pendulum* of my Clock which vibrated *seconds*, needed to be made shorter than it had been in *England* by a very sensible space, (but which at that time I neglected to observe accurately) before it would keep time; and since the like Observations has been made by the *French Observers* near the *Equinoctial:* Yet I dare not affirm that in mine it proceeded from any other Cause, than the great height of my place of Observation above the *Surface* of the *Sea*, whereby the *Gravity* being diminished, the length of the *Pendulum* vibrating *seconds*, is proportionably shortned.

The "Problem of great use in Gunnery," which Halley solves, is one which we now enunciate thus: To determine the direction in which a body must be projected from a given point with a given velocity, so as to hit a given point. Halley considers his solution superior to those which had been previously given; he says the problem was "first Solved by Mr *Anderson*, in his Book of the Genuine use and effects of the *Gunn*, Printed in the year 1674."

Halley observes, that for a given horizontal range the velocity is least when the angle of projection is 45°. He says:

This Rule may be of good use to all *Bombardiers* and *Gunners*, not only that they may use no more Powder than is necessary, to cast their *Bombs* into the place assigned, but that they may shoot with much more certainty, for that a small Error committed in the *Elevation* of the *Piece*, will produce no sensible difference in the fall of the Shot: For which Reasons the *French* Engineers in their late Sieges have used Morter-pieces inclined constantly to the *Elevation* of 45, proportioning their Charge of Powder according to the distance of the *Object* they intend to strike on the *Horizon*.

According to theory the horizontal ranges should be equal for two different angles of projection, one as much below 45° as the other is above 45°; and Halley states that experiments shew there is little difference in the ranges, especially for large shot: see his page 20.

73. Thomas Burnet, master of the Charter-house, published towards the end of the seventeenth century his *Sacred Theory of the Earth*, first in Latin and afterwards in English. The work related to geology and the Mosaic cosmogony, and naturally gave rise to much controversy. I shall, however, not attempt to follow the details of this controversy, as it is but slightly connected with our subject; but content myself with noticing the contributions of one writer, Keill, whose name is not unknown in the history of mathematical science.

The work of Keill now to be considered is entitled *An Examination of Dr Burnet's Theory of the Earth. Together with some remarks on Mr. Whiston's New Theory of the Earth.* By Jo. Keill, A.M. Coll. Ball. Ox. 1698. The book contains 224 pages in octavo, besides the title-page and the dedication "to the Reverend Dr Mander, the worthy master of Balliol College in Oxford."

74. The part of the work which most concerns us is Chapter VI., *Of the Figure of the Earth*, which occupies pages 101...143.

Burnet maintained that the Earth was not oblate but oblong. Keill says on his page 107:

I come now to examin the Theorists reasons by which he proves the Earth to be of an Oblong Spheroidical figure. He tells us that the fluid under the æquator being much more agitated than that which is towards the Poles which describes in its diurnal motions lesser arches, and because it cannot get quite off and fly away by reason of the Air which every way presses upon it, it could no other wayes free it self than by flowing towards the sides, and consequently form the Earth into an Oval figure.

Keill maintains, on the contrary, the oblateness of the Earth; he gives substantially the two investigations of the ratios of the axes which were then known, namely, that of Huygens, which assumed the resultant attraction to be constant, and that of Newton, which assumed the attraction between particles to vary inversely as the square of the distance. Keill also gives, after Huygens, a very clear account of the effect of centrifugal force on the position of a pendulum, and on the weight of a body. Keill does not refer to the work of Huygens, from which he must have obtained a large part of his Chapter VI., namely, the *Discours...de la Pesanteur;* but other works by Huygens are cited.

75. There is nothing new on our subject in Keill's work; he merely reproduces what had been given by Newton and by Huygens. There are, however, some incidental mistakes which we should scarcely have expected from a distinguished member of a distinguished college.

On his page 41 he says, "by calculation it will follow that a body would run down four thousand miles in the space of twenty-three seconds, abstracting from the resistance of the air." He must mean twenty-three *minutes.*

On his page 150 he says, "for the ninty ninth power of 2 is a number which if written at length would consist of a hundred Figures." We know that 2^{99} consists of 30 figures.

On his page 156 he has an angle of which the tangent is $\frac{1}{50}$; he makes the sine 19594, which is ten times too great: by correcting the error his own argument is much strengthened.

On his pages 160 and 161 he has some calculations, which he begins by stating that a perch is 10 feet, and which he continues on the supposition that a perch is 20 feet.

76. Keill's most serious mistake is one which it is very natural to make; but, unfortunately, he is extremely incautious in drawing attention to it. He says on his pages 138 and 139:

Now tho' I have already determined the Earths Figure from other Principles; Yet to comply with the Theorist in this point, I will give him an account of a Book whose extract I have seen in the *Acta Eruditorum Lipsiæ publicata* for the year 1691. written by one *Joh. Casp Eisenschmidt, a German* who calls himself Doctor of Philosophy and Physick. The Title of the Book is, *Diatribe de Figura Telluris Elliptico-Sphæroide.* And it is Printed at *Strasburg* in the year 1691....

Keill then proceeds to give some account of the book. According to Eisenschmidt, the measurements hitherto made of the length of a degree of the meridian in various latitudes shewed that the length decreased as the latitude increased; granting this to be the case, Eisenschmidt inferred quite correctly that the Earth was of an oblong form. But Keill says on his pages 140 and 141:

None but a man of prodigious stupidity and carelessness could reason at this rate! If he had asserted that the Earth was of an Oval Figure because Grass grows or Houses stand upon it, it had been something excusable; for that Argument tho it did not infer the conclusion, yet it could never have proved the contradictory to be true. But to bring an Argument which does evidently prove that the Earth has a Figure directly contrary to that which he would prove it has, is an intolerable and an unpardonable blunder...

Keill's error consisted of course in misunderstanding what was meant by a degree of the meridian. Keill supposed that the difference of latitude of two places on the same meridian is the angle between straight lines drawn from these points to the centre of the Earth; whereas in this subject, the difference of latitude means the angle between the vertical directions at the two places.

77. Keill's Chapter IV. is entitled, *Of the Perpendicular position of the Axis of the Earth to the plane of the Ecliptick.* This Chapter contains some interesting matter; though it is not connected with our subject.

Burnet held that in the primitive Earth, the axis of the Earth's rotation was perpendicular to the plane of the ecliptic. Keill is thus led to consider the advantages resulting from the inclined position which we know the axis actually has. He infers by calculation, that places whose latitude exceeds 45° receive more heat from the Sun than they would do if the axis of the Earth's rotation were perpendicular to the ecliptic; while other places receive less heat. Keill derives his method, and some of his results, from a paper by Halley in the *Philosophical Transactions*, Number 203.

On Keill's page 75, the first and second entries with respect to the Sun in Cancer ought to change places; Halley is correct.

78. Keill on his page 70 charges Dr. Bentley with error for saying that "tho the *axis* had been perpendicular, yet take the whole year about we should have had the same measure of heat we have now." But it is obvious that Bentley is right in a certain sense; namely, that the whole heat received by the Earth is the same in the two cases. I am sorry to see that Keill goes on to shew that Bentley is to be numbered among the advocates of an error which has at all times been popular; according to Keill, page 70,

...in the same Lecture, he confidently saies, that *'tis matter of fact and experience that the Moon alwaies shews the same Face to us, not once wheeling about her own Centre,* whereas 'tis evident to any one who thinks, that the Moon shews the same face to us for this very reason, because she does turn once, in the time of her period, about her own Centre.

The Lecture to which Keill alludes is the "last Lecture for the *Confutation of Atheism.*"

79. Keill published another work on the same subject as the former; it is entitled *An Examination of the Reflections on the Theory of the Earth. Together with a Defence of the Remarks on Mr Whiston's New Theory.* The book contains 208 pages in

octavo, besides the title-page. It furnishes nothing connected with our subject except another reference to Dr Eisenschmidt. Keill seems determined to remain unconvinced of his error; he says on his page 100:

Our *Defender* tells us, that Dr. *Eisenschmidt* supposes the Vertical Lines or Lines of Gravity, to be drawn at right Angles to the Tangent of each respective Horizon. What Dr. *Eisenschmidt* does really suppose I know not, but I am sure he cannot suppose a thing more absurd than what our Author makes him suppose in this place. For that the Line of direction of heavy Bodies is at right Angles with the Tangent of the Horizon, is to me such an incomprehensible supposition, that I shall excuse my self from considering of it, till the *Defender* (who I suppose would have us think he understands it) is at leisure to explain it.

Keill was subsequently appointed Savilian Professor of Astronomy at Oxford: let us hope that before that time he understood this simple matter which had perplexed him.

80. In the Paris *Mémoires* for 1700, published in 1703, we have two articles bearing on our subject: both occur in the historical portion of the volume.

On pages 114...116, there is a notice of some observations of the length of the seconds pendulum made by Couplet in 1697 at Lisbon, and in 1698 at Parayba in Brazil.

On pages 120...124, there is a brief account of the operations up to the current date connected with the French arc of the meridian.

81. In the Paris *Mémoires* for 1701, published in 1704, we have on page 111 of the historical portion of the volume some pendulum observations made by Des Hayes in 1699 and 1700: Newton states the results in the third edition of the *Principia*.

In the same volume, there is a memoir entitled *De la Meridienne de l'Observatoire Royal prolongée jusqu'aux Pyrenées. Par M. Cassini.* The memoir occupies pages 169...182 of the volume.

After noticing what had been done by the ancients as to the measurement of the Earth, the memoir gives an account of the operations in France. The substance of the memoir is reproduced in the first eighteen pages of the work *De la Grandeur et de la Figure de la Terre.*

There is an account of the memoir in pages 96 and 97 of the historical portion of the volume. Here we have the error which Keill adopted, as we saw in Art. 76:

> Mais en supposant, comme il est fort vraisemblable, que cette diminution de la valeur terrestre d'un degré, continue toûjours de l'Equateur vers le Pole, et en conservant d'ailleurs les hypotheses communes, on voit d'abord qu'un Meridien doit être plus petit que l'Equateur, et par consequent que la Terre est un Globe aplati vers les Poles.

The passage was changed in another edition: see La Lande's *Astronomie,* third edition, Vol. III. page 24.

82. In the Paris *Mémoires* for 1702, published in 1704, we have a memoir entitled *Reflexions sur la mesure de la Terre, rapportée par Snellius dans son Livre intitulé, Eratosthenes Batavus. Par M. Cassini le fils.* The memoir occupies pages 61...66 of the volume: see also page 82 of the historical portion of the volume. Cassini shews that Snell's result was quite unsatisfactory. The memoir is substantially reproduced with additions in pages 287...296 of the work *De la Grandeur et de la Figure de la Terre.*

83. In the Paris *Mémoires* for 1703, published in 1705, we have a memoir entitled *Remarques sur les Inégalités du Mouvement des Horloges à Pendule. Par M. De La Hire.* The memoir occupies pages 285...299 of the volume: there is an account of it on pages 130...135 of the historical portion of the volume.

84. David Gregory, Savilian Professor of Astronomy at Oxford, published there in 1702 his *Astronomiæ Physicæ et Geometricæ Elementa:* it is a folio volume containing 494 pages, besides the Title, Dedication, Preface, and Index. The work was reprinted in two quarto volumes at Geneva in 1726, with some additions by an editor who signs himself C. Huart, M. and P. S.

A section of the work is devoted to the Figure of the Sun and the Planets: this section occurs on pages 268...272 of the original edition, and on pages 408...414 of the reprint.

David Gregory contributes nothing new to our subject. He repeats two mistakes from Newton, with rather increased emphasis. One mistake is the assertion that *gravity* at the surface varies inversely as the radius, instead of *gravity resolved along the radius*: see Art. 33. The other mistake is the assertion that if instead of being homogeneous, the central portion is denser than the rest, then the ellipticity is increased: see Art. 30, and Clairaut's *Figure de la Terre*, page 254.

On the hypotheses that the figure is an oblatum, and that gravity varies inversely as the radius; David Gregory gives a good geometrical demonstration of the theorem, that the increase of gravity in proceeding from the equator to the pole varies as the square of the sine of the latitude.

On page 37 of his own edition, David Gregory stated the oblateness of the Earth as a fact. This is the only point at which the editor of the reprint ventures to correct the original author; and on page 51 of the reprint we have this unfortunate note:

Constat ex celeberrimorum Geometrarum observationibus, experimentis et argumentis, Terram quidem Sphæroidem esse, sed oblongam non verò depressam versus Polos, contra quod affirmat Autor noster. Verùm circa hanc quæstionem consulantur Historia et Commentarii Regiæ Scientiarum Academiæ anni præsertim 1720.

Keill very naturally praised the work of his predecessor in the Savilian chair; though with some extravagance of language. The following words occur in the *Ricerche sopra diversi punti*...of Gregory Fontana, Pavia, 1793, pages 93 and 94:

Il famoso *David Gregori* nella sua elegantissima opera intitolata *Astronomiæ Physicæ et Geometricæ Elementa*, che dal celebre Giovanni Keil nella Prefazione della sua Introduzione alla vera Fisica ed Astronomia viene caratterizzata col pomposo elogio di *opus cum sole et luna duraturum*....

85. A memoir by Keill is given on pages 97...110 of Number 315 of the *Philosophical Transactions*. The Number is for the months of May and June, 1708; it forms part of Volume XXVI. which is for the years 1708 and 1709, and is dated 1710.

The memoir is entitled *Joannis Keill ex Æde Christi Oxon. A.M. Epistola ad Cl. virum Gulielmum Cockburn, Medicinæ Doctorem. In qua Leges Attractionis aliaque Physices Principia traduntur.*

The memoir is reprinted at the end of the edition of Keill's *Introductiones ad veram Physicam*...published at Leyden in 1739.

86. The memoir consists of thirty theorems; many of them are merely enunciated; others are supported by a short commentary.

They are but little connected with our subject, being experimental rather than mathematical, and bearing on what we should call molecular attraction.

87. Keill speaks of Newton as

> Vir ingenio pene supra humanam sortem admirabili, dignusque cujus fama per omnes terras pervagata, cœli quos descripsit meatibus permaneat coæva.

The immensity of space and of time with which Astronomy is concerned may cause but can scarcely justify the exorbitant language in which the achievements of those who cultivate the science are sometimes described. The expressions of Keill with respect to Newton may be compared with those which Arago uses when noticing Poisson's famous memoir on the permanence of the solar system:

> Il aura établi qu'à ce point de vue, le seul dont Newton et Euler se fussent préoccupés, les géomètres, ses successeurs, liront encore son beau Mémoire dans plusieurs millions d'années. *Œuvres complétes de François Arago*, vol. II. page 654.
>
> S'il en était besoin, le magnifique Mémoire sur l'invariabilité des grands axes, prouverait que Poisson avait un intérêt personnel à porter ses regards, ses pensées, sur des siecles si éloignés. *The same volume*, page 696.

88. Keill says that he had thought about applying a principle similar to Newton's attraction for the explanation of terrestrial phenomena; and had tested the notion by experiments. He adds:

Meaque hac de re cogitata, abhinc quinquennio, Domino Newtono indicavi; ex eo autem intellexi, eadem fere, quæ ipse investigaveram, sibi diu ante animadversa fuisse.

89. Almost the only passage in the memoir which directly concerns us presents a difficulty. Keill's Theorem XV. asserts that the attractive forces of perfectly solid particles depend much on their figures. He proceeds thus:

Nam si parva aliqua materiæ particula in laminam circularem indefinite exiguæ crassitudinis formetur, et corpusculum in rectâ per centrum transeunte et ad planum circuli Normali locetur; sitque distantia corpusculi æqualis decimæ parti semidiametri circuli: vis qua urgetur corpusculum tricesies minor erit, quam si materia attrahens coalesceret in Sphæram, et virtus totius particulæ ex uno quasi puncto Physico diffunderetur.

Let M denote the mass of the particle, c the distance from the centre of the lamina of the attracted corpuscle, b the radius of the lamina. Then by the ordinary formula we have the attraction

$$= \frac{2M}{b^2}\left\{1 - \frac{c}{\sqrt{(c^2+b^2)}}\right\}.$$

In the case of the sphere the attraction $= \frac{M}{c^2}$.

The ratio of the former to the latter is

$$\frac{2c^2}{b^2}\left\{1 - \frac{c}{\sqrt{(c^2+b^2)}}\right\}.$$

Since $b = 10c$, this ratio $= \frac{1}{50}\left(1 - \frac{1}{\sqrt{101}}\right)$

$$= \frac{2}{101+\sqrt{101}} = \frac{1}{55} \text{ very nearly.}$$

I presume *tricesies* is intended for *thirty times*, though it is not contained in the dictionaries. Hence Keill has $\frac{1}{30}$ instead of $\frac{1}{55}$.

The formulæ for the attraction of a circular lamina and of a sphere are implicitly given by Newton; so that there is no reason for the error.

90. The Paris *Mémoires* for 1708, published in 1709, contain observations of the length of the seconds pendulum made by Feuillée in 1704 at Porto Bello and somewhat later at Martinique: see pages 8 and 16 of the volume. The anomalous results obtained were noticed by Newton in the second and third editions of the *Principia*.

91. We have next to advert to a paper published in pages 330...342 of Number 331 of the *Philosophical Transactions*, which is for the months of July, August, and September, 1711. The number forms part of Volume XXVII. which is for the years 1710...1712, and is dated 1712.

The title of the paper is *Johannis Freind, M.D. Oxon. Prælectionum Chymicarum Vindiciæ, in quibus Objectiones, in Actis Lipsiensibus Anno* 1710. *Mense Septembri, contra Vim materiæ Attractricem allatæ, diluuntur.*

The paper is not mathematical. Freind had published a work on Chemistry, and the editors of the Leipsic *Acta* found fault with the use he made of the principle of Attraction. In this paper Freind maintains the truth and the importance of the principle.

92. In the Paris *Mémoires* for 1713, published in 1716, there is a memoir entitled *De la Figure de la Terre. Par M. Cassini.* The memoir occupies pages 188...200 of the volume.

The arc of the meridian measured from Paris to the south of France, compared with the arc measured northwards, seemed to indicate that the length of a degree of the meridian decreased from the equator to the pole. This result suggested that the Earth is an oblongum. Accordingly Cassini so considers it; and assuming that the excentricity of the generating ellipse is about $\frac{1}{11}$ he calculates a table of the length of a degree of the meridian for every degree of latitude. The memoir is substantially reproduced in pages 237...245 of the work *De la Grandeur et de la Figure de la Terre;* but the table is there calculated for the excentricity $\frac{1}{7}$.

Some introductory matter given in the memoir is not reproduced in the work just cited. This matter contains short accounts of the opinions of Newton and of Huygens in favour of the oblate form of the earth. Then a contrary opinion is noticed at greater length, beginning thus: Tout au contraire, M. Einsenschmid célebre Mathematicien de Strasbourg... We have already learned the nature of this opinion: see Art. 76.

There is an account of the memoir on pages 62...66 of the historical portion of the volume. It is there remarked that supposing the length of a degree of the meridian to decrease from the equator to the pole, it would not follow, as had been erroneously suggested in the historical portion of the *Mémoires* for 1701, that the Earth is flattened at the poles: see Art. 81.

93. James Hermann published at Amsterdam in 1716 a quarto volume, entitled *Phoronomia, sive de Viribus et Motibus corporum et fluidorum libri duo.*

We are concerned only with pages 361...371 of the work.

94. Hermann solves Huygens's problem of the relative equilibrium of rotating fluid under the action of a constant force directed to a point on the axis of rotation. Hermann gives two solutions; one on Newton's principle of columns balancing at the centre, the other on Huygens's principle of the plumb-line.

95. Hermann also solves by both principles the problem in which the central force, instead of being constant, varies as the distance; in this case he shews that the figure is an oblatum. This is the first appearance of the problem and its solution. For the case of the Earth the ratio of the axes would be nearly as $\sqrt{288}$ is to $\sqrt{289}$, that is, approximately as 577 is to 578.

Hermann's investigations of both problems are correct and satisfactory. There is, however, a curious circumstance connected with his second problem. He notices that the result differs very much from that which Newton had obtained for the ratio of the axes of the Earth; he does not expressly say that Newton was wrong, but he seems to imply that his own was the correct result. He observes that neither Newton nor David Gregory had determined what the figure must be for equilibrium; and this is certainly true. See, however, Boscovich, *De Litteraria Expeditione*, pages 442...446.

96. In Newton's fluid mass, assumed to be an oblatum, *so long as we keep to the same radius vector*, the attraction varies as the distance from the centre, and so also does the gravity. And at the surface the gravity *resolved along the radius vector* varies inversely as the length of the radius vector. Now Hermann notices these results; though he seems to pay no attention to the limiting clauses which I have printed in Italics. Both results hold for Hermann's own fluid mass. Moreover, Hermann demonstrates a proposition which we may enunciate thus: Suppose a fluid mass in relative equilibrium under a centrifugal force and a central force to some point of the axis of rotation; then if at the surface the gravity resolved along the radius vector varies inversely as the length of the radius vector, the attraction at the surface varies as the distance from the centre.

Perhaps, from seeing that his fluid mass and Newton's had similar properties, Hermann inferred that Newton's figure and his own ought to be identical. But it is sufficient to observe that Newton's problem and Hermann's are essentially different. Newton does not assume attraction to a fixed centre varying as the distance; he assumes that every particle attracts every other according to the law of the inverse square of the distance. It should have been a caution to Hermann that his own problem and Huygens s led to approximately the same result for the ratio of the axes, though the laws of force were very different; thus from partial agreement he ought not to have expected universal agreement.

97. Hermann seems to have been much surprised at the proposition which, as we have said in the preceding Article, he demonstrates. He observes on his page 369:

Hac verò proprietate posita, quod scilicet solicitationes gravitatis acceleratrices...distantiis à centro...reciproce proportionales sunt, quis crediderit gravitates absolutas corporum in iisdem punctis...eorum distantiis...directe proportionales esse?

Boscovich, nearly forty years later, expressed his surprise at the same result: see his *De Litteraria Expeditione*....page 403, where he says:

gravitates residuæ erunt accuratè in superficie ejus solidi in ratione reciproca distantiarum a centro, quod sane mirum videri possit, cum gravitates primitivæ ibidem sint in ratione directa distantiarum earundem.

It will be observed that what I call *attraction* Hermann calls *gravitas absoluta*, and Boscovich *gravitas primitiva*; what I call *gravity* Hermann calls *solicitatio gravitatis acceleratrix*, and Boscovich *gravitas residua*. See Art. 25.

98. On his page 372, Hermann discusses a problem about rotating fluid, which does not concern our subject. Here he falls into an error, which was pointed out by Clairaut in page 55 of his *Figure de la Terre*.

99. In the Paris *Mémoires* for 1718, published in 1719, we have a memoir entitled *De la Grandeur de la Terre et de sa Figure. Par M. Cassini.* It occupies pages 189...196 of the volume; there is an account of it on pages 64...66 of the historical portion of the volume.

The memoir contains a notice of the labours of the ancients on the subject, and of the recent operations in France. It is substantially reproduced in the work *De la Grandeur et de la Figure de la Terre*, pages 12...18 and 189...196.

100. We have now to consider the account of the measurement of an arc of the meridian through France, which is contained in the work *De la Grandeur et de la Figure de la Terre*; the work has also the title *Suite des Mémoires de l'Academie Royale des Sciences, Année* 1718. The date of publication is 1720.

The volume is in quarto. It contains Title, Half-title and Table of Contents on 6 pages, and 306 pages of text. There is a small map of France, and 4 large maps shewing the meridian line of Paris traced through the kingdom; there are also 15 plates. A list of the misprints in the work is given in the Paris *Mémoires* for 1732, pages 512 and 513.

101. The volume is divided into two parts; in the first part the operations are described which relate to the arc extending from Paris southwards to the Pyrenees, and in the second part the operations are described which relate to the arc extending from Paris northwards to Dunkirk. The author's name is not given explicitly; but we learn incidentally that it was J. Cassini: see pages 5, 10, 193, 302, 303, 304, 305.

The operations which the volume records are the most accurate and important which had as yet been performed in connection with the Figure of the Earth; and the account given of them is interesting and satisfactory. The instruments and the methods of using them are fully and clearly described, and the calculations exhibited in such a manner that they can be easily tested.

102. The determination of an arc of the meridian we are now considering is a continuation of the work commenced by Picard in 1669. Picard measured a base of 5663 toises near Paris; then by a series of triangles he found the distance between the parallels of Malvoisine and Amiens to be 78850 toises, corresponding to a difference of 1° 22′ 55″ in latitude: hence he adopted 57060 as the number of toises in a degree. See pages 273, 256, 281.

It was afterwards proposed to extend Picard's arc through France; and the work was committed to D. Cassini and others: but it was interrupted in 1683. The work was resumed by D. Cassini, J. Cassini, and others in 1700, and the arc was extended southwards to the Pyrenees. In 1718 the extension of the arc northwards to Dunkirk was commenced. See pages 4, 5, 191. In this extension many of Picard's triangles were employed: see pages 191, 255.

103. All the triangles were calculated in succession from Picard's original base, which was not re-measured. Two bases of verification were measured, one near the Pyrenees, and the other

near Dunkirk. The difference between the measured and the calculated length was three toises in the former case; but this was reduced by some necessary corrections of the angles: the difference in the latter case was about a toise See pages 104 and 221. Both bases of verification were measured by wooden rods. In the former case four rods each of two toises in length were joined together, two and two, so as to make two rods each of four toises in length; in the latter case three rods each of three toises in length were used: the lengths of the wooden rods were determined in both cases by the aid of the *same* iron rule, four feet long. See pages 99 and 219.

Picard's original base had been measured by four rods each of two toises in length, which were joined together two and two, so as to make two rods each of four toises in length. See page 255.

104. The general result obtained is the following: from the southern arc which extended over nearly $6^0 19'$, the length of a degree was found to be 57097 toises; from the northern arc which extended over rather more than $2^0 12'$, the length of a degree was found to be 56960 toises. This was considered to make it sufficiently evident that the length of a degree of the meridian must diminish from the equator to the pole. Assuming then that the earth is an oblongum, the ellipticity is found to be $\frac{1}{95}$. See pages 148, 237, 243. A table is given of the length of a degree of the meridian in different latitudes on the Cassinian hypothesis: see Arts. 39 and 92.

It is now well known that the length of a degree of the meridian *increases* from the equator to the pole; the contrary opinion however, maintained by J. Cassini, found advocates for some years after the publication of the work we are now considering As we shall see, the erroneous determination deduced from the French arc was finally corrected by fresh operations.

105. Pages 255...287 of the volume are devoted to the subject of Picard's measure of the Earth. As Picard's book was scarce, large extracts are given from it; a few remarks are made which do not substantially affect Picard's accuracy.

Pages 287...306 of the volume are devoted to the measure by Snell and the measure by Riccioli; the value of both is quite demolished: see Art. 82.

106. A few remarks may be made on some incidental points. I offer with hesitation an opinion as to instruments; but from the descriptions given it seems to me very unlikely that either the geodetical or the astronomical angles could have been observed accurately to *seconds* as is professed. The astronomical instruments used at the north and south extremities of the arc were different; the former had an error of 3 seconds in a degree from false centering. See pages 142, 223, 233.

On pages 225...230 we have an account and an explanation of a fact stated to be then observed for the first time, which gave much trouble until it was understood. The fact is this in modern language: any star which is not an equatorial star does not strictly run along the horizontal wire of a transit instrument as it crosses the meridian of the observer; thus in determining the zenith distance from observations of the star when it is not accurately on the meridian, it is necessary to allow for the curvature of the path.

Speaking of the distinction of the regions of the Earth into East, West, North and South, our author gives a paragraph which I quote for the sake of its last example; see his pages 20, 21.

Cette même distinction des régions fut observée dans la construction du Temple de Jérusalem. Nous voyons aussi qu'elle a été imitée dans la construction des premiers Temples Chrétiens, quand on l'a pû faire commodément, et même dans la situation de la Maison de Notre-Dame de Lorette, comme nous l'avons observé nous-mêmes après plusieurs autres Mathématiciens.

Much importance was attached to the precaution of taking the observations of stars at the same season of the year: see pages 144 and 231. It seems to have been made out even then that the altitudes of the stars varied at different seasons. We know now that the Aberration of Light would certainly cause such variations.

Speaking of the largest Egyptian pyramid our author says on his page 154:

Il y a lieu de s'étonner, que M. Graves Mathématicien Anglois, dans sa Pyramidographie, ait trouvé la base de cette Pyramide, mesurée par les Triangles, de 693 pieds de Londres....

The error is certainly large; for according to trustworthy statements the base was originally 764 feet, and is now 746 feet: see Herschel's *Familiar Lectures on Scientific subjects*, page 427. The inaccurate measurer was John Greaves, Savilian Professor of Astronomy at Oxford.

107. We may state here, though a little out of chronological order, that a German translation of the *De la Grandeur et de la Figure de la Terre* was published in 1741 at Arnstadt and Leipzig. This is entitled *Mathematische und genaue Abhandlung von der Figur und Grösse der Erden.* There is a preface by J. A. Klimmen, from which we learn that the translator, whose name is not stated, did not live beyond the commencement of the printing.

The translation is in a small octavo form; there are no maps, but the other plates of the original are copied, on a diminished scale. The misprints pointed out in the Paris *Mémoires* for 1732 are corrected.

It seems strange that a translation should have been published when the original work was just about to be superseded. In 1739 astronomical observations had been made by Maupertuis, Clairaut, Camus and Le Monnier, in order to determine anew the length of a degree between Paris and Amiens; and in 1740 Picard's base was remeasured: in 1744 the work entitled *La Meridienne de Paris verifiée* appeared.

108. An account of the work *De la Grandeur et de la Figure de la Terre* is given in the Paris *Mémoires* for 1721, published in 1723. The account is on pages 66...77 of the historical portion of the volume: it furnishes references to preceding volumes of the *Mémoires* in which the subject had been noticed. There is nothing of importance in the account.

The following sentence, so far as it is intelligible, suggests a proceeding which may very naturally have been adopted; but I do not know what authority there is for the statement.

En tirant d'un Lieu une perpendiculaire sur la Méridienne, pour avoir la distance de ce Lieu par rapport à elle, on a considéré s'il en étoit proche, ou s'il ne l'étoit pas. Dans le premier cas la perpendiculaire étoit la distance asses juste, mais dans le second, cette perpendiculaire representoit un petit arc de Cercle, et l'on avoit égard à la différence de l'arc et de la Corde, qui étoit la distance cherchée.

Page 146 of the work seems to approach nearest to the latter part of the above statement.

109. We have now to consider a memoir by Mairan, entitled *Recherches Géométriques sur la diminution des Degrés terrestres, en allant de l'Equateur vers les Poles: Où l'on examine les conséquences qui en résultent, tant à l'égard de la figure de la Terre, que de la pesanteur des corps, et de l'accourcissement du Pendule.*

This is contained in the Paris *Mémoires* for 1720, published in 1722. The memoir occupies pages 231...277 of the volume.

The memoir may be described generally as consisting of misapplied mathematics. Mairan was a Cartesian and a Cassinian; so that he upheld the system of vortices, and the oblong form of the Earth. There is an account of the memoir in pages 65...79 of the historical portion of the volume; this is I presume by Fontenelle, who was then Secretary of the Paris Academy of Sciences: Mairan's opinions seem here to be accepted without hesitation.

110. Mairan shews that if the length of a degree of the meridian decreases from the equator to the pole, the polar diameter must be the longest. He compares the effect produced by centrifugal force at a place in the same latitude on the surface of a sphere, an oblong body, and an oblate body; the latitude being determined in each case by the angle between the normal to the surface and the plane of the equator. Part of his page 244 is unsatisfactory, but it can be easily corrected.

111. Mairan supposes that the Earth was originally of an elongated form, and that the amount of elongation was diminished by the centrifugal force, but not entirely destroyed. See Bailly's *Histoire de l'Astronomie Moderne*, Vol. II., page 641.

Mairan's Proposition VIII. on page 253 is a striking example of the vagueness of the mechanical language of the period. He speaks about the centre of the Earth sustaining a part of the effort of gravity: it is difficult to attach any meaning to such an expression.

112. Mairan has a long discussion on the direction of gravity at different points of the interior of the Earth. Suppose that through any point of the interior a surface is drawn, similar, similarly situated, and concentric with the external surface; Mairan takes the normal to this surface at the point for the direction of gravity. Then, to determine the lines of direction of gravity, he solves what we call a problem of orthogonal trajectories; the curves which are cut at right angles being ellipses, similar, similarly situated, and concentric. Thus his result coincides with what we should obtain in seeking the *lines of force* inside a homogeneous mass of rotating fluid, supposing it in relative equilibrium. Mairan seems to attach great importance to the matter; he thinks his lines of direction may extend beyond the Earth to the boundary of the terrestrial vortex; he admits however that there is little prospect of verifying his result by observation: see his page 263.

113. But the most extraordinary part of the memoir is that which treats of the variation of gravity at the surface of the Earth. Newtonians and Cassinians agreed in admitting, as a result of observation, the diminution of gravity in passing from the pole to the equator. Huygens's notion that the resultant attraction is constant at all distances from the Earth's centre would not reconcile this fact with an oblong form of the Earth. Newton's law of attraction according to the inverse square of the distance directly contradicted the oblong form. Accordingly, Mairan had to invent a law; he suggests and rejects various other absurdities before he produces that which he adopts: we will describe this in modern language.

Mairan holds that at every point of the surface of a body of revolution the force of attraction would vary inversely as the product of the two principal radii of curvature at the point. His reason for this assumption depends on the fact that adjacent normals to the surface, taken in the plane of the meridian, intersect at one centre of curvature, while adjacent normals to the surface, taken in the plane at right angles to the meridian, intersect at the other centre of curvature.

With this arbitrary law, Mairan triumphantly shews that the oblong form makes gravity decrease from the pole to the equator, which agrees with observation; while the oblate form makes gravity increase from the pole to the equator. He prudently abstains from numerical calculation which would test the extent of his agreement with observation. If we take an oblongum, we find that Mairan's law makes the attraction at the pole bear to the attraction at the equator the ratio of the fourth power of the polar diameter to the fourth power of the equatorial diameter; thus, assuming with J. Cassini and Mairan, the ellipticity to be about $\frac{1}{95}$, the diminution of gravity in passing from the pole to the equator would be about $\frac{1}{24}$ of the gravity at the pole, besides that caused by the centrifugal force: this is extravagantly greater than observation suggested.

It would be difficult to find a more striking example of misplaced ingenuity than the pages 264...276 of the memoir, which are devoted to Mairan's arbitrary law.

114. With respect to the equation which Huygens obtained, as we stated in Art. 55, Mairan says on his page 253:

M. *Huguens* a donné l'Equation algébrique de la courbe génératrice du sphéroide applati, par rapport à la Terre supposée primitivement sphérique; et M. *Hermann*, qui avoit trouvé la même courbe par le calcul intégral, dans sa réponse à M. *Nieuwentiit*, l'a encore donnée par synthèse, et avec la construction, dans sa *Phoronomie.*

I have not seen the first production of Hermann, to which Mairan refers: I have noticed the second in Arts. 93...98.

115. The writers who have appeared before us in the present Chapter added nothing to Newton's investigations on Attraction and on the Figure of the Earth; while under the powerful influence of D. Cassini and J. Cassini doubts had arisen as to the real shape of the Earth. But the true theory ultimately gained the support of decisive researches and measurements.

The next three Chapters will be devoted to three eminent mathematicians who all contributed essentially to the advancement of our subject. Maupertuis adopted and explained Newton's propositions on Attraction and on the Figure of the Earth; and he conducted an expedition to Lapland, for the measurement of an arc of the meridian, the result of which was fatal to the Cassinian hypothesis. James Stirling enunciated without demonstration approximate propositions respecting the magnitude and the direction of the attraction of a homogeneous oblatum at its surface; and he implicitly established Newton's *postulate*. see Art. 44. Clairaut produced several valuable memoirs; in particular, during his stay in Lapland, he found leisure to compose one on the same subject as Stirling's: another memoir led the way to the investigations of the Figure of the Earth, supposed heterogeneous. These two memoirs were subsequently embodied by Clairaut in a work of enduring interest and importance.

CHAPTER IV.

MAUPERTUIS.

116. We shall notice in this Chapter the various memoirs which Maupertuis contributed to our subject.

117. A memoir is given on pages 240...256 of Number 422 of the *Philosophical Transactions.* The Number is for the months of January, February, and March, 1732; it forms part of Volume XXXVII. which is for the years 1731 and 1732, and is dated 1733.

The memoir is entitled *De Figuris quas Fluida rotata induere possunt, Problemata duo; cum conjectura de Stellis quæ aliquando prodeunt vel deficiunt; et de Annulo Saturni. Authore Petro Ludovico De Maupertuis, Regiæ Societatis Londinensis, et Academiæ Scientiarum Parisiensis Socio.*

118. In the first problem, fluid is supposed to rotate with uniform angular velocity round a fixed axis, and to be attracted to a fixed point in the axis by a force which varies as any power of the distance. Maupertuis uses Newton's principle of balancing columns, and investigates the equation which determines the form of the surface for relative equilibrium. He restricts himself, as we should say, to *space of two dimensions*; but a modern reader will have no difficulty in solving the problem generally, and the result will coincide with that of Maupertuis.

119. The second problem is enunciated thus:

Posito quod materia fluens circa axem extra fluentum sumtum, attrahatur versus centrum in hoc axe positum vi alicui distantiæ à centro dignitati proportionali; dum interea propter fluenti partium attractionem mutuam, sit altera attractio versus aliud centrum intra fluentum sumtum, quæ in quavis sectione fluenti revolutionis perpendiculariter per centrum exterius facta, sit alicui distantiæ a centro interiori dignitati proportionalis: invenire figuram quam fluentum induet.

In the solution of this problem also, Maupertuis restricts himself to space of two dimensions; but it may be shewn by a more general process that his result is correct.

Take the axis of z for that of rotation; let ω be the angular velocity; and (x, y, z) any point of the fluid. Then in the usual way, we may suppose the system reduced to rest, if we impress forces $\omega^2 x$ and $\omega^2 y$ parallel to the axes of x and y respectively.

Let there be a force directed to the origin, denoted by λr^m, where $r = \sqrt{(x^2 + y^2 + z^2)}$. Besides this there is to be a force of a certain kind, arising from the attraction of the mass itself. This mass is supposed to form a symmetrical ring-shaped body. Hence it is obvious that its action at any point (x, y, z) will lie in the plane which passes through this point and the axis of z. It is *assumed* that while we keep to the same plane, this action will pass through a fixed point; so that, denoting the co-ordinates of this point by ξ, η, 0, we have

$$\frac{\xi}{x} = \frac{\eta}{y} = \frac{c}{r_1};$$

where c is a constant quantity, and equal to $\sqrt{(\xi^2 + \eta^2)}$, and r_1 stands for $\sqrt{(x^2 + y^2)}$.

Put s for $\sqrt{\{(x - \xi)^2 + (y - \eta)^2 + z^2\}}$, and denote the action of the mass by μs^n.

Then, with the usual notation,

$$X = \omega^2 x - \frac{x}{r}\lambda r^m - \frac{x - \xi}{s}\mu s^n;$$

and Y and Z can be similarly expressed.

Now $$\frac{x-\xi}{s}\mu s^n = \mu\left(x - \frac{cx}{r_1}\right)s^{n-1}$$

$$= \mu\left(x - \frac{cx}{r_1}\right)\{(r_1 - c)^2 + z^2\}^{\frac{n-1}{2}} = \mu\left(x - \frac{cx}{r_1}\right)\{r^2 - 2r_1c + c^2\}^{\frac{n-1}{2}};$$

and $$\left(x - \frac{cx}{r_1}\right)dx + \left(y - \frac{cy}{r_1}\right)dy + zdz = rdr - cdr_1.$$

Thus, finally, the equation to the surface of relative equilibrium is

$$\frac{\omega^2}{2}(x^2+y^2) - \frac{\lambda r^{m+1}}{m+1} - \frac{\mu(r^2 - 2r_1c + c^2)^{\frac{n+1}{2}}}{n+1} = \text{constant},$$

that is, $$\frac{\omega^2}{2}(x^2+y^2) - \frac{\lambda r^{m+1}}{m+1} - \frac{\mu s^{n+1}}{n+1} = \text{constant}.$$

120. Maupertuis himself gives two investigations, one for the part of the mass which is between the axis of rotation and the point $(\xi, \eta, 0)$, and the other for the part which is beyond this point; but this is unnecessary: a single investigation with proper generality in the symbols applies to the whole mass.

The second problem includes the first as a particular case; we have only to suppose $\mu = 0$. Maupertuis himself makes this remark: see his page 253.

Maupertuis suggests, that the constants may happen to be so adjusted, that what we may call the generating curve of the ring will consist of *two* ovals; so that, in fact, there will be two rings. This is conceivable, but he is wrong in implying that it is possible when $m = 1$, and $n = 1$; for then the generating curve must consist of only a single ellipse.

121. The solutions here given by Maupertuis are reproduced by him in his *Figure des Astres;* and also, though with less detail, in his memoir, which is published in the Paris *Mémoires* for 1734. The problems, though rather theoretical than practical, were doubtless a valuable contribution to the science of Hydrostatics of the period.

As to the popular part of the memoir, we shall say a word hereafter: see Art. 127.

122. We have next to consider the work published by Maupertuis, under the title of *Discours sur les différentes figures des Astres*...Paris, 1732. I have seen only the copy in the library of the Royal Society, which is marked *Ex dono Auctoris.* The volume is in octavo, and contains 83 pages, besides the Title and Table of Contents, on four pages.

The mathematical part of the volume consists of the same problems in French as were given in Latin in the *Philosophical Transactions*, and which we have already noticed. Besides this, we have Chapters of a popular character, which contain general reflexions on the figure of the Earth, a metaphysical discussion on attraction, and explanations of the motions of the planets on the system of vortices, and on the system of gravitation.

123. In his first Chapter, Maupertuis adverts to the researches of Huygens on the figure of the Earth, and afterwards to those of Newton. By taking this order, a reader might be led to suppose that Huygens preceded Newton in this subject; but, as we have already pointed out, Newton was the first: see Art. 65.

124. There is a note on page 44 which presents a difficulty. Suppose a sphere, the radius of which is one foot, and its density the mean density of the Earth. The attraction which this sphere would exert on a particle at its surface, is a very small fraction of the attraction which the Earth would exert on a particle at the surface of the Earth; the numerator of the fraction would be unity, and the denominator the number of feet in the Earth's radius. This substantially agrees with Maupertuis. Then he proceeds thus: "Deux Spheres semblables, placées à la distance d'un quart de pouce dans le vuide, employeroient un mois à se joindre." I suppose the spheres to be such as have been just mentioned, namely, each of a foot radius and of the mean density of the Earth; and that they are to be placed so that their surfaces may be a quarter of an inch apart. But then instead of a month the spheres would require only a few minutes to arrive at contact. Thus I am quite at a loss as to his meaning.

125. A second edition of the *Figure des Astres* was published, which I have not seen. Clairaut refers to it on pages 19 and 59

of his *Figure de la Terre*; see also D'Alembert's *Opuscules Mathématiques*, Vol. VI. page 358. The work seems to have been translated into English.

126. There is an account of Maupertuis's *Figure des Astres* on pages 85...93 of the historical portion of the volume of the Paris *Mémoires* for 1732. Centrifugal Force has puzzled the writer of the account; he says on page 86, "...les directions de la Force centrifuge sont à chaque instant les Tangentes de chaque point...." Of course instead of *tangents* we ought to read *normals*.

127. The popular part of the *Figure des Astres* is reproduced in the collected edition of the works of Maupertuis, published in four volumes at Lyons in 1756; it occupies pages 81...170 of the first volume. The mathematical investigations are not reproduced.

Maupertuis suggests that the variable brightness of certain stars may be explained by supposing that these stars are very much flattened, and that, owing to different positions assumed by their axes of rotation, we sometimes have a much larger disc turned towards us than at other times. He considers that the nebulæ are really suns or planets, of figures more or less deviating from spheres.

He suggests that the ring of Saturn may have been formed out of the tail of a comet which Saturn by the aid of his attraction has appropriated.

128. A memoir by Maupertuis, entitled *Sur les loix de l'Attraction*, is contained in the volume for 1732 of the Paris *Mémoires*, published in 1735. The memoir occupies pages 343...362 of the volume. There is an account of the memoir on pages 112...117 of the historical portion of the volume; this account, like many other attempts to give a translation of mathematical processes into ordinary language, is scarcely intelligible.

The memoir, according to Bailly, is the first example of the adoption of the principle of attraction by French mathematicians: see *Histoire de l'Astronomie Moderne*, Vol. III. page 7.

The memoir may be described as an analytical investigation of most of the results contained in Newton's two sections on Attraction; adding, however, nothing of importance to them. The methods employed are simple and interesting.

129. We will notice the method by which Maupertuis finds the attraction of a spherical shell. Suppose the law of attraction that of the inverse n^{th} power of the distance. Proceeding as in Art. 4 we obtain for the attraction of an element of the shell $\frac{2\pi k\rho y ds}{r^n}\cos\theta$. Now it will be found that $\frac{ds}{dr}=\frac{a}{c\sin\theta}$, and $y=r\sin\theta$. Thus the expression becomes $\frac{2\pi k\rho a dr}{cr^{n-1}}\cos\theta$; and $\cos\theta=\frac{r^2+c^2-a^2}{2cr}$, so that finally we have $\frac{2\pi k\rho a dr}{r^n}\,\frac{r^2+c^2-a^2}{2c^2}$; which is immediately integrable.

This is substantially the method of Maupertuis; the chief part of it consists in making r the independent variable. The method is, in fact, that which Laplace adopted for finding the attraction of a spherical shell; and it has passed into the elementary text-books on the subject: see *Statics*, Chapter XIII.

It will be noticed that in this process Maupertuis made the easy extension which arises from taking the inverse n^{th} power of the distance; while Newton, in the corresponding place, used only the inverse square of the distance.

130. Some incidental statements made in the memoir may be noticed.

Maupertuis says on page 343, that a homogeneous fluid mass which has no motion of rotation, but is left to the influence of its own attraction, will necessarily assume a spherical form: "car il est facile de voir qu'il n'y a que cette figure dans laquelle toutes les parties puissent demeurer en équilibre." The belief here expressed was doubtless held by many of the earlier writers on the subject; but the belief was not founded on evidence. It is observed by Poisson that it has not been demonstrated that the sphere is the only figure which can be taken by a fluid at rest under the mutual attractions of its particles, however natural that may appear. *Traité de Mécanique*, Vol. II. page 543. See also Résal, *Traité élémentaire de Mécanique Celeste*, page 198.

Maupertuis says on page 346, that if a homogeneous fluid rotates round an axis, and its particles are attracted towards a centre by a force which varies as the distance, the form assumed is such that the meridians are ellipses: this we know to be true, with the condition, however, that the centre of force must be at some point of the axis of rotation. He adds with respect to the fluid mass: "Et si elle circule autour d'un axe pris au dehors d'elle, elle forme un anneau dont les sections sont encore des ellipses." This passage taken alone would not be intelligible, but from another memoir we know all that Maupertuis can have intended to say; namely, that relative equilibrium will subsist under a certain peculiar assumption: see Art. 119.

Maupertuis offers some remarks on his pages 347 and 348, commencing with the following sentence: "Supposé que Dieu eût voulu établir dans la matiére quelque loi d'Attraction, toutes ces loix ne devoient pas lui paroître égales." Maupertuis holds that the ordinary law has, as it were, a reason for preference, because it leads to the result that a sphere will attract as if it were a particle collected in its own centre. To this Stay alludes in his *Philosophiæ Recentioris*, Lib. IV. v. 1582...1584:

Scrutantes quidam; quid Mundi illexerit ipsum
Artificem, legem ut voluisset materiai
Ponere, quam doceo;

Boscovich in his note dissents from Maupertuis. See also Bailly, *Histoire de l'Astronomie Moderne*, Vol. III. page 7.

Maupertuis refers on page 361 to thirty propositions relating to attractions, given at the end of Keill's works; and on page 362 he says that Keill and many English philosophers believed precipitations, coagulations, crystallizations, and a multitude of other phenomena to arise from an attraction very powerful at contact, but insensible at great distances. He adds: "Enfin M. Friend a donné une Chimie, toute déduite de ce principe."

131. In the Paris *Mémoires* for 1733, published in 1735, we have a memoir by Maupertuis, entitled *Sur la Figure de la Terre, et sur les moyens que l'Astronomie et la Géographie fournissent pour la déterminer*. The memoir occupies pages 153...164 of the volume.

Maupertuis gives analytical investigations of the length of a degree of longitude and of a degree of meridian on the Earth, supposed to be an ellipsoid of revolution; and he shews how the axes of the ellipsoid may be deduced from lengths of degrees determined by measurement.

Maupertuis refers to Huygens, Newton, Cassini, Mairan, and M. des Aiguiliers; the last is usually written Desaguliers.

Maupertuis also quotes a passage from a letter written by Poleni: we shall notice the letter in Chapter VIII.

132. In the Paris *Mémoires* for 1734, published in 1736, we have a memoir by Maupertuis, entitled *Sur les Figures des Corps Célestes.* The memoir occupies pages 55...109 of the volume; there is an account of it on pages 88...94 of the historical portion of the volume.

The memoir may be regarded as a development of the *Figure des Astres;* for Maupertuis says on page 56:

Je reviens à examiner les figures que les loix de la Statique et de l'Hydrostatique doivent donner aux Corps célestes, et j'entrerai sur cette matiére dans un plus grand détail que je n'ai fait dans le Discours sur la figure des Astres.

The memoir is divided into four parts.

133. The first part of the memoir treats on a subject which Bouguer discussed in the same volume; and adds nothing fresh. Maupertuis shews, as Bouguer did, that if the force on a fluid is always directed to a fixed point, the principles of Newton and of Huygens lead to the same form for equilibrium, provided the force be a function of the distance from the fixed point; but they do not lead to the same form if the expression for the force be the product of a function of the distance into a function of the angle which determines the position of the distance.

134. Maupertuis gives an extract of a letter sent to Fermat by Pascal and Roberval, in order to shew that the idea of attraction had occurred to the writers before Newton proposed it. But we have here only a vague idea, not any suggestion of the law of the inverse square; and of course no pretence at demonstration.

135. In the second part of the memoir we have the problems already given in the *Philosophical Transactions;* though they are here treated with less detail: see Art. 121.

For a particular case of the second problem, Maupertuis supposes that the force which is directed to a fixed point in the axis of rotation varies inversely as the square of the distance, and that the other force vanishes. His result then, expressed in modern notation, becomes $\frac{\mu}{r} + \frac{\omega^2}{2} r^2 \cos^2\theta = \text{constant}$.

This is in fact the equation which is now obtained in investigating the form of the atmosphere. Maupertuis does not discuss the equation; but he implies that it would give him an oval curve about some point not coinciding with the pole from which r is measured. This, however, is not the case; that is to say, the equation does not correspond to the diagram he supplies, and has no application to such an object as Saturn's ring, which he has in view.

136. In the third part of the memoir, Maupertuis refers to certain celestial phænomena which he considers support his theory; such as nebulæ and variable stars.

137. The fourth part of the memoir relates to the figure of the Earth supposed fluid, and taking the ordinary law of attraction.

This may be described as a commentary on Newton's theory of the Figure of the Earth. Newton's process is developed clearly and correctly; with the exception of one slight mistake. In Art. 20, we have stated that the attraction of a certain oblatum is approximately a mean proportional between the attractions of a certain sphere and a certain oblongum. Maupertuis incautiously says that the attractions of these bodies *are as their masses*, and therefore the result which Newton affirms is true. We have already drawn attention to this mistake: see Art. 22.

138. Maupertuis obtains, as Newton did, the value $\frac{1}{9\frac{1}{5}}$ for the ratio of the difference of the axes to the minor axis in the case of Jupiter; see Art. 29. Then Maupertuis says on his page 96:

Comme cette différence est beaucoup plus grande que celle qui résulte des observations de M. Cassini, et que celle qui résulte des observations de M. Pound, M. Newton conjecture que Jupiter est plus dense vers le plan de son équateur que vers les poles. Cet excès de

densité feroit que la colomne qui est dans le plan de l'équateur, pour être en équilibre avec celle qui repond au pole, doit être plus courte que cette Théorie ne la détermine, et par conséquent le diametre de l'équateur différeroit moins de l'axe, et son rapport à l'axe approcheroit plus du rapport observé.

This extract shews in what sense Maupertuis understood a rather obscure passage in Newton; but of course the explanation is not very satisfactory. If the fluid is not homogeneous, the whole investigation must be revised; and it will not be sufficient to consider merely the equilibrium of the polar and the equatorial columns.

This passage in Newton seems to have been considered rather important by Maupertuis, for he had previously noticed it, namely, on his page 73. But this reference was not very appropriate; because Maupertuis is there using, not the law of attraction of nature, but the hypothesis of a force directed to a fixed point.

139. On the whole, it does not seem to me that this long memoir by Maupertuis added anything to the current knowledge of the subject; the commentary on Newton was perhaps the most valuable part.

140. In the Paris *Mémoires* for 1735, published in 1738, we have a memoir by Maupertuis, entitled *Sur la Figure de la Terre.* The memoir occupies pages 98...105 of the volume.

Maupertuis investigates the expression for the radius of curvature of an ellipse in terms of the inclination to the major axis; namely, in modern notation, $\frac{a(1-e^2)}{(1-e^2\sin^2\lambda)^{\frac{3}{2}}}$. This furnishes a very approximate expression for the length of a degree of the meridian: see his page 99.

Maupertuis also solves a problem which we may thus enunciate: find at what point the change in the length of a degree of the meridian is most rapid.

Let σ be the measured length of a degree in the latitude ϕ, and let ρ be the radius of curvature; then we take $\frac{\sigma}{\rho}=\frac{\pi}{180}$, so that $\sigma = \frac{\pi}{180}\rho$. Therefore $\frac{d\sigma}{d\phi} = \frac{\pi}{180}\frac{d\rho}{d\phi}$ Hence $\frac{d\rho}{d\phi}$ measures the rate of increase of the length of a degree; and so we have to make $\frac{d\rho}{d\phi}$ a maximum. This is substantially the process of Maupertuis; see his page 105. The result is that ϕ must be found from the equation $3e^2 \sin^4\phi - (4e^2 - 2)\sin^2\phi - 1 = 0$.

If e is very small, we have approximately $\phi = \frac{\pi}{4}$.

Maupertuis makes some simple remarks on the important subject of comparing the measured lengths of degrees of the meridian in the most advantageous manner, so as to render the gradual change in the length decidedly obvious in spite of the unavoidable errors of observations. See his pages 101...104.

141. In the Paris *Mémoires* for 1736, published in 1739, we have a memoir by Maupertuis, entitled *Sur la Figure de la Terre;* the memoir occupies pages 302...312 of the volume.

Maupertuis suggests the following operation. Take two stars which have about the same right ascension and a difference of one degree in declination. Find two places A and B on the Earth's surface, such that one of these stars passes over the zenith at A, and the other over the zenith at B. Then determine by measurement the place C on the Earth's surface, which is on the arc AB, and equally distant from A and B; and observe at C the zenith distances of the two stars. If the Earth is a sphere these zenith distances ought to be equal; if the zenith distances are not found to be equal, we have evidence that the form is not spherical, and we have information as to whether it is oblate or oblong.

Maupertuis also considers an important point in connexion with a trigonometrical survey; namely, the ultimate effect of a constant cause of error by which each side of the triangles employed in succession to produce the required result is rendered greater than it should be. Then, combining this with the error which may be expected to arise from the astronomical observations for finding the amplitude of the arc, he determines what he considers to be the most advantageous number of triangles to be employed.

Maupertuis refers to this memoir in his account of the operations in Lapland; for there the conditions which, according to the memoir, are most advantageous were reasonably satisfied. See his work *La Figure de la Terre déterminée*, page 35.

142. The Paris *Mémoires* for 1737, published in 1740, contain on pages 389...466 a memoir by Maupertuis, entitled *La Figure de la Terre déterminée...*; the memoir describes the operations in Lapland which established the oblate form. There is an account of the memoir on pages 90...96 of the historical portion of the volume.

The memoir is embodied in the book which Maupertuis published in 1738 under the same title: we shall notice this hereafter.

143. A book was published in 1738, entitled *Examen désintéressé des différens ouvrages qui ont été faits pour déterminer la figure de la terre.* See La Lande's *Bibliographie Astronomique*, page 406.

La Lande says that this book is marked *Oldenbourg*, but was printed at *Paris:* he adds, that owing to the censorship of the press a book was often marked with the name of some supposed place where the press was free, as London or Amsterdam.

I have not seen this edition.

La Lande on his next page gives the title of another work published in 1738 and also marked Oldenbourg, namely, *Examen des trois dissertations que M. Desaguliers a publiées sur la figure de la terre, dans les Transactions Philosophiques, Nos.* 386, 387 *et* 388.

I have not seen this edition.

144. The two works appear together in one volume which is dated 1741, and marked Amsterdam; this volume I will now describe.

The volume is in octavo; there are forty-six unnumbered pages, followed by 160 which are numbered. The *Examen désintéressé* extends to page 104, and the rest of the volume is devoted to the *Examen des trois dissertations.*

145. I begin with the *Examen désintéressé.* The title-page says that it is the second edition, augmented by the history of the book. The title-page has the motto, "Et mundum tradidit disputationi eorum. Eccles. cap. III. v. 11."

146. The work is anonymous; but La Lande says that it was written by Maupertuis. This is also clear from other sources. See Bouguer's *Figure de la Terre*, pages 174 and 175, and his *Lettre...Astronomique Pratique*, pages 6, 7, 9, and 10; also La Condamine's *Réponse*...page 5. It affects to be very impartial, and is certainly very clever and amusing; but it contributes nothing new to the knowledge of the subject. The work seems to have attracted great attention at the time; and, as we learn from the Introduction, it was attributed to Mairan and to Fontenelle, although they were opposed to the opinion of Maupertuis. In fact, as La Lande remarks, the smart bantering tone of the work might easily deceive a reader and leave him doubtful whether the author was in favour of the oblate or oblong form.

147. Thirty-six of the unnumbered pages are devoted to the *Histoire du Livre*; these pages constitute an outline of the contents of the work. But one matter here considered is not included in the work; it had, I presume, happened since the publication of the first edition. A distinguished Danish astronomer, named P. Horrebow, had written a work on the Theory of the Earth, and well-feigned surprise is expressed at his rashness in declaring for the oblate form. See Petri Horrebowii *Opera*, 1740, Vol. I. page 381.

148. In the first part of the work the writer speaks of the important measurements which had been made, namely, that at the polar circle which favoured the oblate form, and five operations by Cassini which favoured the oblong form. The measurement at the polar circle will be discussed in Chapter VII.

In noticing the operations at the polar circle the writer puts the amplitude of the arc at 57′ 25″, omitting the correction for Aberration which he says is not yet allowed by all the world. By omitting the Aberration the two determinations of the amplitude, by two different stars, agree to a second. The length of the degree is first stated as 57437 toises; but this is the length which Maupertuis really obtained by allowing for Aberration, and is, I presume, a misprint. Afterwards, the number is given as 57497: and this is what it should be if we neglect Aberration.

The five operations by Cassini are those which are described in the Paris *Mémoires* for 1701, 1713, 1718, 1733 and 1734. The writer says in his usual jesting manner that since all these operations were in favour of the oblong form he is astonished that any more should be sought; and he often recalled the saying of an ancient, that if ignorance is the punishment of too little study, uncertainty is often the reward of too much.

149. In the second part of the work, we have notices of the authors who had discussed the theory of the Figure of the Earth. For the oblate form Huygens, Newton, David Gregory, and Hermann are brought forward. For the oblong form the far less eminent names of Childrey, Burnet, Eisenschmidt and Mairan are brought forward. Childrey seems to have been the author of a description of England; the others we have already mentioned.

150. Let us now turn to that part of the volume which is devoted to the consideration of the dissertations published by Desaguliers.

The title-page has the motto:

Magnus sine viribus ignis
Incassum furit.

VIRG. *Georg.* Lib. III. v. 99, 100.

There is no statement that this is a second edition; it is dated 1741.

After the title, we have a notice by the bookseller; he ascribes the work to a learned friend to whom he had shewn the former work. La Lande does not say by whom it was written, but I presume that the whole volume is really by the same author, that is, by Maupertuis.

The work shews that some of the objections which Desaguliers had brought against Cassini were really unfounded; especially those in the first of the three dissertations. It will be made clear hereafter, that Desaguliers was not judicious in his criticisms: see Chapter VIII.

CHAPTER V.

STIRLING.

151. STIRLING was the first person who turned his attention to the important point which had been assumed by Newton in his theory of the Figure of the Earth; see Art. 44. The memoir which we shall now notice is entitled, *Of the Figure of the Earth, and the Variation of Gravity on the Surface. By Mr. James Stirling, F.R.S.*

The memoir occupies pages 98...105 of Number 438 of the *Philosophical Transactions*, which is for the months July, August and September, 1735. The Number forms part of Vol. XXXIX. which is for the years 1735, 1736, and is dated 1738.

152. Stirling begins thus:

The Centrifugal Force, arising from the Diurnal Rotation of the Earth, depresseth it at the Poles, and renders it protuberant at the Equator; as has been lately advanced by Sir *Isaac Newton*, and long ago by *Polybius*, according to *Strabo* in the Second Book of his *Geography*. But although it be of an oblate spheriodical Shape, yet the kind of that Spheroid is not yet discovered; and therefore I shall suppose it to be the common Spheroid generated by the Rotation of an Ellipsis about its lesser Axis; although I find by Computation, that it is only nearly, and not accurately such. I shall also suppose the Density to be every where the same, from the Center to the Surface, and the mutual Gravitation of the Particles towards one another to decrease in the duplicate Ratio of their Distances.

The late Sir J. W. Lubbock says in the Preface to his *Account of the "Traité sur le Flux et Réflux de la Mer" of Daniel Bernoulli:*

I have searched in Strabo in vain for the remarkable passage alluded to by Stirling; but at all events the glory of the discovery of the true figure of the Earth belongs to Newton.

Perhaps Sir J. W. Lubbock expected too much. Strabo certainly says that Polybius supposed the equatorial regions to be elevated. See page 97, near the bottom, of Casaubon's edition of Strabo, the paging of which is given in the margins of other editions. See also a note on page 254 of Vol. I. of the French translation of Strabo by De la Porte du Theil and Coray.

153. Stirling states without demonstration approximate results respecting a homogeneous oblatum. He gives the direction and the magnitude of the action which the oblatum exerts on a particle at its surface, both when the oblatum does not revolve, and when it does. The approximations are true to the order of the square of the excentricity of the generating ellipse.

Let P denote any point on the generating ellipse; let CA and CB be the semi-axes. Let PG be the normal at P, meeting the greater axis at G. Take $CH = \frac{3}{5} CG$.

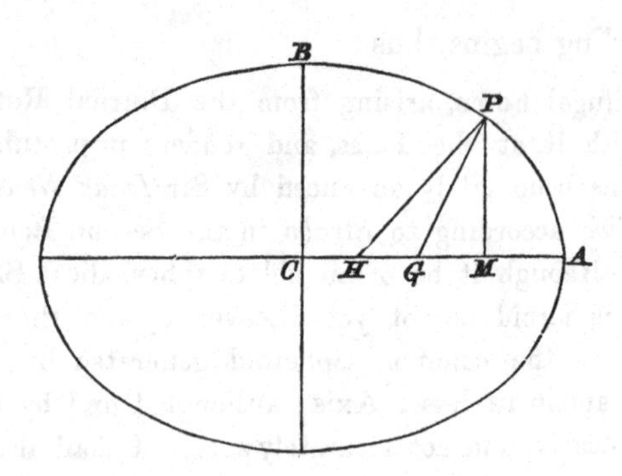

Then Stirling says, when there is no rotation PH is the direction of gravity and proportional to the value of it.

Draw PM perpendicular to CA; let $CM = x$, and $PM = y$; let X and Y denote the attractions at P parallel to CA and CB respectively. Then if ρ denote the density, and e the excentricity, we have by the modern theory

$$X = 2\pi\rho (1 - e^2)\, x \int_0^{\frac{\pi}{2}} \sin^3\theta\, (1 - e^2 \sin^2\theta)^{-1}\, d\theta,$$

$$Y = 4\pi\rho y \int_0^{\frac{\pi}{2}} \cos^2\theta \sin\theta \,(1 - e^2 \sin^2\theta)^{-1}\, d\theta\,;$$

see *Statics*, Chapter XIII.

If we neglect e^4 and higher powers of e we shall obtain

$$X = \frac{4\pi\rho}{3} x \left(1 - \frac{e^2}{5}\right), \quad Y = \frac{4\pi\rho}{3} y \left(1 + \frac{2e^2}{5}\right);$$

thus
$$\frac{Y}{X} = \frac{y\left(1 + \frac{2e^2}{5}\right)}{x\left(1 - \frac{e^2}{5}\right)} = \frac{y}{x\left(1 - \frac{3e^2}{5}\right)}$$

approximately.

But by the nature of the ellipse $CG = e^2x$, so that

$$CH = \frac{3e^2x}{5}, \text{ and } HM = x\left(1 - \frac{3e^2}{5}\right).$$

Thus the component attractions may be represented by PM and MH in magnitude and direction; and therefore the resultant may be represented by PH in magnitude and direction.

When there is rotation and relative equilibrium PG represents the resultant action in magnitude and direction. Stirling does not make any distinction in language corresponding to the fact that this statement is *exact* while the former is *approximate*. We know that for the equilibrium of a fluid, the resultant action must be normal to the surface, so that PG is exactly the direction of this action. Now take the expressions given for X and Y, and introduce the centrifugal force; then the actions at P parallel to the axis of y and x respectively will be μy and λx, where μ and λ are constants: so that these actions are proportional to y and $\frac{\lambda}{\mu} x$ respectively. But as we know that PG is the direction of the resultant, the components must be proportional to PM and MG, respectively; hence MG must be equal to $\frac{\lambda}{\mu} x$, and PG will represent the resultant in magnitude and direction.

This simple process does not occur very often in works on the subject: it is given on page 113 of Laplace's *Théorie...de la Figure elliptique des Planetes*, at least substantially.

154. Let λ denote the latitude of P, that is, the angle PGM: this will of course be very nearly equal to the angle PCM. We can express PH and PG in terms of λ and the elements of the ellipse; thus we obtain the following approximate results which in effect Stirling gives: first suppose no rotation, then if F denote the attraction at the pole, the attraction at P is $F\left(1 - \frac{e^2}{10}\cos^2\lambda\right)$; next suppose rotation, then if G denote the gravity at the pole, the gravity at P is $G\left(1 - \frac{e^2}{2}\cos^2\lambda\right)$.

In the diagram of Art. 153, the attraction at P is denoted by PH, and the gravity at P by PG: thus, as Stirling remarks, HG represents the centrifugal force at P.

It is easy to give *exact* statements of the nature of Stirling's *approximations*; this, as we shall see hereafter, was done by Thomas Simpson.

155. Stirling applies the expression for the value of gravity at any point of the surface to some observations respecting the relative number of vibrations of the seconds pendulum at London and at Jamaica; he deduces from these observations $\frac{1}{191}$ as the ellipticity: but he goes on to shew that this value is inadmissible.

Stirling makes the following remark respecting pendulum observations:

From all the Experiments made with Pendulums, it appears that the Theory makes them longer in Islands, than they are found in fact...This Defect of Gravity in Islands is very probably occasioned by the Vicinity of a great Quantity of Water, which being specifically lighter than Land, attracts less in Proportion to its Bulk.

Modern writers however appear to suggest that gravity may be *greater* on islands than on continents: see Airy's *Figure of the Earth* in the *Encyclopædia Metropolitana*, page 230, and Stokes's *Variation of Gravity at the Surface of the Earth* in the *Cambridge Philosophical Transactions*, Vol. VIII.

156. We have seen in Art. 44, that Newton assumed without demonstration an oblatum as a possible form of relative equilibrium for a mass of revolving fluid. Laplace asserts that the

defect was first supplied by Clairaut in the *Philosophical Transactions* for 1737; see the *Mécanique Céleste,* Vol. v. page 6. But perhaps we may consider that Stirling had already obtained this result. The main thing to be proved was that the resultant action at any point of the surface would be normal to the surface, when a proper relation was established between the ellipticity and the ratio of the centrifugal force to the attraction. The relation, in the notation we have used, is that $\frac{\lambda}{\mu} = \frac{GM}{CM}$; that is, $\frac{\lambda}{\mu} = 1 - e^2$. I do not say that Stirling gives this relation *explicitly*; but it seems to me *implied* in his remarks. Such too appears to have been the opinion formed at the time; as we may infer from a passage in the *Philosophical Transactions,* Vol. XL. page 278, which will be quoted in Art. 168. See also Lubbock, *Account of the Traité...,* page vi. However, Stirling's results were given without demonstration; moreover, we find from the passage in the *Philosophical Transactions,* to which reference has just been made, that they could not have been known to Clairaut when he wrote his first paper on the subject; so that Clairaut's merits remain undiminished.

157. I find it difficult to ascertain what opinion Stirling held as to the agreement of the theory with facts. He says, as we have seen, in his commencement referring to the Earth's elliptic figure, "that it is only nearly, and not accurately such." But further on he says very positively:

And whereas the Earth could not be of an oblate spheroidical Figure, unless it turned round its Axis; nor could it turn round its Axis, without putting on that Figure...

Moreover he compares his theory with observation in the case of Jupiter, and finds them to agree nicely; then he says:

And if this Theory agrees so well with Observations in *Jupiter*, there is no doubt but it will be more exact in the Earth, whose Diameters are much nearer to Equality.

After he has made the suggestions respecting pendulum observations on islands, which we have quoted, he gives the following statements:

And I find by Computation, that the Odds in the Pendulums betwixt Theory and Practice is not greater than what may be accounted for on that Supposition. I shall also observe, that although the Matter of the Earth were entirely uniform, yet the Hypothesis of its being a true Spheroid is not near enough the Truth to give the Number of Vibrations which a Pendulum makes in twenty-four Hours.

He concludes thus:

But after the *French* Gentlemen who are now about measuring a Degree, and making Experiments with Pendulums in the North and South, shall have finished their Design, we may expect new Light in this Matter.

158. Stirling's mathematical powers were highly esteemed by his contemporaries. Clairaut calls him "one of the greatest Geometricians I know in Europe." *Philosophical Transactions*, Vol. XL. page 278. See also Maclaurin's *Fluxions*, page 691; Todhunter's *History of the Theory of Probability*, pages 188, 190.

Stirling's name seems to be omitted in the ordinary biographical dictionaries. The *Abridgement of the Philosophical Transactions* by Hutton, Shaw, and Pearson, contains some notices entitled *Biography; or, Account of Authors*. All that is there recorded of Stirling is in Vol. VI. page 428, where we read: "This very respectable mathematician was agent for the Scotch Mine Company, Leadhills. He died the 5th of December, 1770." Sir John Leslie gives an interesting notice of Stirling in the *Dissertation on the Progress of Mathematical and Physical Science*, which forms part of the *Encyclopædia Britannica*: see page 711 in the eighth edition of the *Encyclopædia*.

CHAPTER VI.

CLAIRAUT.

159. In this Chapter we shall give an account of certain memoirs by Clairaut; these exhibit the high mathematical power of their author, and form the origin of the researches afterwards embodied by him in his great work entitled *Théorie de la Figure de la Terre.*

160. In the Paris *Mémoires* for 1733, published in 1735, we have a memoir by Clairaut, entitled *Détermination géométrique de la Perpendiculaire à la Méridienne tracée par M. Cassini; avec plusieurs Méthodes d'en tirer la grandeur et la figure de la Terre.* The memoir occupies pages 406...416 of the volume.

Clairaut shews that by such a process as Cassini adopted, the curve of minimum length between its extreme points on the surface of the Earth is obtained; and this curve is not in general a plane curve, unless the Earth is a sphere.

Clairaut then proceeds to investigations respecting curves of minimum length. For a surface of revolution he obtains the property, now well known, that the sine of the angle made by the curve at any point with the meridian varies inversely as the length of the perpendicular from the point on the axis of revolution. He gives special attention to the case in which the surface is an ellipsoid of revolution.

A mistake occurs on page 414, which also influences page 416. Clairaut says that if m is greater than unity $u^2 - 1 + \frac{u^2}{p^2}$ is obviously greater than $\frac{u^2-1}{m^2}+\frac{u^2}{p^2}$; but u^2-1 is a negative quantity, and so his statement is wrong.

161. In the Paris *Mémoires* for 1735, published in 1738, we have a memoir by Clairaut entitled *Sur la nouvelle Méthode de M. Cassini, pour connoître la Figure de la Terre.* The memoir occupies pages 117...122 of the volume.

This memoir consists of simple and interesting investigations of the geometrical theorems involved in the application of Cassini's method.

An important proposition in solid geometry occurs here, perhaps for the first time. At any point, M, of a surface of revolution, let a normal section be made at right angles to the plane of the meridian; then the radius of curvature of this section at M is the length of the normal between M and the axis of revolution. Clairaut's demonstration is sound; but he leaves to his readers the trouble of constructing a diagram without any directions.

162. In the Paris *Mémoires* for 1736, published in 1739, we have a memoir by Clairaut, entitled *Sur la Mesure de la Terre par plusieurs Arcs de Méridien pris à différentes Latitudes.* The memoir occupies pages 111...120 of the volume.

Let x be the abscissa and y the ordinate of any point on a curve; and suppose that the radius of curvature is equal to $a+bA+cA^2+\ldots$, where A is the angle whose tangent is $\frac{dx}{dy}$, and $a, b, c,\ldots$ are constants. Then Clairaut shews how we may express x and y in terms of z, which denotes the sine of A.

He practically confines himself to the case in which the above series contains only the three terms explicitly given; and for this case he calculates some numerical results which might be useful for application to the arcs about to be measured in Lapland and Peru, compared with that measured in France.

Let m denote the excess of the radius of curvature at the equator above that at latitude 45°, and let p denote the excess of the radius of curvature at latitude 45° above that at 67°; then Clairaut finds $a - \frac{2551m + 2904p}{3283}$ for the equatorial semi-diameter, and $a + \frac{200m - 2263p}{3283}$ for the polar semi-diameter. I have corrected a sign in the former value. On the Cassinian hypothesis m and p will both be positive, on the Newtonian hypothesis they will both be negative.

163. We have next to consider a memoir by Clairaut entitled *Investigationes aliquot, ex quibus probetur Terræ figuram secundum Leges attractionis in ratione inversâ quadrati distantiarum maximè ad Ellipsin accedere debere, per Dn. Alexin Clairaut, Reg. Societ. Lond. et Reg. Scient. Acad. Paris. Soc.*

This memoir occupies pages 19...25 of Number 445 of the *Philosophical Transactions;* which is for the months January... June, 1737. The Number forms part of Vol. XL. which is for the years 1737, 1738, and is dated 1741.

The object of the memoir is to demonstrate Newton's postulate; see Art. 44. Clairaut obtains an approximate expression for the attraction of an oblatum at any point of its surface; and thus shews, that with a suitable value of the ellipticity the resultant of the attraction and centrifugal force at any point of the surface will be normal to the surface at that point.

164. In Clairaut's work on the *Figure of the Earth* he did not reproduce this approximate solution of the problem of the *homogeneous* oblatum; for Maclaurin had in the meantime given an exact determination of the attraction of such a body, and so Clairaut followed him and exhibited an exact solution: see Clairaut's *Figure de la Terre,* page 157. But the method used in this memoir for the homogeneous oblatum is used in the work for the heterogeneous oblatum: pages 233...243 of the work reproduce the essence of this very ingenious method.

165. In this memoir we have for the first time the approximate method of determining the attraction of an oblatum on a

particle at its pole, which still retains a place in elementary works: see *Statics*, Art. 217. The method occurs in the *Figure de la Terre*, pages 239...243, where it is used for a particle situated at any point of the polar axis produced.

166. We may observe that Clairaut's memoir begins rather inauspiciously by apparently adopting the error we have noticed in Newton and David Gregory: see Arts. 33 and 84. However, as we proceed we find that Clairaut really understood the theorem correctly: see especially page 24 of the memoir, and also pages 188...190 of the *Figure de la Terre.*

167. The next memoir is entitled, *An Inquiry concerning the Figure of such Planets as revolve about an Axis, supposing the Density continually to vary, from the Centre towards the Surface; by Mr. Alexis Clairaut,* F.R.S. *and Member of the Royal Academy of Sciences at Paris. Translated from the French by the Rev. John Colson, Lucas. Prof. Math. Cantab. and F.R.S.*

This memoir occupies pages 277...306 of Number 449 of the *Philosophical Transactions,* which is for the months August and September, 1738. The Number forms part of Vol. XL.

168. Clairaut begins by adverting to Newton's researches on the Figure of the Earth, and especially to his important postulate; see Art. 44. Clairaut says:

What at first seem'd to me worth examining, when I apply'd myself to this Subject, was to know why Sir *Isaac* assumed the Conical Ellipsis for the Figure of the Earth, when he was to determine its Axis......

I began then with convincing myself by Calculation, that the Meridian of the Earth, and of the other Planets, is a Curve very nearly approaching to an Ellipsis; so that no sensible Error could ensue by supposing it really such. I had the Honour of communicating my Demonstration of this to the ROYAL SOCIETY, at the Beginning of the last Year; and I have since been inform'd, that Mr *Stirling*, one of the greatest Geometricians I know in *Europe*, had inserted a Discourse in the *Philosophical Transactions*, No. 438. wherein he had found the same thing before me, but without giving his Demonstration. When I sent that Paper to *London*, I was in *Lapland*, within the frigid *Zone*, where I could have no Recourse to Mr *Stirling*'s Discourse, so that I could not take any Notice of it.

Of course Clairaut did not demonstrate, as he says, that the meridian *is* nearly an ellipse, but only that an ellipse is *an* approximate solution. As we have stated in Art. 130, the earlier writers often *assumed* that a fluid mass, if acted on by no external force, would *necessarily* assume a spherical form. In like manner, when Newton's postulate had been established, it was often assumed, as here implicitly by Clairaut, that a fluid mass rotating with uniform angular velocity, and in relative equilibrium, would *necessarily* assume the form of an oblatum.

169. The first part of the present memoir determines the attraction at any point of an ellipsoid of revolution, supposing it to be composed of similar strata varying in density. The investigations are only approximate, extending to the first power of the ellipticity.

All that this part of the memoir contains is included in Clairaut's *Figure de la Terre;* but in the work there is a gain both as to simplicity and to generality. Problem I. of the memoir corresponds to Section 45 on pages 239...243 of the work. Problem II. and Problem III. of the memoir are included in Section 46 on pages 243...247 of the work. The Theorem on page 282 of the memoir corresponds to Section 44 on pages 236...239 of the work. Problem IV. of the memoir corresponds to Sections 24 and 25 on pages 200...202 of the work. Problem V. of the memoir corresponds to Section 26 on pages 203...208 of the work; the investigation is given at full in the work, but only the result in the memoir. Problem VI. and Problem VII. of the memoir are included in Section 29 on pages 209...218 of the work.

The work is more general than the memoir. In the memoir it is assumed that the strata are similar, so that the ellipticity is the same for all the strata; in the work this is not assumed. In the work the formulæ contain a general symbol to represent the density; in the memoir a law of density is assumed, the density being denoted by $fr^p + gr^q$, where f, g, p, q are constants, and r is the variable polar semi-axis of the strata: the integrations are effected in the memoir, but the formulæ are thus rendered less simple in appearance than they are in the work.

170. The second part of the memoir contains the application of the first part, to find the figure of a nearly spherical fluid mass which rotates about an axis.

This part is unsatisfactory, because the only condition of equilibrium which Clairaut regards is, that the resultant action at every point of the free surface shall be normal to the surface at the point. This is not *sufficient* for the equilibrium of a heterogeneous fluid mass. Clairaut discovered his error, and acknowledged it; see page 155 of his *Figure de la Terre:* here he allows that his investigations in the memoir are untenable, except on the supposition that the interior parts of the Earth had been originally solid. In the Sections 37 and 39, on pages 225, 226, 228, and 229 of the work, we have an equivalent for pages 288...294 of the memoir, but expressed more accurately.

171. On page 294 of the memoir, we have the first appearance of the theorem which is now known as *Clairaut's Theorem:* see Section 49, on pages 249, 250 of the *Figure de la Terre.* We will state the theorem. From the value of gravity at the pole subtract the value of gravity at the equator, and divide the remainder by the value of gravity at the equator; this fraction we shall call *Clairaut's fraction.* Then Clairaut's Theorem asserts that *the sum of the ellipticity of the surface and Clairaut's fraction is equal to twice the ellipticity of the Earth considered as a homogeneous fluid.* We shall defer the demonstration of the theorem until we give an account of Clairaut's *Figure de la Terre.*

172. Clairaut deduces from his theorem a result contrary to a statement made by Newton; see Art. 30.

Clairaut, speaking of Newton, says:

He affirms, that the Earth is denser towards the Centre than at the Superficies, and more depress'd than his Spheroid requires. But by the foregoing Theory we may easily perceive, that if the Density of the Earth diminishes from the Centre towards the Superficies, the Diminution of Gravity from the Pole towards the Equator will be greater than according to Sir *Isaac's* Table; but at the same time the Earth will not be so much depress'd as his Spheroid requires, instead of being more so, as he affirms.

The two statements made by Clairaut are connected by his Theorem, so that one will follow from the other. In the Section 38, on pages 226, 227 of his work, he shews that if the density diminishes from the centre to the surface, the ellipticity is *in general* less than for the homogeneous body: the condition which prevents the statement from being universally true is there given.

Clairaut proceeds to say:

Yet I would not by any means be understood to decide against Sir *Isaac*'s Determination, because I cannot be assured of his Meaning, when he tells us, that the Density of the Earth diminishes from the Centre towards the Circumference. He does not explain this, and perhaps instead of the Earth's being compos'd of parallel Beds or *Strata*, its Parts may be conceived to be otherwise arranged and disposed, so as that the Proposition of Sir *Isaac* shall be agreeable to the Truth.

In his *Figure de la Terre*, however, Clairaut does not hesitate to decide against Newton: see Art. 30.

173. As an example, Clairaut takes the following case:

Setting aside all Attraction of the Parts of Matter, if the Action of Gravity is directed towards a Centre, and is in the reciprocal Ratio of the Squares of the Distances, the Ratio of the Axes of the Spheroid will then be that of 576 to 577 And the Gravity at the Pole is greater than at the Equator by $\frac{1}{144}$th Part, or thereabouts. Which may be a Confirmation of what is here advanced, especially to such as will not be at the Pains of going through the foregoing Calculations. For we may consider the Spheroid now mention'd, in which Gravity acts in a reciprocal Ratio of the Squares of the Distances, as composed of Matter of such Rarity, in respect of that at the Centre, that the Gravity is produced only by the Attraction of the Centre or Nucleus.

This is the first appearance of a problem which may be described as a companion to that discussed by Huygens; and which has sometimes been erroneously ascribed to Huygens: see Art. 64.

174. Clairaut makes some remarks on the two principles which were then in use for determining the form of a fluid in equilibrium, namely, Newton's principle of balancing columns and Huygens's principle of the plumb-line: he states the reasons which induced him to adopt the latter principle. He proceeds to examine whether the solution which he has obtained does make the polar and equatorial columns balance; he finds that, in order to secure this, a certain relation must hold among the constants which enter into the expression for the density. In fact, as we have already stated, Clairaut's solution in the memoir did not satisfy all the necessary conditions: see Art. 170.

175. Clairaut demonstrates a result on pages 302...304 of the memoir, which though quite obvious on the modern theory of fluid equilibrium must have appeared remarkable at the time. We will state the general proposition of which his result is a particular case. Suppose a solid, not necessarily homogeneous, covered with a stratum of homogeneous fluid which is in equilibrium; then if a fine channel be made in the body from one point of the fluid to another, and be filled with the fluid, the fluid in the channel will remain in equilibrium. In fact, the pressure p at any point of the channel of fluid can theoretically be found so as to satisfy the necessary conditions.

176. The memoir closes with some reference to the results obtained by observations. Clairaut admits that those furnished by the expedition to Lapland do not agree well with the theory; for, according to these, each of the two fractions which occurs in *Clairaut's Theorem* is greater than $\frac{1}{230}$. However he will wait for the observations made in Peru.

177. In the Paris *Mémoires* for 1739, published in 1741, there is a memoir by Clairaut, entitled *Suite d'un Mémoire donné en* 1733, *qui a pour titre: Détermination Géométrique de la Perpendiculaire à la Méridienne, &c.* The memoir occupies pages 83...96 of the volume.

In modern language we should say that this memoir relates to geodesic curves on the surface of an ellipsoid of revolution.

The investigations are approximate, extending to the first power of the ellipticity.

It may be interesting to give a specimen of Clairaut's investigations.

Let the polar semi-diameter be taken for unity, and let $\frac{1}{m}$ denote the equatorial semi-diameter. Let x denote the longitude of any point in a geodesic curve, measured from the meridian which the geodesic curve crosses at right angles; let t denote the cotangent of the latitude of this point; let p denote the value of t when $x = 0$; then

$$\frac{dx}{dt} = \frac{pm\sqrt{(1+t^2)}}{t\sqrt{(t^2+m^2)}\sqrt{(t^2-p^2)}}.$$

Clairaut established this formula in his memoir of 1733; and it may be easily obtained from well-known works on solid geometry.

Now put $m = 1 - \alpha$, and suppose α so small that its square may be neglected; thus we get

$$\frac{dx}{dt} = \frac{p}{t\sqrt{(t^2-p^2)}} - \frac{\alpha pt}{(1+t^2)\sqrt{(t^2-p^2)}}:$$

hence $$x = \sin^{-1}\frac{\sqrt{(t^2-p^2)}}{t} - \frac{\alpha p}{\sqrt{(1+p^2)}}\sin^{-1}\frac{\sqrt{(t^2-p^2)}}{\sqrt{(1+t^2)}} \quad\text{.............(1).}$$

Clairaut does not use the symbol $\sin^{-1}$; but he proposes the symbol As to denote what we denote by $\sin^{-1} s$.

The equation (1) determines x when t is known. Now Clairaut proceeds to determine t from it when x is known; and for this he employs a special process, which we will now explain.

Suppose that $t = \tau + \Delta\tau$, where τ is the value of t which would correspond to the known value of x when α is zero, and so $\Delta\tau$ is very small. Hence from (1) we get

$$x = \sin^{-1}\frac{\sqrt{(\tau^2-p^2)}}{\tau} + \frac{p}{\tau\sqrt{(\tau^2-p^2)}}\Delta\tau - \frac{\alpha p}{\sqrt{(1+p^2)}}\sin^{-1}\frac{\sqrt{(\tau^2-p^2)}}{\sqrt{(1+\tau^2)}}$$
$$- \frac{\alpha p\tau\Delta\tau}{(1+\tau^2)\sqrt{(\tau^2-p^2)}} \quad\text{.............(2).}$$

But by supposition $x = \sin^{-1} \frac{\sqrt{(\tau^2 - p^2)}}{\tau}$.

Hence, neglecting the term which involves the product of α and $\Delta\tau$, we have from (2)

$$\frac{p}{\tau\sqrt{(\tau^2 - p^2)}} \Delta\tau = \frac{\alpha p}{\sqrt{(1 + p^2)}} \sin^{-1} \frac{\sqrt{(\tau^2 - p^2)}}{\sqrt{(1 + \tau^2)}}.$$

This furnishes the correction $\Delta\tau$, which will be required in the cotangent of the latitude when calculated for a sphere, in order to obtain the value for the ellipsoid of revolution.

Clairaut himself uses t for our τ, and dt for our $\Delta\tau$.

Clairaut's memoir consists of the solution of four problems; the other three resemble that which we have taken as a specimen. They are illustrated by numerical application to an oblatum in which $\alpha = \frac{1}{100}$; this value Clairaut says does not differ much from that obtained by means of the degree of the meridian measured at the polar circle.

This memoir is the last of the series of Clairaut's contributions to our subject before the publication of his work entitled *Théorie de la Figure de la Terre*, which we shall examine in Chapter XI.: we now proceed to give an account of the measurement in Lapland, to which allusion has just been made.

CHAPTER VII.

ARC OF THE MERIDIAN MEASURED IN LAPLAND.

178. THE Academy of Sciences at Paris seems to have selected the problem of the Figure of the Earth as peculiarly its own. But the success hitherto attained scarcely corresponded to the labour which had been expended; partly perhaps owing to the fact that the able observers, trained by the astronomers who bore the justly celebrated name of Cassini, had adopted the oblong form and maintained it firmly.

In order to settle the question in dispute between the Cassinians and the Newtonians, the scheme was seriously proposed in 1733 of measuring an arc of the meridian near the equator, in order to compare the corresponding length of a degree with that which had been obtained from the French arc by Picard and by J. Cassini. The task was entrusted to three members of the Academy, Bouguer, La Condamine, and Godin, who started in May, 1735. Two Spanish naval officers, Juan and Ulloa, assisted in the work.

179. After this expedition had started for Peru it was resolved to measure also an arc as near as possible to the pole: see La Condamine, *Journal du Voyage...* page 1. This task was entrusted to four members of the Academy, Maupertuis, Clairaut, Camus, and Le Monnier; moreover l'Abbé Outhier, who was a correspondent of the Academy, and Celsius, who was professor of Astronomy at Upsal, were associated with the Academicians.

180. The Arctic expedition seems to me to have been stronger than the Equatorial. The genius of Clairaut outshone that of the

whole Academy, which was not yet adorned by the rising splendour of D'Alembert. But even if we leave out of consideration this transcendant name the superiority remains, I think, still with the Arctic party. I should place Maupertuis, Camus, and Le Monnier, above Bouguer, La Condamine, and Godin; while the priest and the professor who accompanied the former are at least equal to the two sailors who assisted the latter.

The two operations were conducted on different principles. The members of the Arctic expedition worked in harmony under the general direction of Maupertuis. La Condamine calls Maupertuis, the senior (*l'ancien*) of the party, *Journal du Voyage...* page iii.; and Maupertuis is called *Chef de l'entreprise du Nord* in the *Histoire de l'Académie*...for 1737, page 96. There was but little cordiality in the Equatorial party; and the three Academicians performed much of their work separately. Thus in the former case we find friendship and subordination; and in the latter case isolation and independence. On a purely scientific estimate it may be maintained that there are advantages in each course which the other does not secure.

We are here concerned only with the Arctic party which left Paris on the 20th of April, 1736. Two narratives of the proceedings were printed; we will now describe these works.

181. Maupertuis published *La Figure de la Terre déterminée par les observations...au cercle polaire.* Paris, 1738. This is an octavo volume; the Title, Preface, and Table of Contents, occupy xxviii. pages; the text occupies 184 pages; there are 9 plates besides a map.

In the historical portion of the Paris *Mémoires* for 1737, pages 90...96 relate to the Arctic expedition: the date of publication is 1740. Moreover, in this volume, pages 1...130 of Maupertuis's work are reprinted; they occupy pages 389...465 of the volume. Maupertuis here says there have been too many editions of his book in various languages to render it necessary to repeat the other observations made in the North: he contents himself with referring to the observations on the force of gravity, and reproduces the Table which occurs on page 181 of his book.

It is stated by La Condamine that Maupertuis's work was translated into all the languages of Europe: *Journal du Voyage...* page iii. I have seen a German translation and a Latin translation. The German translation was published at Zurich in 1741; it contains also a dedication to Frederic III. of Prussia, by Samuel König, an introduction by the translator, and a memoir by Celsius on Cassini's work *De la Grandeur et de la Figure de la Terre.* The Latin translation was published at Leipsic in 1742; it contains also an introduction by the translator, Alaricus Zeller: he says on the third page of his introduction that he has preserved the paging of the Amsterdam edition in his margin. This translator's introduction contains some criticisms which are not devoid of interest; they do not however practically affect the determination of the length of the degree of the meridian, but relate to incidental matters, such as refraction. There are also a few notes to the translation, which supply corrections of slight misprints or mistakes.

There is an English translation which I have not seen.

182. Outhier published *Journal d'un Voyage au Nord...*, Paris, 1744. This is a quarto volume; the Half-title, Title, Dedication, and Preface, are on eight pages; the text occupies 238 pages, followed by two pages which contain an *Extrait des Registres de l'Académie...*, and the *Privilege du Roi.* According to the *Table des Figures* on page 238, there ought to be 18 plates. But in the single copy which I have seen there are only 16 plates. The plate which is marked 15 in the list does not occur; there is only one plate corresponding to the two which are marked 9, 10 on the list; and there are only two plates corresponding to the three which are marked 6, 7, 8 on the list. On the other hand, there is a *Veüe de la Montagne de Niemi, du côté du Midy,* which is not named in the list.

Outhier's work seems never to have attracted much attention and to be now scarce.

183. The calculations and the theoretical deductions are given most fully by Maupertuis; the details of the daily occupations of the party, and the peculiarities of the country and of the inhabit-

ants, are given most fully by Outhier. I shall refer to the pages of Maupertuis in the original French edition, and distinguish them by the letter M. I shall refer to Outhier's work by the letter O.

184. Maupertuis was for a long time in doubt whether he should go to Iceland, to Norway, or to the Gulf of Bothnia; he decided for the last, intending to carry on his operations among the islands along the shores of the Gulf. O. 3. But on examination these islands were found to be too low, and too near the shore, to form advantageous stations; and after some consideration Maupertuis resolved to proceed to the mountains north of Tornea, which is at the head of the Gulf. M. 11; O. 52.

Finally Tornea was taken as the most Southern station, and Kittis as the most Northern; both are on the river Tornea, and nearly on the same meridian. The other stations were mountains not far from the river. The base which was to be measured was chosen about midway between Tornea and Kittis, and the extremities denoted by signals. M. 29; O. 86.

All the geodetical angles were observed in the space of about two months, between the beginning of July and the beginning of September, 1736. The observations were made with a quadrant of two feet radius. M. 33, 79; O. 204...219.

185. The next step was to determine the difference of latitude of the extreme points of the arc. The star δ Draconis was selected which passed the meridian very near to the zenith; observations of this star were made at Kittis on the 4th, 5th, 6th, 8th, and 10th of October; and at Tornea on the first five days of November. The difference of zenith-distance was found to be $57' 25''.55$. M. 104.

The instrument used for determining this difference of zenith-distance was a zenith-sector made by Graham at London; instrument resembled that used by Bradley in the observations which established the aberration of light. M. 38. A copper telescope-tube of nine feet long formed one radius of the sector; the extent of the arc of the sector was $5^{\circ}\frac{1}{2}$, graduated at every $7'\frac{1}{2}$. At the focus of the telescope were fixed two wires at right angles.

The telescope and the arc formed one instrument. A large pyramid of wood 12 feet high served as the support of the instrument. M. 38, 94. The instrument could turn freely round a horizontal axis; it was moved by a micrometer screw acting in opposition to a weight. A plumb-line was suspended from the centre of motion, and marked on the graduated arc the angle through which the instrument had been turned. The absolute zenith-distance of a star at a given place was not determined by the French observers, but only the *difference* of zenith-distance at two given places.

186. The base was measured on the frozen surface of the river Tornea, very nearly in the direction of the stream; the extremities of the base were on the land. The measurement was begun on December 21st, and occupied a week. Eight rods of fir were employed, each five toises long; the correct length of these rods was determined by the aid of an iron toise which had been carefully adjusted to the length of the standard toise at Paris. O. 137. This iron toise is known henceforth in the history of the subject as the *Toise du Nord*. A similar iron toise had been taken by the Equatorial expedition, which is known as the *Toise du Pérou*. Neither Maupertuis nor Outhier records the fact that these two toises were made at the same time and by the same artist, Langlois; this we learn from La Condamine: see the Paris *Mémoires* for 1772, Part II. pages 482...501.

187. The measurers of the base divided themselves into two bands; each band had four of the fir rods, and measured independently: the length of the base was found to be 7406 toises 5 feet 4 inches by one band, and 7406 toises 5 feet by the other band. After the measurement was finished three of the party verified that no error could have arisen in counting the hundreds, by using a cord 50 toises long over the whole base. O. 144.

The sun scarcely rose above the horizon, but the twilight, the white snow, and the Aurora Borealis supplied enough light for four or five hours work daily. M. 51.

188. It followed from the length of the base that the length of the arc of the meridian intercepted between the parallels of

Tornea and Kittis was $55023\frac{1}{2}$ toises; and that the length of a degree of the meridian at the Arctic circle was nearly 1000 toises greater than the length calculated according to the Cassinian theory in the book *De la Grandeur et de la Figure de la Terre.* M. 58.

The party then went to Tornea and remained shut up in their chambers in a kind of inaction until March. The difference between their result and that of the Cassinian theory was so great that it astonished them; and although they considered their operations to be incontestable, yet they resolved to execute some rigorous verifications. M. 63. We read in the Paris *Mémoires* for 1737, page 94 of the historical portion:

On la tint fort secrette, tant pour se donner le loisir de la réflexion sur une chose peu attendue, que pour avoir le plaisir d'en apporter à Paris la premiére nouvelle.

189. The angles of the triangles were supposed to admit of no doubt; these angles had been observed many times by various persons; and the three angles of every triangle had been observed. The calculations were verified by combining the triangles in a different series; and also by assuming that errors had arisen in measuring the angles, which all tended to make the length greater than it should have been. But the length of the arc of the meridian still remained without any very decided diminution. M. 63...65.

The measurement of the base was considered to be also above suspicion; thus there remained only the very important point of the difference in latitude of the extreme stations; and accordingly this was redetermined. The star α Draconis was now selected; observations of this star on the meridian were made with the zenith-sector at Tornea on the 17th, 18th, and 19th of March, 1737, and at Kittis on the 4th, 5th, and 6th of April: the difference of zenith-distance was found to be 57′ 25″. 85. M. 115.

The reason given in the Paris *Mémoires* for 1737, on page 95 of the historical portion, for going over the astronomical part of the work again is that it could be done much more easily than the other parts.

190. The observations for determining the difference of latitude required corrections for aberration, for precession, and for

a third inequality which had been recently discovered by Bradley, and which is called nutation. No correction was applied for refraction. M. 125. See Bouguer's *Figure de la Terre,* page 290.

Thus, finally, the amplitude of the arc of the meridian was 57′ 26″.93 by the star δ Draconis, and 57′ 30″.42 by the star α Draconis; the difference is 3″.49. Maupertuis considered that 0″.95 of this difference was owing to an inequality in the graduation of the sector, which was discovered by careful scrutiny. M. 124.

Maupertuis took the mean of the two results, 57′ 28″.67 for the amplitude; and from this he deduced that the length of the degree of the meridian which is bisected by the Arctic circle is 57437.9 toises.

191. Important pendulum experiments were made at Pello, which is close to Kittis. The result is that a pendulum which oscillates in a second at Paris will make 59 more oscillations in 24 hours at Pello than at Paris. M. 172.

192. The Academicians endured great hardships during their operations. The severe cold of the winter months must have been anticipated; and the precautions which the natives had learned from experience would afford some mitigation of this evil. But the most painful period of the survey seems to have been that which was spent among the mountains in observing the geodetical angles: in one instance they remained for ten days on a mountain. M. 21. The exposure to extremes of heat and of cold, the excessive rains, and the want of proper food, all contributed to the sufferings of the party. But the worst torment seems to have been that inflicted by insects. Maupertuis calls them flies, and says they were of different kinds. M. 14, 16, 22. Outhier calls them by various names; flies, gnats, midges: thus *cousins* 55, 57, 58, 59, 63, 64, 74, 82; *moucherons* 64, 65, 75, 79, 82; *mouches* 57, 58, 64. Le Monnier fell very ill. M. 24; O. 75, 79, 81. According to Hutton's *Mathematical Dictionary* the health of Maupertuis was permanently impaired by the hardships he underwent.

The Academicians left Tornea in June, 1737, and reached Paris in August.

193. The measurement of the arc of the meridian by the French in Lapland is historically the most important of all such operations. The question as to the oblate or oblong form of the Earth was decisively settled.

Two generations of the best astronomical observers formed in the school of the Cassinis had struggled in vain against the authority and the reasoning of Newton.

194. Some incidental matters may now be noticed which present themselves in studying the narratives.

Maupertuis says on his page xii.:

Sur des routes de 100 degrés en Longitude, on commettroit des erreurs de plus de 2 degrés, si naviguant sur le Sphéroïde de M. Newton, on se croyoit sur celui du Livre de la Grandeur et Figure de la Terre.

I cannot understand this. Nothing is said about the *latitude;* but the amount of error in a course of 100 degrees of longitude will depend mainly on the latitude.

In the life of Maupertuis in the *Biographie Universelle,* which is partly by Delambre, reference is made to the exaggerations of Maupertuis on this point.

Clairaut is the mode of spelling which the bearer of this distinguished name himself adopted: Outhier, however, generally uses Clairaux; once he has Clairault. O. 25.

Maupertuis, in returning to France, was shipwrecked in the Gulf of Bothnia; he merely alludes to this misfortune himself: but we find from Outhier that the instruments were immersed, and were cleaned rather more than a month after the accident. M. 78; O. 169, 189.

195. The success of the Arctic expedition may be fairly ascribed in great measure to the skill and energy of Maupertuis: and his fame was widely celebrated. The engravings of the period represent him in the costume of a Lapland Hercules, having a fur cap over his eyes; with one hand he holds a club, and with the other he compresses a terrestrial globe. Voltaire, then his friend, congratulated him warmly for having "aplati les pôles et les Cassini." See articles entitled *Histoire des Sciences* in the *Revue*

des deux Mondes, Jan. and Nov., 1869. Readers of Carlyle's *History of Frederick the Great* will remember the allusions to the Earth-flattener.

196. Although the measurement of the Lapland arc settled the question as to the oblate or oblong form of the Earth, yet it introduced a great difficulty; for by comparing the result with that obtained from the French arc the ellipticity of the Earth appeared to be about $\frac{1}{178}$. This was greater than had been expected, and greater than subsequent operations, such as that in Peru, furnished. From our present knowledge it is certain that this value of the ellipticity is far too large.

We have seen indeed, in Art. 177, that Clairaut assigned $\frac{1}{100}$ as the ellipticity furnished by the Lapland arc; this must have been obtained by using for the French arc a certain value obtained by Maupertuis in his *Figure de la Terre*, page 126; but this value of the French arc was soon afterwards found to be too small.

197. According to La Lande, Maupertuis himself was not satisfied with his operations. We read in the *Bibliographie Astronomique*:

> je sais que Maupertuis n'en était pas lui-même très-content. Page 407.
>
> ... Au reste, on m'écrit de Suède que Maupertuis s'était proposé de recommencer la mesure à ses dépens; ce qui prouve qu'il n'en était pas très-content. Page 811.

It is well known that the Lapland arc was remeasured at the beginning of the present century by Svanberg and others under the direction of the Stockholm Academy of Sciences. La Lande alludes to the early stages of this operation; see the *Bibliographie Astronomique*, pages 811, 837, 857. Svanberg obtained a decidedly shorter length for a degree of the meridian than that of Maupertuis, namely, 57196.159 toises instead of 57437.9 toises; but the middle points of the two degrees are not quite identical.

198. We may just notice the memoir by Celsius, which is contained in the German translation of Maupertuis's *Figure de la*

Terre: see Art. 181. This is probably a translation of one which was originally published at Upsal in 1738 under the title of *De observationibus pro figurâ telluris determinandâ in Galliâ habitis disquisitio,* according to La Lande's *Bibliographie Astronomique,* page 406.

In the translation Celsius first defends the astronomical operations in Lapland from an objection which had been urged against them by J. Cassini before the Paris Academy, because the sector had not been reversed at each place of observation. Celsius maintains that this was unnecessary for the purpose of the observers, especially considering the excellence of Graham's sector. Then Celsius proceeds to criticise the French operations recorded in the work *De la Grandeur et de la Figure de la Terre;* and he considers that he shews both the astronomical and geodetical parts to be untrustworthy. These operations indeed were just about to be given up and replaced by the more accurate determinations recorded in the work *La Meridienne de Paris verifiée.*

199. For further information respecting the Lapland arc of the meridian, I may refer to my memoir on the subject published in the *Cambridge Philosophical Transactions,* Vol. XII.; I have there corrected the numerous and serious errors which have been made by distinguished astronomers in their account of this remarkable measurement.

CHAPTER VIII.

MISCELLANEOUS INVESTIGATIONS BETWEEN THE YEARS 1721 AND 1740.

200. WE have first to consider a production to which allusion has been made in Arts. 143 and 150. It is entitled *A Dissertation concerning the Figure of the Earth, by the Reverend John Theophilus Desaguliers,* L.L.D. F.R.S. This is contained in Vol. XXXIII. of the *Philosophical Transactions:* the volume is for 1724, 1725; and is dated 1726.

The dissertation consists of four parts.

201. The first part occupies pages 201...222 of the volume. This part criticises the conclusions at which J. Cassini had arrived as to the form of the Earth in his *De la Grandeur et de la Figure de la Terre,* of which we have given an account in Arts. 100...108.

Desaguliers endeavours to shew that the Cassinian figure is impossible, because it would lead to a deviation of the plumb-line, from the direction which is at right angles to the surface of water, to the amount of five minutes: but the process is unsound. We know now that under certain hypotheses as to the form of the solid nucleus, the outer surface of the fluid might be an oblongum: see Clairaut's *Figure de la Terre,* page 224.

Desaguliers maintains that the latitudes in the French survey of the meridian cannot be relied on as sufficiently accurate to establish the oblong figure of the Earth; and he is not satisfied that the heights of the mountains were properly determined. Desagulier's criticisms have perhaps some foundation; but like many controversialists he seems disposed to be unfair. For instance, he considers that the height of one mountain was over-estimated, and the height of another under-estimated; and thus, he says, we must add 20 toises to the length of the 44th degree of latitude, and take away 30 toises from the length of the 45th degree of latitude. But even admitting these corrections to be necessary, they tend to balance each other; and they produce no perceptible effect on the definite result obtained by Cassini, namely, that the whole southern arc from Paris to the Pyrenees gives a longer average length of a degree than the whole northern arc from Paris to Dunkirk.

Strictly speaking, what Desaguliers calls the 44th degree of latitude should be the 45th; and what he calls the 45th should be the 46th.

Desaguliers assigns one reason which may have induced Cassini to make the Earth oblong, in these words: "especially because in this Hypothesis, the Degrees differ most in Length from one another about the 45th Degree." But this is quite unsatisfactory. For if we suppose the Earth to be nearly spherical, then whether it be oblate or oblong the degrees will differ most in length at about the 45th degree: see Art. 140.

202. The second part of the dissertation occupies pages 239...255 of the volume. The object of this part is to shew "How the Figure of the Earth is deduc'd from the Laws of Gravity and Centrifugal Force." Instead of giving anything of his own, Desaguliers transcribes a long extract from Keill's book against Burnet; the extract consists of that matter which Keill took substantially from Huygens: see Art. 74.

Desaguliers says:

I own indeed that he has made a Mistake in that Book concerning the Measure of the Degrees of an Ellipse; but I find that all that relates to the oblate Spheroidical Figure of the Earth is right....

The mistake of course is that which we have noticed in Art. 76. Desaguliers would probably have thought it unnecessary to warrant the accuracy of the matter which he transcribed, if he had known that it was substantially all due to Huygens.

203. The third part of the dissertation occupies pages 277...304 of the volume. This part is chiefly a criticism of the memoir by Mairan which we have examined in Arts. 109...114. Much of what Desaguliers says, though quite true, would have failed to produce any effect on Mairan. For instance, according to Mairan, Paris is more distant from the centre of the Earth than a place at the equator is; *hence the attraction at Paris will be less than* it is at the equator; hence, although the centrifugal force at the equator is greater than at Paris, we may have gravity at Paris less than gravity at the equator: and this is contrary to observation. But Mairan would have declined to admit the statement in Italics; he had invented a law of attraction for himself which made the attraction greater at Paris than at the equator.

Of course the assailable part of Mairan's memoir was the arbitrary law of attraction which he had invented; and against this Desaguliers directs a decisive argument. He finds that, taking Mairan's law, and allowing for centrifugal force, the Paris seconds pendulum would have to be shortened at the equator nearly an inch. He says: "But this being about five Times more than agrees with Observation; what proves too much, proves nothing at all." See Art. 52.

Desaguliers finds, that on Mairan's law the polar and equatorial columns of fluid would not balance; but Mairan might have replied that the Earth was solid, and for this reason he might have declined to admit the principle of balancing columns.

204. Desaguliers in the third part of his dissertation returns to the subject of the French arc. He arranges a table which gives the observed latitudes of successive stations on the meridian, and also the distance from Paris in toises. He shews that there is not a constant decrease in the length of a degree in passing from the southern extremity of the arc to the northern. But the objection is of no value; because the French observers did not require, and did not attempt to find, the latitudes of intermediate stations with the same accuracy as the latitudes of Paris and of the two extremities of the arc.

Desaguliers says on page 303:

To conclude, I will propose a Method of observing the Figure of the Shadow of the Earth in Lunar Eclipses, whereby the Difference between the Diameters in the oblong spheroidical Figure, if there be such an one as Mons. *Cassini* affirms (*viz.* of 96 to 95), may be discover'd.

But the method has, I believe, no practical value.

205. The fourth part of the dissertation occupies pages 344, 345 of the volume. It consists of an account of an experiment to "illustrate" what had been said in the preceding parts. The essence of the experiment may be thus described. Take a hoop of thin elastic steel; let it revolve round a diameter as axis, the axis passing freely through the steel: then the greater the angular velocity the more will the hoop bulge out into an oblate form. The toy with which Desaguliers amused himself of course proved nothing to the point; however, he boldly asserts that from this experiment, compared with what had been said, "it will appear that the Earth cannot preserve its Figure, unless it be an oblate Spheroid."

206. There are some incidental matters of interest in the dissertation which may be noticed.

Desaguliers suggests on page 209, that

...... a Degree of Latitude shou'd be measur'd at the Æquator, and a Degree of Longitude likewise measur'd there; and a Degree very northerly, as for Example, a whole Degree might be actually measur'd upon the *Baltick* Sea, when frozen, in the Latitude of sixty Degrees.

We read on pages 219, 220:

when once an Hypothesis is set on Foot, we are too apt to draw in Circumstances to confirm it; tho', perhaps, when examin'd impartially, they may rather weaken, than strengthen our Hypothesis; otherwise, the Author of the History of the *Royal Academy*, for the Year 1713, wou'd not have alledg'd, that *the late Mons.* Cassini *observ'd* Jupiter *to be oval*, as a Proof of young Mons. *Cassini*'s Hypothesis; because *Jupiter* is oval the other Way, that is, an oblate Spheroid flatted at the Poles...

But I cannot find anything in the volume which justifies this remark by Desaguliers.

The only reference to Jupiter occurs after a notice of the fact that the Earth deviates but little from a sphere; then we read:

Si Jupiter est ovale, comme il l'a paru quelquefois à feu M. Cassini, il faut qu'il le soit bien davantage pour le parôitre de si loin.

It is obvious that these words do not bear any such meaning as Desaguliers suggests.

Desaguliers refers to the opinion of Dr Burnet, which we have noticed in Art. 74. Desaguliers says on his page 221: "But Dr. *Burnet*, afterwards, alter'd his Opinion, as I am credibly inform'd."

Desaguliers asserts "That a fluid Substance, of any Figure, will by the Gravity of its Parts become spherical, ..." He gives what he calls a demonstration of this on his pages 278, 279; but, as might be expected, his demonstration is quite inconclusive. See Art. 130.

Desaguliers adopts on his page 280 the erroneous notion that by increasing the density of the central part of the Earth, the ellipticity is also increased; see Arts. 30, 84 and 172. Newton and David Gregory do not state whether they suppose the central part still to remain fluid or to become solid. Desaguliers, however, says distinctly, "Then if, when the Central Parts are fix'd, and the superficial *Strata* are still fluid, ..."

To shew that Desaguliers is wrong, we have only to put $\alpha = 0$ on page 219 of Clairaut's *Figure de la Terre;* then we find that δ is less than $\frac{5\phi}{4}$ Or see Simpson's *Mathematical Dissertations,* page 30.

A paragraph which occurs on pages 280 and 281 is to be cancelled, according to an Advertisement by Desaguliers at the end of Number 399 of the *Philosophical Transactions.*

207. Desaguliers, on his page 285, deviates from accuracy in saying that "on different Parts of the Surface of the Earth (in the Condition it is now) the Gravity on Bodies is reciprocally as their Distance from the Centre of the Earth." I have already stated that this proposition should be enunciated thus: Gravity

resolved along the radius-vector varies inversely as the radius; see Art. 33. Desaguliers omits the resolution along the radius-vector. Moreover, I think from his context, and from a calculation on his page 287, that he made another mistake, and supposed that the *attraction* along the radius-vector varied inversely as the radius; that is, I think, he neglected the distinction between *attraction* and *gravity*. On his pages 286 and 287 he assumes that for an oblongum the gravity will vary inversely as the radius-vector; and by gravity he means here *attraction* alone, for he proceeds to allow separately for the centrifugal force. The assumption is unjustifiable, and seems to have arisen from the confusion of gravity with attraction in the case of the oblatum.

208. Desaguliers obtained from a friend a "Philosophical Argument" against Mairan; it is thus stated on his page 298:

If the Earth was of an oblong spheroidical Figure, higher at the Poles than the Æquator; the Axis of its Revolution, wou'd either go thro' one of its short Diameters, or be continually changing unless the said Axis did exactly coincide with the Axis of the Figure.

These words themselves are true; they are, however, applicable to the oblatum if we change *short* into *long*. The so-called *demonstration* which follows shews that Desaguliers and his friend were wrong in their notions on the subject. In modern language these notions amount to considering that the rotation of an oblongum round its axis of figure is *unstable*. The mechanical knowledge of the period was inadequate to the discussion of a difficult problem in Rigid Dynamics.

209. A work was published at Padua in 1728, entitled *Joannis Poleni...Epistolarum Mathematicarum Fasciculus*. The work is in quarto; the pages are not numbered.

One of the letters relates to the Figure of the Earth; it is addressed "Viro celeberrimo Abbati Gui. Grando." This letter occupies eleven pages; it is of little importance.

Since some persons maintained that the Earth was oblate, and others that it was oblong, Poleni considers it safer to adopt the spherical form as a compromise between the two extremes. He suggests, however, that by measuring an arc of longitude, say in latitude 48°, a test might be obtained as to the two extreme hypotheses. For, assuming the same perimeter of the meridian in the two cases, the arc of longitude would be much shorter if the figure be an oblongum than if it be an oblatum. Poleni states that for an arc of one degree of longitude, the difference would be about 777 toises. See Art. 215.

He considers that the spherical form may be reconciled with the existence of centrifugal force, by supposing the Earth not to be homogeneous.

210. Some pendulum observations were made at Archangel in 1728 by L. Delisle de la Croyere. They are recorded in the *Commentarii Academiæ...Petropolitanæ*, Vol. IV. which is for 1729, and was published in 1735: see pages 322...328 of the volume.

211. In the Paris *Mémoires* for 1732, published in 1735, there is a memoir entitled *Réponse aux Remarques qui ont été faites dans le Journal Historique de la République des Lettres sur le Traité De la Grandeur et de la Figure de la Terre. Par M. Cassini.* The memoir occupies pages 497...513 of the volume.

In the *Journal Historique de la République des Lettres* for January and February, 1733, some extracts were given from several printed letters of the Marquis Poleni; among these letters one related to the Figure of the Earth: see Art. 209. The editor of the Journal added some remarks impugning the accuracy of the observations and the soundness of the results given in the work *De la Grandeur et de la Figure de la Terre.* J. Cassini replies to the remarks.

The chief point urged in the remarks seems to be that *some* of the observations of latitudes recorded in the work differ considerably from the latitudes finally adopted; the chief point urged in the reply seems to be that observations made with less care and with small instruments were rejected in favour of observations made with more care and with larger instruments.

The reply seems to me temperate and able.

There is on pages 512, 513 a list of the misprints which had been detected in the work *De la Grandeur et de la Figure de la Terre.*

The following succinct account of the French survey of the meridian is given on page 498:

Cet ouvrage fut proposé par mon Pere, et prolongé en 1684 jusqu'au delà de Bourges vers le Midi, pendant que M. de la Hire y travailloit du côté du Nord. Je l'ai continué avec mon Pere et M. Maraldi, depuis Bourges jusqu' à Collioure en 1700 et 1701, et après l'avoir achevé entierement en 1718 avec M^rs. de la Hire le fils et Maraldi, en le prolongeant jusqu'à l'extrémité septentrionale du Royaume, j'en ai donné le résultat au Public; ainsi c'est à moi à en prendre la défense.

212. In the Paris *Mémoires* for 1733, published in 1735, there are five memoirs which are connected more or less closely with our subject. A connected account of them is given in pages 46...63 of the historical portion of the volume.

The first memoir is by Maupertuis; we have noticed it in Art. 131.

213. The next memoir is entitled *Méthode pratique de tracer sur Terre un Parallele par un degré de latitude donné; et du rapport du même Parallele dans le Sphéroïde oblong, et dans le Sphéroïde applati. Par M. Godin.* The memoir occupies pages 223...232 of the volume.

The memoir shews that for various reasons the accurate determination of the latitude of a place is not an easy problem in practical astronomy. Nevertheless it is maintained that an arc of longitude may be traced without much difficulty; and the best way of conducting the operation is explained.

Some numerical results are given as to the length of a degree of longitude; and remarks are made on the letter of Poleni which we have noticed in Art. 209.

Godin finishes with determining the arcs common to an oblatum and an oblongum which have the same centre, and their axes in the same straight line. The matter is very simple, but the account which is given of it in page 53 of the historical portion of the volume is not altogether intelligible.

214. The next of these memoirs is entitled *Description d'un Instrument qui peut servir à déterminer, sur la surface de la Terre, tous les points d'un Cercle parallele à l'Equateur. Par M. De La Condamine.* The memoir occupies pages 294...301 of the volume.

The instrument is intended to facilitate the operation described in Godin's memoir; but it does not seem to me that it would be of any practical value.

An extract of a letter written from Quito by La Condamine is given in the volume of *Mémoires* for 1734, which shews that he had himself discovered grave faults in the memoir, and requested that it might not be printed.

215. The next of these memoirs is entitled *De la Carte de la France, et de la Perpendiculaire à la Méridienne de Paris. Par M. Cassini.* The memoir occupies pages 389...405 of the volume.

The memoir gives an interesting account of the operations in tracing a line perpendicular to the meridian of Paris westwards to the coast of Normandy.

Cassini finds that the length of a degree of longitude in the parallel of St Malo is 36670 toises; and he says that on the supposition of the spherical form of the Earth it should be 37707 toises. Hence he infers that the Earth must be of an oblong form. It will be observed that the discrepancy here is very wide; and a less extravagant result was obtained by Cassini in the *Mémoires* for 1734: see Art. 220. Results much more moderate than this were obtained by Cassini de Thury in the *Mémoires* for 1735 and 1736: see Arts. 224 and 226.

It will be convenient to place here the formulæ relating to this matter.

Let λ denote the latitude, ρ the corresponding radius of curvature of the meridian, r the radius of the section parallel to the equator. If the earth were spherical, we should have $r = \rho \cos \lambda$.

If the earth is an oblatum, a denoting the semi-axis major, and e the excentricity of the generating ellipse, we have

$$\rho = \frac{a(1-e^2)}{(1-e^2\sin^2\lambda)^{\frac{3}{2}}}$$

and

$$r = \frac{a\cos\lambda}{(1-e^2\sin^2\lambda)^{\frac{1}{2}}}.$$

Thus it is obvious that r is now greater than $\rho\cos\lambda$.

If therefore it appeared by observation and measurement that r is less than $\rho \cos \lambda$, it would follow that the Earth could not be an oblatum.

The values of ρ and r in the case of the oblatum are often required in our subject.

216. It was found that the distances between places determined by the trigonometrical operations in France were in many cases less than had been previously supposed; and Cassini makes the following obvious remark:

...... ce qui vient apparemment des grands détours que l'on est obligé de faire pour chercher des routes praticables, joint à ce que les mauvais chemins paroissent toûjours plus longs qu'ils ne le sont réellement.

The operations terminated at Bayeux; Cassini says, after speaking of St Malo:

Nous allâmes de-là à Bayeux où nous fîmes diverses observations de hauteurs du Soleil, d'Etoiles fixes, et principalement de l'Etoile polaire, dans le Palais épiscopal qui joint à la Cathédrale, et où M. l'Evêque de Bayeux a fait tracer dans sa bibliotheque une grande Méridienne, avec des lignes qui marquent les heures avant et après midi, de cinq en cinq minutes, par M. l'Abbé Outhier qui a travaillé avec nous à la description de la Perpendiculaire depuis Caen jusqu'à St Malo.

The last of the five memoirs is by Clairaut; we have noticed it in Art. 160.

217. We have a memoir on pendulum observations in pages 302...314 of Number 432 of the *Philosophical Transactions*. The Number is for the months of April, May, and June, 1734, and forms part of Vol. XXXVIII. which is for the years 1733, 1734, and is dated 1735. The memoir is entitled *An Account of some Observations made in London, by Mr. George Graham, F.R.S. and at Black-River in Jamaica, by Colin Campbell, Esq.; F.R.S. concerning the Going of a Clock; in order to determine the Difference between the Lengths of Isochronal Pendulums in those Places. Communicated by J. Bradley, M.A. Astr. Prof. Savill. Oxon. F.R.S.*

The observations were made during 10 days in England, and during 26 days in Jamaica. Bradley deduced from them that the seconds pendulum of London lost 1 minute 58 seconds in a day at Jamaica; and from this he obtained for the ellipticity of the Earth the value $\frac{1}{190}$.

Bradley gives the reasons which led him to "esteem Mr. *Campbell's* Experiment to be the most accurate of all that have hitherto been made..."

This memoir is referred to by Stirling in the *Philosophical Transactions,* Vol. XXXIX. page 103; by Clairaut in the *Philosophical Transactions,* Vol. XL. page 291; and by Maclaurin in his *Fluxions,* Art. 664.

218. In the Paris *Mémoires* for 1734, published in 1736, there are four memoirs which are connected more or less closely with our subject.

The first of these memoirs is entitled *Méthode de vérifier la Figure de la Terre par Parallaxes de la Lune. Par M. Manfredi.* The memoir occupies pages 1...20 of the volume; there is an account of it on pages 59...63 of the historical portion of the volume.

Supposing the Earth not to be spherical, the parallax of the Moon will be different at different places on the Earth's surface, even when all other circumstances are alike. Manfredi suggests that observations of the Moon taken at two distant places, nearly on the same meridian, would therefore supply information as to the figure of the Earth. In spite of the errors to which such observations might be liable, he maintains that it would be possible to decide in this way the question as to the oblate or oblong form of the Earth.

219. The next of these memoirs is entitled *Comparaison des deux Loix que la Terre et les autres Planetes doivent observer dans la figure que la pesanteur leur fait prendre. Par M. Bouguer.* The memoir occupies pages 21...40 of the volume; there is an account of it on pages 83...87 of the historical portion of the volume.

This memoir is important in the history of Hydrostatics. The two principles to which it refers, are Newton's principle of balancing columns and Huygens's principle of the plumb-line. Bouguer's object is to shew that under certain conceivable laws of force either principle might be satisfied, while the other was not; and then there could not be equilibrium. The whole matter is now well understood; and it is admitted that for equilibrium the forces acting must satisfy a certain condition, namely, in ordinary notation, supposing the fluid homogeneous, $Xdx + Ydy + Zdz$ must be a perfect differential; and it is known that this condition is satisfied for such forces as occur in nature.

Bouguer says on his first page:

...... Entre plusieurs Mathématiciens d'un grand nom qui ont tourné leur vûë vers cette matiére, M. Huguens et M. Herman sont les seuls qui ont appliqué en même temps les deux loix; ils ont trouvé qu'elles s'accordoient à donner à la Terre une même figure dans les suppositions particuliéres d'une pesanteur originairement constante, et d'une pesanteur proportionnelle aux distances au centre.

This statement is correct with respect to Hermann; but there seems no authority for it with respect to Huygens. Hermann did consider both principles and both the laws of attraction: see Arts. 94 and 95. Huygens confined himself to the use of Newton's principle, and to the supposition of a constant attraction: see Arts. 54 and 55.

In his investigations, Bouguer, as we should now say, considered only forces in one plane. He supposes the direction of the force to be always perpendicular to a given curve. This hypothesis was afterwards discussed by Clairaut in pages 63...77 of his *Figure de la Terre.* Clairaut shews that, in order to render this hypothesis reasonable, we must suppose a solid nucleus to the fluid: see his pages 64 and 74.

Although Bouguer's own examples are not of great value, because they depend on laws of force which can hardly be considered natural, yet the memoir must have been very useful at the time, as it called attention to an important subject, and probably suggested to Clairaut the occasion of his own investigations.

220. The next of these memoirs is by Maupertuis; we have noticed it in Arts. 132...139.

The last of these memoirs is entitled *De la Perpendiculaire à la Méridienne de Paris, prolongée vers l'Orient. Par M. Cassini.*

It occupies pages 434...452 of the volume; there is an account of it on pages 74...77 of the historical portion of the volume.

This memoir contains an account of the operations in tracing a line perpendicular to the meridian of Paris, eastwards to Strasbourg; the operations and the memoir are in continuation of those which we have already noticed: see Art. 215.

Cassini finds that the length of a degree of longitude in the latitude of Strasbourg is 37066 toises; and he says that on the supposition of the spherical form of the Earth the length would be 37745 toises. Hence he infers, as before, that the form of the Earth must be oblong. The result differs very considerably from that given in the *Mémoires* for 1733: see Art. 215. The present result depends of course on the longitude of Strasbourg; and this is determined by the aid of observations formerly made by Eisenschmidt. Cassini assumes credit to himself for taking a mean between three determinations, though less favourable to his theory of an oblong form than the value which Eisenschmidt himself adopted. Thus we read at the close of the account in the historical portion of the volume, with respect to these observations:

> mais enfin ces observations se sont trouvées si favorables au Sphéroïde allongé, que M. Cassini a eu la modération de n'eu pas vouloir tirer tout l'avantage qu'il eût pû à la rigueur, et de s'en retrancher une partie.

221. A double prize was offered by the Paris Academy for the year 1734; the subject related to the inclination of the planes of the orbits of the planets to the plane of the Sun's equator. The prize was divided between John Bernoulli and his son Daniel. The essay by Daniel Bernoulli is memorable in the history of the Mathematical Theory of Probability: see my *History*, page 223.

The essay by John Bernoulli is reprinted in his *Opera Omnia*, Vol. III. pages 261...364, under the title *Essai d'une nouvelle Physique Céleste....* Pages 345...355 relate to the Figure of the Earth; but it would be a waste of time to discuss them. The essay uses a system of vortices; and as those who invented such visionary machinery were guided by no principle and restrained by no law, they could easily arrive at any result they pleased.

John Bernoulli disliked and depreciated Newton, and he was now competing for a prize from the Paris Academy; he had, therefore, a double reason for taking the side of error. This he does much to his own satisfaction, and concludes thus in the language of premature triumph:

> Après cette heureuse conformité de nôtre théorie, avec les observations célestes, peut-on plus long-temps refuser à la Terre la figure de sphéroïde oblong, fondé d'ailleurs sur la dimension des degrés de la méridienne, entreprise et exécutée par le même M. Cassini, avec une exactitude inconcevable?

222. In the Paris *Mémoires* for 1735, published in 1738, we have some memoirs which bear, though slightly, on our subject. An account of them is given on pages 47...65 of the historical portion of the volume; but the last six pages of this account refer to some memoir attributed to Clairaut, which does not seem to have been published. According to this account, an arc of longitude, if measured in a very high latitude, might be expected to yield as good a result as an arc of meridian. Bouguer, however, in an able memoir published in the volume for 1736, shewed that this expectation was quite unfounded.

The first memoir is entitled *Méthode de déterminer si la Terre est Sphérique ou non, et le rapport de ses degrés entr'eux, tant sur les Méridiens que sur l'Equateur et ses Paralleles. Par M. Cassini.* The memoir occupies pages 71...86.

The idea of the memoir can be easily stated. Select a mountain, from which the sea is visible in various directions, and observe the dip of the horizon. If the Earth is spherical, the dip will be the same in all directions. If the Earth is not spherical, the dip will be different in different directions. By observing the dip in the directions of the meridian and of the prime vertical, Cassini shews that a sensible difference ought to be obtained on the two current hypotheses as to the form of the Earth; and that thus the question between the two hypotheses might be settled.

I presume, however, that the method has never been found of any use in practice.

The next memoir is by Maupertuis; we have noticed it in Art. 140. The next to this is by Clairaut; we have noticed it in Art. 161.

223. The next memoir is entitled *Seconde Méthode de déterminer si la Terre est Sphérique ou non, indépendamment des Observations Astronomiques. Par M. Cassini.* The memoir occupies pages 255...261 of the volume.

The idea of the memoir can be easily stated. Take two points A and B on the same meridian; say the summits of two mountains. At A observe the angle which AB makes with the vertical at A; at B observe the angle which BA makes with the vertical at B. Let the verticals at A and B, when produced, meet at O. Let the distance AB be measured. Then by solving the triangle ABO we can find AO, which may be considered as the radius of curvature at A of the arc AB. Take a third point C, which is due East or due West of A. Then in the same way we may determine the radius of curvature at A of the arc AC. If the Earth is a sphere, we ought to obtain the same value of the radius of curvature in the two cases; if the values obtained are different, we have information which may serve to settle whether the form is oblate or oblong.

The method is substantially the same as was used by Riccioli in attempting to find the size of the Earth towards the middle of the seventeenth century. See *De la Grandeur et de la Figure...* pages 296...306. I believe the method is of no practical value.

224. The next memoir is entitled *De la Perpendiculaire à la Méridienne de Paris, décrite à la distance de* 60000 *Toises de l'Observatoire vers le Midi. Par M. De Thury.* The memoir occupies pages 403...413 of the volume.

M. De Thury was a son of Jacques Cassini, and is usually called Cassini de Thury. The perpendicular was traced from the meridian of Paris to the western coast of France. Cassini de Thury finds that the length of a degree of longitude in the parallel of Brest is nearly 300 toises shorter than it should be on the supposition of the spherical form of the Earth. Hence he infers that the Earth must be oblong.

It must however be observed that for Nantes, which has nearly the same latitude, Cassini de Thury obtained a difference of 781 toises. It is surprising that such discordant results were considered to be worth preserving. It is plain that the observations were not good enough to furnish trustworthy inferences.

Cassini de Thury assigns 47° 13′ 8″ for the latitude of Nantes, which agrees with the modern value. But he assigns 47° 13′ 2″ for the latitude of Brest; and the modern value is 48° 23′ 22″ See the table published in the *Connaissance des Temps.* There must of course be some error in his figures.

225. The volume for 1735 contains also some important memoirs on the length of the seconds pendulum.

A memoir by Mairan on pages 153...220 relates to the length at Paris; there is an account of this on pages 81...92 of the historical portion of the volume.

A memoir by Godin relates to the lengths at Paris and at St Domingo.

A memoir by Bouguer relates to the length at St Domingo.

A memoir by La Condamine relates to the length at St Domingo.

These three memoirs will be found on pages 505...544 of the volume.

There is some notice of the memoirs by Godin, Bouguer, and La Condamine on pages 115...117 of the historical portion of the volume for 1736. We are told that these investigators did not arrive in Peru so soon as they had hoped; and it is added: "Mais quoiqu'ils ne pussent pas encore s'occuper du principal objet de leur Voyage, la Nature est par-tout, et ils trouvoient par-tout à observer."

226. In the Paris *Mémoires* for 1736, published in 1739, we have four memoirs bearing on our subject.

The first memoir is by Clairaut; we have noticed it in Art. 162. The next memoir is by Maupertuis; we have noticed it in Art. 141.

The third memoir is entitled *Sur la Perpendiculaire à la Méridienne de l'Observatoire à la distance de* 60000 *toises vers le*

Nord. Par M. Cassini De Thury. The memoir occupies pages 329.. 341 of the volume. There is an account of the memoir in pages 103 and 104 of the historical portion of the volume.

The perpendicular was traced from the meridian of Paris to the western coast of France. According to these operations the length of a degree of longitude in the parallel of Brest is 310 toises shorter than it should have been on the supposition of the spherical form of the Earth. Hence, as before, it is inferred that the Earth must be oblong.

It seems, from what is stated on pages 332 and 333, that in the operations before the present, the angle subtended between two objects had not been distinguished from the projection of the angle on the plane of the horizon.

It was sometimes found necessary to construct scaffolds on the tops of lofty trees; one tree so used was above 100 feet high. Then we read on page 104 of the historical portion of the volume: "Ces édifices hardis demandoient que ceux qui s'en servoient, le fussent aussi."

227. The last memoir is entitled *De la maniere de déterminer la Figure de la Terre par la mesure des degrés de Latitude et de Longitude. Par M. Bouguer.* The memoir occupies pages 443...468 of the volume.

Bouguer obtains expressions for the length of a degree of the meridian and for the length of a degree of longitude, assuming the Earth to be an ellipsoid of revolution. Then from the lengths of two different degrees he deduces the ratio of the axes of the Earth. By the aid of the Differential Calculus he finds the change in this ratio produced by a given small change in one of the elements on which it depends.

Bouguer makes some interesting remarks on what he calls "la différente délicatesse de la vûë des Observateurs," or as we now call it the *personal equation* of observers, see his page 457. He says that if two astronomers have observed several times together and know what we call their personal equation, yet this may be altered by the fatigues of a voyage, by the changes in the body, or by a greater or less density of the atmosphere.

Bouguer's main conclusion is that attention should be given almost exclusively to the measurement of arcs of meridian, since practically arcs of longitude could not be determined with sufficient accuracy to settle the question of the Earth's form.

228. We have next to notice *A Proposal for the Measurement of the Earth in Russia, read at a Meeting of the Academy of Sciences of St Petersbourg, Jan. 21. 1737. by Mr Jos. Nic. de L'isle, first Professor of Astronomy, and F.R.S. Translated from the French printed at St Petersbourg,* 1737. *4to. By T. S. M.D. F.R.S.*

This paper occupies pages 27...49 of Number 449 of the *Philosophical Transactions.* The Number is for the months January...June, 1737, and forms part of Vol. XL. which is for the years 1737, 1738, and is dated 1738.

The paper is very interesting; it gives an account of the history of opinion on the Figure of the Earth. The work of Eisenschmidt is cited, and its full title reproduced, which agrees with that in La Lande's *Bibliographie Astronomique,* page 324: but here it is added *pag.* 54. *cum fig.*

The paper was written after the French expeditions had gone to Peru and to Lapland, but before the results of their measurements were known; however, some pendulum observations reported by both expeditions favoured the oblate form.

An *Extract of a Letter* from Delisle is given on pages 50, 51 of Vol. XL. of the *Philosophical Transactions;* from this it appears that he measured on the ice a base of 74250 English feet, as the commencement of the proposed operations in Russia.

In the work by F. G. W. Struve, entitled *Arc du Méridien de* 25° 20′ *entre le Danube et la Mer Glaciale*...there is a slight notice of Delisle's project: see Vol. I. page viii. The title of the original document is given thus; *Projet de la mesure de la Terre en Russie. Saint-Pétersbourg,* 1737, 4to. It is stated that Delisle himself published no account of the measurement of the base or the angles. His manuscripts were preserved in the Observatory of Paris, and examined in 1844 by M. O. Struve.

Delisle was brother to the person who made the pendulum observations at Archangel in 1728: see Art. 210.

229. We have next to consider a memoir by Euler, entitled *De attractione corporum sphaeroidico-ellipticorum.*

This memoir is contained in the *Commentarii Academiæ... Petropolitanæ*, Vol. x. which is for 1738; the date of publication is 1747. The memoir occupies pages 102...115 of the volume.

The memoir finds expressions in the form of infinite series for the attraction of an oblatum on a particle at the pole, and on a particle at the equator. In the former case the series is not complicated, and converges rapidly; as Euler says *vehementer convergit.* In the latter case the series is very complicated, and this case of the problem cannot be considered to be really solved.

We are not told at what date the memoir was read to the Academy; so that there may have been merit and value in it at the time; but before the volume was published the solution of the problem by Maclaurin and by Simpson had appeared, in which the results were expressed in exact finite forms, so that Euler's memoir was completely superseded.

I have not verified all the work in this memoir. I will give some indication of Euler's method.

Required the attraction of an elliptic lamina on a point directly over the centre of the lamina.

Let c denote the distance of the point, δc the thickness of the lamina. Then the attraction is

$$\delta c \iint \frac{c\,dx\,dy}{(c^2+x^2+y^2)^{\frac{3}{2}}},$$

where the integration is to extend over the whole area of the ellipse

$$\frac{x^2}{a^2}+\frac{y^2}{b^2}=1.$$

Integrate first for y; thus we find that the attraction is equal to

$$4bc\,\delta c \int_0^a \frac{(a^2-x^2)^{\frac{1}{2}}\,dx}{(c^2+x^2)\,\{a^2(b^2+c^2)+x^2(a^2-b^2)\}^{\frac{1}{2}}}$$

By expanding, this becomes

$$\frac{4bc\,\delta c}{a\,(b^2+c^2)^{\frac{1}{2}}}\int_0^a \frac{(a^2-x^2)^{\frac{1}{2}}}{c^2+x^2}\left\{1-\frac{1}{2}z^2+\frac{1\,.\,3}{2\,.\,4}z^4-\frac{1\,.\,3\,.\,5}{2\,.\,4\,.\,6}z^6+\dots\right\}dx,$$

where z^2 stands for $\dfrac{x^2\,(a^2-b^2)}{a^2\,(b^2+c^2)}$

This expansion will not give a convergent series throughout the range of integration unless $a^2 - b^2$ is less than $b^2 + c^2$. Euler, however, does not pay any attention to this point. Moreover, he also expands $\frac{1}{c^2 + x^2}$ in ascending powers of x^2 before the integration, so that this expansion is really not permissible if a is greater than c.

However, Euler evaluates in this way the expression

$$\int_0^a \frac{(a^2 - x^2)^{\frac{1}{2}}}{c^2 + x^2} dx,$$

namely, by expanding the denominator, integrating each term separately, and then summing the infinite series which arises. We should now of course avoid the expansion. By putting $a \sin \theta$ for x, the expression becomes

$$\int_0^{\frac{\pi}{2}} \frac{a^2 \cos^2 \theta \, d\theta}{c^2 + a^2 \sin^2 \theta}, \quad \text{that is} \quad \int_0^{\frac{\pi}{2}} \frac{c^2 + a^2}{c^2 + a^2 \sin^2 \theta} d\theta - \int_0^{\frac{\pi}{2}} \frac{c^2 + a^2 \sin^2 \theta}{c^2 + a^2 \sin^2 \theta} d\theta,$$

that is
$$\frac{\pi}{2} \left\{ \frac{\sqrt{(a^2 + c^2)}}{c} - 1 \right\}.$$

Hence in the required attraction we have the terms

$$\frac{2\pi b \sqrt{(a^2 + c^2)}}{a \sqrt{(b^2 + c^2)}} \delta c - \frac{2\pi bc}{a \sqrt{(b^2 + c^2)}} \delta c.$$

Next consider the term which arises from z^2. We may proceed thus without expansion:

$$\int \frac{(a^2 - x^2)^{\frac{1}{2}} x^2 dx}{c^2 + x^2} = \int \left\{ \frac{c^2 + x^2}{c^2 + x^2} - \frac{c^2}{c^2 + x^2} \right\} (a^2 - x^2)^{\frac{1}{2}} dx.$$

Then taking the integrals between the limits 0 and a, we obtain

$$\frac{\pi a^2}{4} - \frac{\pi c^2}{2} \left\{ \frac{\sqrt{(a^2 + c^2)}}{c} - 1 \right\}$$

Hence in the required attraction we have the terms

$$-\frac{1}{2} \frac{a^2 - b^2}{a^2 (b^2 + c^2)} \frac{4bc\delta c}{a \sqrt{(b^2 + c^2)}} \left\{ \frac{\pi a^2}{4} - \frac{\pi c^2}{2} \left(\frac{\sqrt{(a^2 + c^2)}}{c} - 1 \right) \right\},$$

or $$\frac{\pi bc^2 (a^2 - b^2) \sqrt{(a^2 + c^2)}\, \delta c}{a^3 (b^2 + c^2)^{\frac{3}{2}}} - \frac{\pi bc^3 (a^2 - b^2)\, \delta c}{a^3 (b^2 + c^2)^{\frac{3}{2}}} - \frac{\pi bc (a^2 - b^2)\, \delta c}{2a (b^2 + c^2)^{\frac{3}{2}}}.$$

Similarly we might proceed with the term which arises from z^4, which will introduce $(a^2 - b^2)^2$; and so on.

The result of course will be very complicated. Euler seems to me to increase the complication by putting

$$\sqrt{(a^2 + c^2)} = \sqrt{(b^2 + c^2 + a^2 - b^2)},$$

and expanding the latter in powers of $a^2 - b^2$. He offers a reason for this which I do not quite comprehend. "Vel cum ad applicationem ad computum expediat ipsas series retinere, quo singulorum terminorum integralia algebraice exhiberi queant..."

Euler's approximate values for the attraction at the pole and at the equator are respectively $4\pi b \left(\frac{1}{3} + \frac{4\epsilon}{15} - \frac{2\epsilon^2}{21}\right)$, and $4\pi b \left(\frac{1}{3} + \frac{\epsilon}{5} - \frac{3\epsilon^2}{35}\right)$, where b is the polar semi-axis, and $b(1 + \epsilon)$ is the equatorial semi-axis. It will be found on examination that these are correct: see Art. 153.

Euler applies his results to determine the ratio of the axes in order that a rotating fluid oblatum may be in relative equilibrium; he obtains a value for the ellipticity, which is sensibly the same as Newton's in the case of the Earth.

230. A few words may be given to the treatise published by Daniel Bernoulli at Strasbourg in 1738 under the title of *Hydrodynamica*, although it is rather beyond our subject.

On pages 244 and 245 Daniel Bernoulli solves the problem of determining the form for relative equilibrium of the free surface when fluid in a cylinder rotates round a vertical axis; the angular velocity is not assumed to be the same throughout the mass. The solution is correct, and is recognised as such by Clairaut in his *Figure de la Terre*, page 55.

Daniel Bernoulli however proceeds on page 246 to make some unsatisfactory remarks on vortices. He begins by saying that he thinks the fluid cannot continue permanently in its state if the centrifugal force *increases* from the axis to the circumference: the context seems to shew that instead of *increases* he meant *de-*

creases. But it is plain from his remarks that the subject was not understood at the time.

Daniel Bernoulli criticises implicitly Propositions 51 and 52 of Book II. of the *Principia,* which he considers do not both correspond to possible cases.

231. The volume of the Paris *Mémoires* for 1739 was published in 1741. On page 30 of the historical portion there is a short notice of a memoir communicated to the Academy by D'Alembert. The memoir does not bear on our subject, but it is interesting to observe the early appearance of a writer with whom we shall be much occupied hereafter. We are told that: "On a trouvé dans M. d'Alembert beaucoup de capacité et d'exactitude." The later writings of D'Alembert do not in general seem to me to deserve the praise of *exactness.*

A memoir by Clairaut occurs in the volume; of this we have given an account in Art. 177.

There is a memoir entitled *Sur les Opérations Géométriques faites en France dans les années* 1737 *et* 1738. *Par M. Cassini De Thury.* The memoir occupies pages 119...134 of the volume.

The operations were chiefly directed to surveying parts of the coast of France, with the view of rectifying the maps. Some observations as to the velocity of sound are recorded.

232. The Academy of Sciences at Paris proposed *The Tides* as the subject for a prize essay in 1740. Four essays were published in consequence at Paris. One essay was by a Jesuit named Cavallieri; this adopted the Cartesian system of vortices. The other essays were by Daniel Bernoulli, Maclaurin, and Euler; these are reprinted in the Jesuits' edition of the *Principia,* and it is stated that many errors in the original impression have been corrected. I have used the reprint in consulting these Essays.

It will be convenient to postpone an account of Maclaurin's essay until we have examined the part of his *Treatise of Fluxions* which relates to our subject; for this contains all that was in the essay with great additions and improvements.

233. The second chapter of Daniel Bernoulli's essay contains some lemmas relating to the Attraction of Bodies. The result

may be summed up thus: he determines the attraction at any superficial or internal point of an ellipsoid of revolution which is nearly spherical, neglecting powers of the ellipticity beyond the first. The method used consists in finding accurately the attraction of a sphere, and then approximately the attraction of the difference between the sphere and the ellipsoid on a particle at the pole or at the equator; as we have stated in Art. 165 this method had been previously used by Clairaut. But Daniel Bernoulli seems to claim the method as his own; he says at the end of his second Chapter:

Ceux qui voudront employer l'analyse pure pour la solution de nos deux derniers Problêmes, se plongeront dans des calculs extrêmement pénibles, et verront par là l'avantage de notre méthode.

Although Daniel Bernoulli employed attraction for the purpose of his essay, yet he seems to have had but a weak faith in the principle: see his Chap. I., Art. 6, and his Chap. II., Art. 1.

Daniel Bernoulli added nothing to our subject; all his results respecting Attraction are included in the formulæ given by Clairaut in 1737. But his theory of the Tides is very important in the history of that subject, though it would be out of place for us to discuss it here.

An account of Daniel Bernoulli's essay was published in 1830 by the late Sir J. W. Lubbock; it is in octavo, entitled *Account of the "Traité sur le Flux et Réflux de la Mer" of Daniel Bernoulli; and a Treatise on the Attraction of Ellipsoids,* pages vii. + 47.

234. Euler's essay on the Tides contains scarcely anything that concerns us. He finds the attraction of a spherical shell on an internal particle in his Art. 20. The results in his Art. 30 are interesting as examples: we will state them. The attraction of the Sun, or of the Moon, at the surface of the Earth, is of course not strictly the same as the attraction at the centre; hence arises a *disturbing attraction* as it may be called, which at a given place will depend on the zenith-distance of the attracting body. Euler finds that the number of oscillations made by a pendulum when the Sun and the Moon are together in the zenith is to the number made in the same time by the same pendulum when the Sun

and the Moon are together in the horizon as 4666666 is to 4666667. Also if the Sun and the Moon are together at 45° from the zenith, first on one side and then on the other side, in the same great circle, the plumb-line on the whole experiences a deviation of less than $\frac{1}{12}$ of a second. These results are obtained of course by using the values then adopted for the masses and the distances of the Sun and the Moon.

The following passage occurs at the beginning of Euler's Article 12:

Explosis hoc saltem tempore qualitatibus occultis missâque Anglorum quorumdam renovatâ attractione......

At first sight this looks as if Euler intended to reject the principle of attraction; but we find on examination that he practically adopts the principle, after assuming the existence of a subtle fluid in order to account for it to his own satisfaction.

235. A work entitled *Degré du Méridien entre Paris et Amiens*...was published in 1740. I have not seen the original but only a German translation published at Zurich in 1742: I must assume therefore that the translation corresponds to the original. Maupertuis and his companions in the polar expedition were charged with the business of verifying the length of a degree of the meridian assigned by Picard. They assumed the accuracy of Picard's terrestrial measurement, but determined the amplitude of the arc afresh. The observations were made in the latter half of the year 1739; the instrument employed was the same zenith-sector as had been employed in Lapland.

The book contains a description of the sector and an account of the observations made with it. More than half the volume however is a reprint of Picard's account of his own operations. Some observations are also given relating to Aberration.

236. In the Paris *Mémoires* for 1740, published in 1742, we have a Memoir entitled *De la Méridienne de Paris, prolongée vers le Nord, et des Observations qui ont été faites pour décrire les frontieres du Royaume. Par M. Cassini De Thury.* The memoir

occupies pages 276...292 of the volume. There is an account of it on pages 69...75 of the historical portion of the volume. The memoir is very important in the history of the subject. Hitherto the accuracy of Picard's base had not been questioned; but now it was resolved to examine this point. A base not quite coincident with Picard's, but very near to it, was measured five times; by the aid of a certain length deduced from this it was found that Picard had ascribed to his base a length nearly 6 toises greater than it should have had. In order to leave no doubt on the point, the last measurement was made in the presence of Commissioners from the Academy, at the request of Cassini de Thury himself. These Commissioners were Clairaut, Camus, and Le Monnier. See *La Meridienne de Paris verifiée*, page 36.

Bailly implies that Picard's actual base was remeasured, which as we see was not the case. Moreover, he erroneously states that *all the five* measurements were made in the presence of the Commissioners from the Academy. *Histoire de l'Astronomie Moderne*, Vol. III. page 35.

It will be convenient to bring together the various lengths assigned to the degree of the meridian between Paris and Amiens.

Picard himself in 1671 adopted 57060 toises; see *De la Grandeur et de la Figure de la Terre*, page 281.

Maupertuis in 1738 by correcting Picard's observations for aberration arrived at 56926 toises. *Figure de la Terre*, page 126.

Maupertuis and his companions in 1740 by new astronomical observations obtained 57183 toises. *Degré du Méridien*...First Part, Chapter VIII.

Cassini de Thury, after the remeasurement of Picard's base, using the amplitude determined by Maupertuis and his companions, gave 57074 toises. Paris *Mémoires* for 1740, page 289. The errors made by Picard in his astronomical and geodetical work had by accident almost balanced each other. The subject is discussed by La Condamine in his *Mesure des trois premiers degrés*, pages 239...258.

237. We return to the memoir by Cassini de Thury. The memoir is remarkable for being, I presume, the first since the

discussion had arisen as to the form of the Earth in which a member of the family of Cassini recognised the oblateness. We learn from page 288 of the memoir that at the north of France the length of a degree of the meridian was found to be 57081½ toises, and at the south of France 57048 toises.

Then Cassini de Thury adds:

...ainsi, suivant ces observations, les degrés vont en diminuant en s'approchant de l'Equateur, ce qui est favorable à l'hypothese de l'applatissement de la Terre vers les Poles.

It may be interesting to compare results given in the present memoir with some given in the earlier work.

According to the *De la Grandeur et de la Figure de la Terre*, page 148, the distance between the parallels of Paris and Collioure is 360614 toises, the amplitude 6° 18′ 57″, and the mean length of a degree 57097 toises. According to the present memoir, the distance between the parallels of Paris and Perpignan is 350142 toises, the amplitude 6° 8′ 17″, and the mean length of a degree 57045 toises.

Again, according to the *De la Grandeur et de la Figure de la Terre*, page 236, the distance between the parallels of Paris and Dunkirk is 125454 toises, the amplitude 2° 12′ 9″.5, and the mean length of a degree 56960 toises. According to the present memoir, for the same arc the corresponding numbers are 125508 toises, 2° 11′ 55″.5, and 57081.5 toises.

238 It must be observed that the error in Picard's base does not account for the apparent diminution in the length of a degree of the meridian in passing from the equator to the pole which the school of Cassini had deduced and maintained. For in the *De la Grandeur et de la Figure de la Terre*, which was the main support of this hypothesis, the lengths are all deduced from that of Picard's base; and so the proportions would not be affected by any error in the base. This remark is necessary because the contrary has been asserted, or obviously implied. Thus Bailly says, "l'erreur de cette mesure étoit le nœud de la difficulté:" *Histoire de l'Astronomie Moderne*, Vol. III. page 38. And on page 169 of the article *Figure of the Earth* in the *Encyclopædia Metropolitana* we read

"On measuring new bases and making new observations of every kind, *the cause of the original difficulty was soon discovered.* The measure of Picard's base was erroneous by about $\frac{1}{1000}$th part of the whole, and *this error had affected one part only of the arc.*" The statements which I have here put in Italics do not seem to me supported by the evidence. It is true that in 1739 and 1740 anomalies were revealed which cast suspicion on Picard's measurement, and which were explained when that measurement was corrected; but these were quite distinct from the original difficulty. See *La Meridienne de Paris verifiée,* page 19.

We perceive from this memoir that in 1740 the oblate form of the Earth was fully established and admitted.

239. An edition of Newton's *Principia* appeared at Geneva in 1739...1742, edited by Thomas Le Seur and Francis Jacquier. The editors are usually styled Jesuits, and the edition is called the Jesuits' edition. I have already referred to this edition: see Arts. 16, 22, and 232.

The commentary on Propositions XVIII., XIX. and XX. of Newton's third Book does not seem to me very successful; there are some serious mistakes in it, which occur chiefly in notes marked with an asterisk. It appears from the *Monitum* and the *Editoris monitum,* prefixed to the third Book, that these are due to J. L. Calandrinus, to whom Le Seur and Jacquier acknowledge great obligations.

I will point out these mistakes.

A curious note is given on the words which I have quoted in Art. 26: "Et propterea dico..." The note in effect states that Newton must have had better reason than appears at once obvious for applying the rule of proportion. The note then proceeds to justify the proportion which Newton uses; but the investigation is unsatisfactory for the reason which often applies to approximations, namely, that the calculations are not carried to the same degree of accuracy throughout. Using the letters as in Art. 20 the note asserts that the ratio of the attraction at Q to the attraction of a sphere having C for centre and CQ for radius, is equal to

$\frac{3 \cdot CA - 2 \cdot CQ}{CA}$; if the ellipticity ϵ be very small, this reduces to $3 - 2(1 - \epsilon)$, that is, to $1 + 2\epsilon$: but, as we have stated in Art. 20, the true value is $1 + \frac{4\epsilon}{5}$

A long note is given on Newton's Proposition XIX., which involves some singular errors; indeed it seems to me quite extraordinary that such a note should have been printed towards the middle of the eighteenth century. The note proposes to investigate the resultant attraction of a homogeneous solid of revolution at the surface; and it begins correctly by observing that if we take a pyramid with an *infinitesimal solid angle*, the attraction exerted by a segment of the pyramid on a particle at the vertex varies as the height of the pyramid.

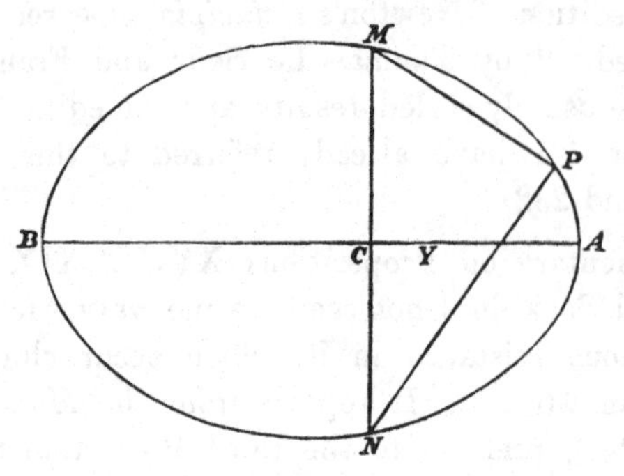

Let AB be the axis of the solid of revolution, P any point at its surface, MCN any double ordinate at right angles to AB. The note supposes P to be the vertex of a system of infinitesimal pyramids, the axes of the pyramids all passing through the circle generated by the revolution of CM round CA. The note concludes that the resultant attraction of these pyramids will be in the direction PC: this conclusion is obtained by taking the pyramids in pairs, so that the bases of a pair may be at the opposite ends of a diameter of the circle; for example, the pyramid which has PM for its axis is combined with that which has PN for its axis. Now it is quite true that such a pair of pyramids will exert a resultant attraction along PC, *provided the two pyramids have equal infinitesimal solid angles:* but this important condition is practically forgotten in the note. A laborious calculation is given for determining the resultant attraction of all the pyramids which have their axes passing through the circle formed by the revolution of CM round CA; but this is of no use, because the bases of all these pyramids will not form a strip of the surface contained between this circle and an adjacent circle in a parallel plane, though the note implicitly assumes that they will.

Again, the language of the note seems to suggest that we are to obtain the attractions exerted on P by all the circular elements like that considered, and *add* them together. This would however be useless; for as these attractions are not all in the same direction they would have to be resolved according to fixed directions, and the resolved parts in the *same direction* added.

Again, we are in effect told to obtain the direction of the resultant attraction of the solid in the following manner: Suppose Y a point in AB such that the attractions on P of the two segments into which the solid is divided by a plane through Y at right angles to AB are equal; then PY is the direction of the resultant. This statement is certainly untrue. For instance, if the solid is a sphere, the resultant attraction passes through the centre; but the two halves formed by cutting the sphere by a plane do not in general exert equal attractions on a particle at the surface.

It is incidentally stated, that in the triangle PMN we have $(PM + PN)\,PN$ greater than MN^2; but this is not necessarily true.

The following extraordinary principle is offered for obtaining the condition of equilibrium of a mass of fluid in the form of a solid of revolution. Let t denote the distance at any point P between the bounding surface and a similar surface indefinitely near, f the attraction at P, y the distance of P from the axis, ds an element of the generating curve at P; then $tfyds$ is to be constant. It is sufficient to observe that in the simplest possible case, that of a sphere, this condition does not hold; for then t and f are constant, but yds is not constant, except by an arbitrary hypothesis.

The commentators notice the inaccuracy of Newton, on which I have remarked in Art. 33; they assert that gravity at different places varies inversely as the radius of curvature: "...gravitates in singulis punctis forent reciproce ut radii osculatores curvæ." This is untrue; it would make the gravity greatest at the equator and least at the poles. The fact is that gravity would vary as the length of the normal between the point and the major-axis.

The commentators having obtained an expression substantially equivalent to the $\frac{a^3-p^3}{a^3p^3}$, which I have given in Art. 35, immediately proceed to take a^3-r^3 for the numerator; but this approximation is not exact to the order which has been retained. I should add, however, that in their next note there is a correct analytical investigation of the matter.

240. We may next advert to a memoir entitled *Determinatio exactior Graduum Parallelorum Æquatoris et Meridiani...Auctore C. N. de Winsheim.* This is contained in the *Commentarii Academiæ...Petropolitanæ,* Vol. XII. which is for 1740; the date of publication is 1750. The memoir occupies pages 222...240 of the volume. Here we have Tables giving the lengths of a degree of the meridian and of a degree of longitude in various latitudes, for a sphere, and for an oblatum in which the ellipticity is $\frac{1}{183}$. This ellipticity is found from the Lapland degree of 57438 toises, and Picard's taken at 57183 toises: see Art. 236. Winsheim ascribes to Euler the rule which he uses for calculating the Tables with respect to the oblatum.

CHAPTER IX.

MACLAURIN.

241. MACLAURIN'S researches on Attractions first appeared in his Essay on the subject of the Tides, which gained a prize from the French Academy in 1740; see Art. 232. These researches are reproduced in an enlarged and improved form, in Maclaurin's work entitled *A Treatise of Fluxions*, Edinburgh, 1742. The work is in two quarto volumes; it contains Title Pages, a Dedication to His Grace the Duke of Argyle and Greenwich on two pages, a Preface on six pages; then the text on 763 pages, and a page of Errata: there are XL Plates.

The *Treatise of Fluxions* embodies much of the analysis and mechanics of the period. Maclaurin touches on the equilibrium of fluids in his pages 409, 410. We may infer that he had a correct idea of what we now call the differential equation to the surface of a homogeneous fluid in equilibrium under given forces.

242. The part of the *Treatise of Fluxions* with which we are concerned, occupies pages 522...566, which are in the second volume.

Maclaurin shews that the attraction of a homogeneous cone with a given infinitesimal solid angle on a particle at the vertex varies as the length of the cone; and that the same result holds for a frustum of the cone; the particle being still supposed at the vertex of the cone. See his Article 628. Then his Article 629 draws an important inference, which Newton had given in the first corollary to his Proposition 87. Maclaurin says:

The forces with which particles similarly situated with respect to similar homogeneous solids gravitate towards these solids are as their Distances from any points similarly situated in the solids, or as any of their homologous sides. For such solids may be conceived to be resolved into similar cones, or frustums of cones, that have always their vertex in the particles, and the gravitation towards these cones, or frustums, will be always in the same ratio.

In future, if nothing is said about the density of the attracting body it is to be understood to be a *homogeneous* body.

243. Maclaurin shews in his Article 630, that a particle will be in equilibrium if it is placed at any point within the hollow part of a shell, the surfaces of which are concentric, similar, and similarly situated ellipsoids of revolution; the demonstration is the same as Newton's: see Art. 13.

244. Let the attraction of an ellipsoid of revolution on any constituent particle be resolved into two components, one perpendicular to the axis, and the other parallel to the axis; then the former component varies as the distance of the particle from the axis, and the latter component varies as the distance from the plane of the equator. Maclaurin demonstrates these theorems, first formally stated by himself, by a beautiful geometrical process: see his Articles 631...634.

Clairaut preserves the essence of Maclaurin's demonstration: he says, "Cette méthode m'a paru si belle et si savante...": see *Figure de la Terre*, pages 157...170.

Suppose that λ denotes the constant coefficient for the component attraction parallel to the axis, and μ the constant coefficient for the component perpendicular to the axis; then, by some general reasoning, Maclaurin arrives at the result that the product of λ into the square of the polar axis is less or greater than the product of μ into the square of the equatorial axis according as the ellipsoid of revolution is oblate or oblong: see his Article 635.

245. Let there be an ellipsoid of revolution; let $2a$ be the equatorial diameter, and $2b$ the polar diameter. Suppose the ellipsoid to be fluid; and besides the mutual attractions let there be at every point any other force perpendicular to the axis varying as the distance from the axis, and any other force parallel to the axis varying as the distance from the plane of the equator:

the necessary and sufficient condition for equilibrium is that a must be to b, as the resultant force at the pole is to the resultant force at the equator. This theorem can be demonstrated immediately by the aid of the well-known equations for the equilibrium of a fluid. Maclaurin, however, was not in possession of these equations; so that he adopted a different method. He says in his Article 636:

To demonstrate this proposition fully, we shall shew, 1. That the force which results from the attraction of the spheroid and those extraneous powers compounded together acts always in a right line perpendicular to the surface of the spheroid. 2. That the columns of the fluid sustain or ballance each other at the center of the spheroid. And 3. That any particle in the spheroid is impelled equally in all directions.

He gives his demonstrations in his Articles 637, 638, 639.

Maclaurin then was in this position: there was as yet no theory of fluid equilibrium which indicated what conditions were *sufficient*, so he shews that all the conditions which had then been recognised as *necessary* for equilibrium would be satisfied in the case supposed. He easily demonstrates the first condition, which, as we know, was given by Huygens: see Art. 53. Maclaurin's second condition is a particular case of his third, and was given by Newton: see Art. 23. The meaning which Maclaurin attaches to his third condition is the following: Take any definite point within the mass; draw from this point a straight line to the surface *in any direction;* let this straight line be the axis of a column of given infinitesimal section: then the attraction on the column resolved along the column, is independent of the direction. Maclaurin, however, only demonstrates this for the case in which the direction is in the *meridian plane* of the definite point; he says that "in like manner, it is shewn" that the result is true for columns not in the meridian plane: but it is not obvious how he would have proceeded. The result can be obtained very easily by modern methods.

Maclaurin's third condition is thus an extension of Newton's principle of balancing columns, *any point* being taken instead of the centre, at which the balancing is to hold. Huygens had briefly alluded to this extension: see Art. 55.

246. This extension of Newton's principle of balancing columns seems to have been considered important at the time. D'Alembert says on page 14 of his *Essai...de la Résistance des Fluides*:

Quoique le Principe de l'équilibre des Canaux rectilignes, soit comme l'on voit, une conséquence très-naturelle de la pression des Fluides en tout sens; cependant je dois reconnoître ici, que feu *M. Maclaurin* est le premier qui ait fait usage de ce Principe, et qui l'ait appliqué à la recherche importante de la Figure de la Terre. Voyez son *Traité des Fluxions*, art. 639, et son Traité *de Causa Fluxûs et Refluxûs maris*, Paris, 1740.

See also D'Alembert's *Traité...des Fluides*, second edition, page 49.

247. In Maclaurin's Article 637, we have the important result which we have noticed in our account of Stirling; namely, that when rotating fluid in the form of an oblatum is in relative equilibrium the gravity at any point of the surface varies *exactly* as the length of the normal between the point and the plane of the equator; see Art. 153. This result had however been communicated to the Royal Society by Simpson, in 1741, before the publication of Maclaurin's *Fluxions:* see the preface to Simpson's *Mathematical Dissertations.* Simpson seems to claim priority for himself; but he overlooks the fact that Maclaurin had previously given the result in his prize essay on the Tides: it is the *Theorema Fundamentale* of the essay.

It follows immediately from conic sections that instead of the gravity varying as the length of the normal between the point and the plane of the equator, we may take the length of the normal between the point and the axis of revolution.

248. Maclaurin, in his Article 640, states the conclusions which he had thus demonstrated respecting the problem of Art. 245. Among them we may observe that he says, surfaces similar, similarly situated, and concentric with the bounding surface "will be level surfaces at all depths."

This is the first mention I find of *level* surfaces; the essential property of a level surface is that the resultant force at any point of the surface is in the direction of the normal to the surface at that point.

D'Alembert in his *Essai...de la Résistance des Fluides*, page 202, says:

... *M. Maclaurin*, le premier qui ait parlé de ces couches......auxquelles la pesanteur est perpendiculaire, et qu'il appelle *surfaces de niveau*....

249. Maclaurin now applies the results obtained for the general problem of Art. 245 to the particular case of the relative equilibrium of a revolving fluid.

He says in his Article 641:

It appears therefore that if the earth, or any other planet, was fluid and of an uniform density, the figure which it would assume in consequence of its diurnal rotation, would be accurately that of an oblate spheroid generated by an ellipsis revolving about its second axis, as Sir ISAAC NEWTON supposed.

Here, Maclaurin says more than he was justified in saying; he had not proved that the planet *would* assume the form of an oblatum, but only that this form is *a* form of relative equilibrium. See Art. 168.

The proposition really investigated was first established *exactly* by Maclaurin; as we have stated, Stirling and Clairaut had given approximate investigations of it: see Arts. 156 and 163.

250. Maclaurin now proposes to calculate the attraction of an ellipsoid of revolution at the pole or at the equator. He begins with a lemma which forms his Article 642. Let a slice of an attracting body be formed by two planes, both containing the attracted particle, and inclined to each other at an infinitesimal angle: then the lemma shews how to calculate the attraction of the slice resolved along a given direction in one of the planes.

251. Before discussing the attraction of an ellipsoid of revolution, Maclaurin considers that of a sphere in his Article 643. The following general result is obtained: Let C be the centre of a circle, P any external point in the plane of the circle. From P draw any straight line cutting the circumference of the circle at L and M; and let a solid be formed by the revolution round PC of the smaller segment of the circle cut off by LM. Then the attraction of this solid on a particle at P varies as $\frac{(LM)^3}{(PC)^2}$.

This may be easily verified by the aid of the general expression given in Art. 4. The formula is very remarkable; it does not involve the radius of the sphere; that is, if LM is constant, we get the attraction constant whatever may be the value of the radius. The result was generalised by Legendre, as we shall see, in his third memoir.

252. Maclaurin then in his Articles 644...647 investigates accurate expressions for the attraction of any ellipsoid of revolution on a particle at the pole or at the equator. The investigations are conducted in the manner of the time by representing the attractions by the areas of certain curves, and finding the areas by the method of fluents. The results agree with those obtained by analysis, and presented in modern works on Statics. Maclaurin's processes are remarkable specimens of ingenuity, considering the date of their publication; but they will not be very interesting to a modern reader.

253. Maclaurin says in his Article 647:

> What has been shown concerning the gravity at the pole...agrees with what was advanced long ago by Sir ISAAC NEWTON and Mr. COTES, who contented themselves with an approximation in determining the gravity at the equator, which is exact enough when the spheroid differs very little from a sphere. The approximations proposed lately for this purpose, *Phil. Trans*. N. 438 and 445. are more accurate; and Mr. STIRLING after determining the gravity at the equator by a converging series, since found that the sum of the series could be assigned from the quadrature of the circle.

I do not know what is intended by the reference to Mr Cotes. Of course Cotes, as editor of the *Principia*, may be supposed to have accepted some of the responsibility which would otherwise have fallen on Newton alone: but Maclaurin's words seem to imply that Cotes had made some investigations of his own. The paper in the *Philosophical Transactions*, Number 438, is that by Stirling, of which we gave an account in Chapter V.; and the paper in the *Philosophical Transactions*, Number 445, is that by Clairaut, of which we gave an account in Arts. 163...166. I do not know what Maclaurin means by the words "and Mr Stirling...circle."

This passage from Maclaurin was quoted, and the difficulty

as to its meaning noticed, by the late Sir J. W. Lubbock: see page 24 of his work cited in Art. 233.

I do not know whether the conjecture may be considered plausible that Maclaurin wrote *Stirling* by mistake for *Simpson.* It appears from the preface to Simpson's *Mathematical Dissertations* that his researches on the Figure of the Earth were read to the Royal Society in March or April, 1741; and what Maclaurin says with respect to Mr Stirling is not unsuitable to the investigation we find in Simpson's work, except that Simpson does not restrict himself to a point at the equator, but takes any point on the surface.

254. Maclaurin proceeds in his Articles 648...652 to one of the most important of his investigations, remarkable as forming a large part of the theorem which now usually bears the name of Ivory, though it was substantially first demonstrated by Laplace. Maclaurin's theorem is as follows in modern language: Let there be two confocal ellipses, and let them both revolve round their major-axes, or round their minor-axes, so as to generate two ellipsoids of revolution: then the attractions of the two ellipsoids on the same particle external to both will be as the volumes, provided the particle be on the prolongation of the axis of revolution, or in the plane of the equator. Two such ellipsoids may be called confocal ellipsoids of revolution. Legendre shewed that the theorem was true for any position of the external particle.

The general theorem demonstrated by Laplace is as follows: If there be two confocal ellipsoids, that is, ellipsoids which have the same foci for their principal sections, their attractions on any particle external to both will be as their volumes, that is, will be the same in direction, and in amount will be as their volumes. The simplest statement in modern language is this: The potentials of confocal ellipsoids on a given external particle are as their volumes.

Maclaurin in a later Article, namely 653, gave so much of this general theorem as consists with the limitation that the particle must be on the prolongation of an axis of the ellipsoids. Ivory merely supplied an improved form of demonstration to Laplace's theorem; and combined it with the fact that inside an ellipsoid, along any radius-vector, the attraction varies as the distance from the centre.

Maclaurin's Articles 648 and 649 contain his demonstration for the case in which the external particle is on the prolongation of the axis of revolution. These Articles may be read without difficulty, apart from Maclaurin's other investigations, by those who are desirous of seeing a specimen of his own processes.

255. It is easy to translate into modern language the essence of Maclaurin's demonstration.

Let $2a$ and $2b$ be the axes of an ellipse; let the ellipse revolve about the axis of length $2a$, and thus generate an ellipsoid of revolution: required the attraction of the ellipsoid on a particle which is on the prolongation of the axis of revolution at a distance c from the centre.

Let r be the distance of the attracted particle from any point of the ellipsoid; let θ be the angle between r and the axis of revolution. We see in the usual way that the attraction is found by integrating with respect to r and θ the expression

$$\frac{2\pi r dr\, r \sin\theta \cos\theta\, d\theta}{r^2}.$$

Integrate with respect to r and we obtain

$$2\pi (r_2 - r_1) \sin\theta \cos\theta\, d\theta,$$

where r_2 and r_1 are respectively the greatest and the least values of the radius-vector drawn from the attracted particle to the ellipsoid at the inclination θ to the axis of revolution.

Hence r_2 and r_1 are the roots of the quadratic equation

$$\frac{(r\cos\theta - c)^2}{a^2} + \frac{r^2 \sin^2\theta}{b^2} = 1,$$

and thus we shall find that

$$r_2 - r_1 = \frac{2ab\sqrt{(b^2\cos^2\theta + a^2\sin^2\theta - c^2\sin^2\theta)}}{b^2\cos^2\theta + a^2\sin^2\theta}$$

$$= \frac{2ab\sqrt{\{b^2 + (a^2 - b^2 - c^2)\sin^2\theta\}}}{b^2 + (a^2 - b^2)\sin^2\theta}.$$

Now let there be a second ellipsoid of revolution, having the foci of its generating ellipse in the same position as before; and let accented letters be used to denote the analogous quantities; so that

$$r_2' - r_1' = \frac{2a'b'\sqrt{\{b'^2 + (a'^2 - b'^2 - c^2)\sin^2\theta'\}}}{b'^2 + (a'^2 - b'^2)\sin^2\theta'}$$

Since the foci of the generating ellipses are coincident, we have $a^2 - b^2 = a'^2 - b'^2$, whether the ellipsoids are oblate or oblong. Assume $\sin\theta' = \frac{b'}{b}\sin\theta$; then we see that

$$\frac{r_2' - r_1'}{r_2 - r_1} = \frac{a'}{a};$$

and therefore $$\frac{(r_2' - r_1')\sin\theta'\cos\theta'\,d\theta'}{(r_2 - r_1)\sin\theta\cos\theta\,d\theta} = \frac{a'b'^2}{ab^2}.$$

Thus the attractions of the corresponding elements of the two ellipsoids resolved along the direction of the axis of revolution are in the same proportion as the volumes of the ellipsoids; and so the resultant attractions of the whole ellipsoids will be in that proportion.

It will be observed that on our assumption $r_2' - r_1'$ and $r_2 - r_1$ *vanish together;* so that our elements *always* correspond. If the density of one ellipsoid is not the same as the density of the other, then the attractions will of course be in the ratio of the *masses* instead of the ratio of the *volumes.* This remark will be obviously applicable in some subsequent Articles.

Maclaurin's own investigation in his Art. 648 applies to his figure 292, which is drawn for an oblatum; but the figure may be drawn for an oblongum, and it will be found that the investigation is equally applicable. In Maclaurin's investigation the point P is *on* the larger ellipsoid; but still this involves the result in as general a form as we have stated it.

256. Maclaurin's Article 650 consists of three sentences; it would have been advantageous, for the sake of clearness, if they had been printed as three distinct paragraphs: the last sentence most certainly should have been separated from the others.

In the first sentence Maclaurin gives an expression for the attraction of an oblatum on an external particle which is situated on the axis of revolution: this follows from his former results, which we have noticed in Arts. 252 and 255.

In the second sentence Maclaurin gives the corresponding expression for the attraction of an oblongum.

The third sentence is very remarkable. It has been shewn that the attraction of a homogeneous ellipsoid of revolution on an external particle which is situated on the axis of revolution, varies as the mass, so long as the generating ellipse keeps its foci fixed; now suppose an ellipsoid of revolution, not homogeneous, but made up of shells, each shell being bounded by confocal ellipsoids of revolution, and the density being uniform throughout each shell, but varying in any manner from shell to shell: then the attraction of this heterogeneous ellipsoid on an external particle situated on the axis of revolution is to the attraction of a homogeneous ellipsoid of the same size as the mass of the former is to the mass of the latter. This is the first appearance of these *confocal* shells, which play an important part in modern works on Attraction.

257. Maclaurin now proceeds in his Articles 651, 652 to the case in which the attracted external particle is in the plane of the equator of the attracting ellipsoid of revolution. He uses a most ingenious artifice by which this case is made to depend on that already considered, in which the attracted particle is on the prolongation of the axis of revolution. We will translate his process into modern language.

Let the equation to one ellipsoid of revolution be $\frac{x^2}{a^2} + \frac{y^2 + z^2}{c^2} = 1$, and the equation to another $\frac{x^2}{a'^2} + \frac{y^2 + z^2}{c'^2} = 1$. Suppose the generating ellipses to have the same foci; then, whether the ellipsoids are oblate or oblong, $a^2 - c^2 = a'^2 - c'^2$.

Suppose the second ellipsoid to be the larger. We propose to compare the attractions of these ellipsoids on a particle which is *on* the equator of the larger ellipsoid; the co-ordinates of the particle may be taken to be 0, 0, c'. We shall shew that the attractions of the ellipsoids are as their volumes.

Let C denote the centre of the ellipsoids, and P the position of the attracted particle.

Let two planes pass through CP, and make with the axis of y the angles θ and $\theta + \delta\theta$, respectively: we will call these planes the *first pair of planes*. Let two other planes pass through CP, and make with the axis of y the angles θ' and $\theta' + \delta\theta'$ respectively: we will call these planes the *second pair of planes*. The volume comprised between the first pair of planes and the first ellipsoid we will call the *element of the first ellipsoid*; the volume comprised between the second pair of planes and the second ellipsoid we will call the *element of the second ellipsoid*: each element then consists of *two* wedge-shaped slices. We shall shew that when a suitable relation is made to hold between θ and θ', the attractions of these elements on the particle at P are as their volumes.

The relation between θ and θ' is found by assuming that the ellipses which form the boundaries of the elements shall be *confocal*. Thus we have $r^2 - c^2 = r'^2 - c'^2$,

where $r^2 = \dfrac{a^2c^2}{a^2\cos^2\theta + c^2\sin^2\theta}$, and $r'^2 = \dfrac{a'^2c'^2}{a'^2\cos^2\theta' + c'^2\sin^2\theta'}$.

Since $a^2 - c^2 = a'^2 - c'^2$, we obtain

$$\frac{c^2\sin^2\theta}{a^2\cos^2\theta + c^2\sin^2\theta} = \frac{c'^2\sin^2\theta'}{a^2\cos^2\theta' + c'^2\sin^2\theta'}:$$

this is the relation between θ and θ'. It is obvious that to the limits 0 and $\frac{\pi}{2}$ for θ correspond the same limits for θ'.

Suppose now that a solid were formed by the revolution round CP of an ellipse having C for centre, $2c$ for the axis of revolution, and $2r$ for the other axis. Let F denote the attraction of this solid on the particle at P. Then it is obvious that ultimately the attraction of the *element of the first ellipsoid* on the particle is $\dfrac{\delta\theta}{\pi}F$.

Also suppose that a solid were formed by the revolution round CP of an ellipse having C for centre, $2c'$ for the axis of revolution, and $2r'$ for the other axis. Let F' denote the attraction of this solid on the particle at P. Then it is obvious that ultimately the attraction of the *element of the second ellipsoid* on the particle is $\frac{\delta\theta'}{\pi} F'$

Therefore if f and f' denote the attractions of the elements, we have
$$\frac{f}{f'} = \frac{F \cdot \delta\theta}{F' \cdot \delta\theta'}.$$

Now, as we have seen in Art. 255, Maclaurin had shewn that
$$\frac{F}{F'} = \frac{r^2 c}{r'^2 c'};$$
therefore
$$\frac{f}{f'} = \frac{r^2 c\,\delta\theta}{r'^2 c'\,\delta\theta'}.$$

But $r^2\delta\theta$ represents the area intercepted by the first pair of planes from the ellipse $\frac{x^2}{a^2} + \frac{y^2}{c^2} = 1$; and $r'^2\delta\theta'$ represents the area intercepted by the second pair of planes from the ellipse $\frac{x^2}{a'^2} + \frac{y^2}{c'^2} = 1$. Thus we see that f is to f' as the volume of the element of the first ellipsoid is to the volume of the element of the second ellipsoid. And as this proportion holds for every corresponding pair of elements it holds for the entire ellipsoids; which is what we had to demonstrate.

258. The process may be easily extended to the case in which the ellipsoids are not of revolution, as Maclaurin himself indicates in his Article 653.

Let the equations to the ellipsoids be
$$\frac{x^2}{a^2} + \frac{y^2}{b^2} + \frac{z^2}{c^2} = 1, \quad \frac{x^2}{a'^2} + \frac{y^2}{b'^2} + \frac{z^2}{c'^2} = 1;$$
and let the principal sections of the ellipsoids be confocal, so that
$$c^2 - a^2 = c'^2 - a'^2, \text{ and } c^2 - b^2 = c'^2 - b'^2.$$

The relation between θ and θ' will then be found from the condition
$$r^2 - c^2 = r'^2 - c'^2,$$
where
$$r^2 = \frac{a^2 b^2}{a^2 \cos^2\theta + b^2 \sin^2\theta}, \text{ and } r'^2 = \frac{a'^2 b'^2}{a'^2 \cos^2\theta' + b'^2 \sin^2\theta'}.$$

As before, we shall find that to the limits 0 and $\frac{\pi}{2}$ for θ correspond the same limits for θ'. Then the investigation and the result will be as in the preceding Article.

259. Thus in the attraction of homogeneous ellipsoids Maclaurin's position was as follows: he solved completely the problem of the attraction of an ellipsoid of revolution on any internal particle; and with respect to an external particle, he obtained for ellipsoids, not necessarily of revolution, the theorem of Laplace, so far as relates to a particle on the prolongation of *an axis* of the ellipsoids. All this was *exactly* demonstrated.

Maclaurin states also something more as approximately true in his Article 654. The statement amounts to this, that the theorem of Art. 254 is true "either accurately or nearly when the spheroids differ little from spheres," when the attracted particle has *any* position. He gives no detail as to the investigation of this result; but merely says it may be deduced from his Article 653. We know now that the theorem is exact and not merely an approximation; and, as we have stated, the demonstration was first given by Legendre, and the theorem is a part of Laplace's general theorem.

260. The extent to which Maclaurin carried his investigations was under-estimated by many of the succeeding writers. He was supposed to have merely *enunciated* the result which we have noticed in Art. 258, whereas he really *demonstrates* it: he says "it will appear in the same manner..." and it is clear from an examination of his context that this is the case. The erroneous account will be found in the following places: D'Alembert, *Opuscules Mathématiques*, Vol. VI. 1773, page 243; Lagrange, Berlin *Mémoires* for 1775, page 279; Laplace, *Théorie...de la Figure elliptique des Planetes*, 1784, page 96; Legendre, *Mémoires ...par divers Savans*, Vol. X. 1785, page 412. Laplace, *Mécanique Céleste*, Vol. V. page 9. Plana in Crelle's *Journal für...Mathematik*, Vol. XX. page 190. According to the catalogues of booksellers, it appears that Maclaurin's *Fluxions* was translated into French, so that there is less excuse for the error. I suppose that D'Alembert went astray, and the others followed in succession without examination. Chasles is correct; he says that Maclaurin

did demonstrate his theorem, and he points out the error in this matter made by D'Alembert, Lagrange, Legendre, and others: see the *Mémoires...par divers Savants*, Vol. IX. 1846, page 632. The error is also noticed by Dr F. Grube in a paper in the *Zeitschrift für Mathematik und Physik*, Vol. XIV. Leipsic, 1869, page 272.

On the other hand, some recent English writers have gone to the opposite extreme, and given to Maclaurin more than his due, by ascribing to him in effect the entire theorem called Ivory's, but more strictly Laplace's; see *Natural Philosophy*, by Thomson and Tait, Vol. I. page 392, and Routh's *Rigid Dynamics*, 2nd edition, page 421.

261. It will be convenient to give the results obtained by Maclaurin as to the attraction of an oblatum on an external particle which is in the plane of the equator, or on the prolongation of the axis of revolution.

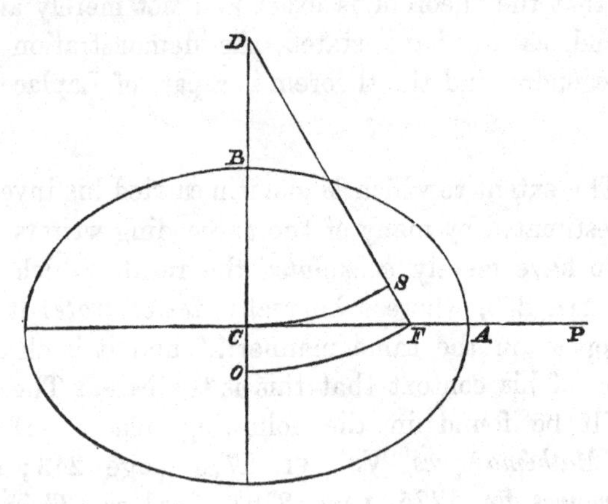

Let P be the position of an external particle which is in the plane of the equator. Let F be the focus of the section of the oblatum made by the plane which contains P and the axis of revolution. Let C be the centre, CA and CB the semi-axes of the section. With F as centre, and a radius equal to CP, describe a circle cutting CB produced at D. With D as centre, and DF as radius, describe the arc FO, and with D as centre and DC as radius, describe the arc CS.

Then Maclaurin obtains for the attraction on a particle at P the expression

$$\frac{2 \,.\, CB \,.\, CA^2}{CF^3} \times \frac{\text{area } FCO}{CP}$$

And for the attraction on a particle at the point D on the prolongation of the axis of revolution, he obtains the expression

$$\frac{2 \,.\, CB \,.\, CA^2}{CF^3} \times (CF - CS).$$

If we multiply these expressions by $2\pi\rho$, where ρ denotes the density, they will be found to agree with those given in modern works on Statics when we suppose P to be on the surface; and the case where P is not on the surface may be deduced from that where P is on the surface, by Maclaurin's theorem of Art. 254. The presence or absence of such a factor as 2π merely depends on the choice we have made of the unit of attraction.

Put a for CA, and ae for CF; also put r for CP in the first expression, and r for CD in the second; then, introducing the factor $2\pi\rho$, our expressions become:

$$\frac{2\pi\rho \sqrt{(1-e^2)}}{e^3} \left\{r \sin^{-1}\frac{ea}{r} - \frac{ae\sqrt{(r^2 - e^2a^2)}}{r}\right\} \ldots\ldots\ldots (1),$$

and

$$\frac{4\pi\rho \sqrt{(1-e^2)}}{e^3} \left\{ea - r \tan^{-1}\frac{ea}{r}\right\} \ldots\ldots\ldots\ldots\ldots (2),$$

so that (1) applies to the particle in the plane of the equator, and (2) to the particle on the prolongation of the axis of revolution.

It will be useful for us to collect here some obvious deductions from (1) and (2).

The attraction at the equator is obtained by putting a for r in (1); and the attraction at the pole is obtained by putting $a\sqrt{(1-e^2)}$ for r in (2).

Let x and y be the co-ordinates of any point *on the surface* of the oblatum, measured from the origin C parallel to CA and CB respectively. Then, by Art. 244, combined with the values of the attraction at the equator and at the pole, to which we have just alluded, we obtain for the attractions at the point (x, y), resolved parallel to CA and CB respectively,

$$\frac{2\pi\rho\sqrt{(1-e^2)}}{e^3}\{\sin^{-1}e - e\sqrt{(1-e^2)}\}\,x,$$

and $$\frac{4\pi\rho}{e^3}\{e-\sqrt{(1-e^2)}\sin^{-1}e\}\,y.$$

If we expand these and neglect e^4 and higher powers of e we obtain respectively

$$\frac{4\pi\rho}{3}\left(1-\frac{e^2}{5}\right)x \text{ and } \frac{4\pi\rho}{3}\left(1+\frac{2e^2}{5}\right)y.$$

By expanding their second factors in powers of e, the expressions (1) and (2) become respectively

$$\frac{4\pi\rho}{3}\sqrt{(1-e^2)}\left\{\frac{a^3}{r^2}+\frac{3e^2}{10}\frac{a^5}{r^4}+\frac{9e^4}{56}\frac{a^7}{r^6}+\ldots\right\},$$

and $$\frac{4\pi\rho}{3}\sqrt{(1-e^2)}\left\{\frac{a^3}{r^2}-\frac{3e^2}{5}\frac{a^5}{r^4}+\frac{3e^4}{7}\frac{a^7}{r^6}+\ldots\right\}.$$

In the expressions (1) and (2) change a into $a+\delta a$, and subtract the original values; thus we obtain the attraction of a shell bounded by similar, similarly situated, and concentric oblata, on an external particle in the plane of the equator or on the prolongation of the axis: supposing δa so small that all powers beyond the first may be neglected, the results are respectively

$$\frac{4\pi\rho\sqrt{(1-e^2)}}{e^3}\,\frac{e^3a^3}{r\sqrt{(r^2-e^2a^2)}}\,e\delta a,$$

and $$\frac{4\pi\rho\sqrt{(1-e^2)}}{e^3}\,\frac{e^3a^3}{r^2+e^2a^2}\,e\delta a.$$

Maclaurin, subsequently, in his Articles 668 and 669, gives without demonstration, in a geometrical form, results which are equivalent to these.

262. Maclaurin, in his Article 655, applies his results to find the condition for the relative equilibrium of an oblatum of fluid rotating round the minor axis. Let a be the semi-axis major, and e the excentricity. Let X denote the attraction at the equator, and Y the attraction at the pole. Then we obtain X by putting a for r in the expression (1) of Art. 261, and we obtain Y by putting $a\sqrt{(1-e^2)}$ for r in the expression (2). Thus we find

$$\frac{Y}{X} = \frac{2\{e - \sqrt{(1-e^2)} \sin^{-1} e\}}{\sin^{-1} e - e\sqrt{(1-e^2)}}$$

Suppose that jX denotes the value of the centrifugal force at the equator; then for relative equilibrium we must have, by Art. 245,

$$\frac{X - jX}{Y} = \sqrt{(1-e^2)};$$

therefore

$$j = \frac{X - Y\sqrt{(1-e^2)}}{X}.$$

Put for $\frac{Y}{X}$ its value, and this becomes

$$j = \frac{3\{\sin^{-1} e - e\sqrt{(1-e^2)}\} - 2e^2 \sin^{-1} e}{\sin^{-1} e - e\sqrt{(1-e^2)}}.$$

These expressions are exact. By approximation we obtain

$$\frac{Y}{X} = \frac{1 + \frac{2}{5}e^2 + \frac{8}{35}e^4 + \frac{16}{105}e^6 + \ldots}{1 + \frac{3}{10}e^2 + \frac{9}{56}e^4 + \frac{5}{48}e^6 + \ldots},$$

$$j = \frac{\frac{2}{5}e^2 + \frac{9}{35}e^4 + \frac{5}{28}e^6 + \ldots.}{1 + \frac{3}{10}e^2 + \frac{9}{56}e^4 + \frac{5}{48}e^6 + \ldots.}.$$

Maclaurin gives these approximations as far as e^4 inclusive.

By reversion of series we obtain

$$e^2 = \frac{5j}{2} - \frac{15j^2}{7} + \ldots;$$

so that when the oblatum differs very little from a sphere we may take

$$e^2 = \frac{\frac{5j}{2}}{1 + \frac{6j}{7}}$$

Maclaurin then says, "in this case the excess of the semi-diameter of the equator above the semiaxis is to the mean semi-diameter nearly as" $5j$ is to $4 - \frac{11j}{7}$. By the *mean semi-diameter* he intends half the sum of the polar and equatorial radii. Taking 1 for the equatorial radius, we have $\sqrt{(1-e^2)}$ for the polar radius; then the ratio of the difference to the half-sum is expressed exactly by $\frac{2\{1-\sqrt{(1-e^2)}\}}{1+\sqrt{(1-e^2)}}$.

If we wish to be correct only to the first power of e^2 this becomes $\frac{e^2}{2}$.

If we wish to be correct to the second power of e^2 this becomes $\frac{e^2\left(1+\frac{e^2}{4}\right)}{2-\frac{e^2}{2}}$. We might use other forms which would coincide with this as far as the second power of e^2. For instance, we have the ratio exactly equal to $\frac{2e^2}{\{1+\sqrt{(1-e^2)}\}^2}$, and thus to the order of e^4 we get $\frac{2e^2}{4-2e^2}$; and then we may put this to the same order in the form $\frac{e^2}{2}\left(1+\frac{e^2}{2}\right)$.

Taking the form $\frac{2e^2}{4-2e^2}$ and putting $\frac{\frac{5j}{2}}{1+\frac{6j}{7}}$ for e^2, we obtain with Maclaurin $\frac{5j}{4-\frac{11j}{7}}$.

263. Maclaurin shews how the value of e^2 for the Earth, supposed homogeneous, may be deduced from the measured length of a degree of the meridian in any latitude, and the measured length of the pendulum which vibrates in a given time in that latitude: see his Articles 656...658. He shews in his Article 657 that the radius of curvature in the ellipse varies as the cube of the length of the normal terminated by the major axis; he was, probably, the first to demonstrate this: see the *Mécanique Céleste*, Vol. v. page 6.

Maclaurin also shews how the value of e^2 may be deduced from the distance and the periodic time of a satellite revolving in the plane of the equator: see his Articles 659 and 660.

Maclaurin in his Articles 661...665 obtains numerical results with respect to the Earth, supposed homogeneous. He does not determine strictly the value of the quantity we denote by j; but he finds $\frac{1}{289.3}$ as the value of the ratio of the centrifugal force at the equator to the force of gravity at Paris, and $\frac{1}{287.8}$ as the value of the ratio of the centrifugal force at the equator to the force of gravity at the Polar circle. For the ratio of the axes of the Earth he obtains a result practically equivalent to Newton's value of 230 to 229.

Maclaurin shews, however, that this result is not consistent with that obtained by means of the observations of pendulums in various latitudes; nor with that obtained from the measured lengths of a degree of the meridian in France and in Lapland: both these methods gave for the ellipticity a larger value than $\frac{1}{230}$. We have now more accurate observations and measurements than those accessible to Maclaurin; and we know that the true value of the ellipticity is about $\frac{1}{300}$.

264. Maclaurin then proposes to treat the Earth as not uniform in density. In his Article 666 he supposes that there is *more* matter at the centre than is consistent with the hypothesis of uniform density; and in his Article 667 he supposes that there is *less* matter at the centre. He concludes that both these suppositions are inadmissible, as not agreeing with facts; for, relying on the French and Lapland arcs, he considered that the ellipticity must be greater than $\frac{1}{229}$.

In his investigations he does not shew that there will be relative equilibrium in the supposed fluid mass; but he shews that if there be relative equilibrium, certain relations will exist between the lengths of the polar and the equatorial diameters.

Maclaurin's investigations do not appear quite satisfactory; let us take his Article 667. With the notation of Art. 262 we have $X-jX$ for the gravity at the equator, and Y for the gravity at the pole. The ratio of the difference to the half-sum is

$$2\frac{Y-X+jX}{Y+X-jX}.$$

Now for relative equilibrium we must have

$$Y\sqrt{(1-e^2)}=X(1-j);$$

substitute, and we find that the above ratio becomes

$$\frac{2\{1-\sqrt{(1-e^2)}\}}{1+\sqrt{(1-e^2)}}.$$

As we have seen in Art. 262, this result can be put in various *approximate* forms.

Now Maclaurin supposes that matter is removed from the centre of the oblatum, so as to diminish the attraction at the equator by a certain fraction of the mean attraction; we shall denote this fraction by λ, and the mean attraction by G. The attraction at the pole will be diminished by $\frac{\lambda G}{1-e^2}$. The ratio of the centrifugal force to the attraction at the equator is supposed to remained unchanged.

Thus the gravity at the equator is $(X-\lambda G)(1-j)$, and at the pole is $Y-\frac{1}{1-e^2}\lambda G$. The ratio of the difference to the half-sum is

$$2\frac{Y-X(1-j)-\lambda G\left\{\frac{1}{1-e^2}-(1-j)\right\}}{Y+X(1-j)-\lambda G\left\{\frac{1}{1-e^2}+1-j\right\}}$$

Maclaurin considers that this is approximately equal to $\frac{5j-14j\lambda}{4-4\lambda+2j\lambda}$; and this is less than $\frac{5j}{4}$ which he takes for the approximate value of the ratio before the matter was removed from the centre.

But these statements are liable to the objection which is fatal to so many approximate calculations; the investigation is not true to the order of the small quantities which are retained. Put $\frac{1}{2}(Y+X)$ for G; and observe that $Y\sqrt{(1-e^2)}=X(1-j)$. Then the ratio after the matter is removed from the centre is accurately

$$\frac{\frac{2}{\sqrt{(1-e^2)}}-2-\lambda\left\{\frac{1}{\sqrt{(1-e^2)}}+\frac{1}{1-j}\right\}\left\{\frac{1}{1-e^2}-(1-j)\right\}}{\frac{1}{\sqrt{(1-e^2)}}+1-\frac{\lambda}{2}\left\{\frac{1}{\sqrt{(1-e^2)}}+\frac{1}{1-j}\right\}\left\{\frac{1}{1-e^2}+1-j\right\}}.$$

If we neglect powers of e^2 and j above the first the numerator of this fraction becomes $e^2-2\lambda(e^2+j)$; and the denominator becomes $2+\frac{e^2}{2}-\frac{\lambda}{2}\left(2+\frac{e^2}{2}+j\right)(2+e^2-j)$, that is, to our order of approximation $2+\frac{e^2}{2}-\lambda\left(2+\frac{3e^2}{2}\right)$. If we now put $\frac{5j}{2}$ for e^2, we obtain for the ratio $\frac{5j-14j\lambda}{4+\frac{5j}{2}-4\lambda-\frac{15}{2}j\lambda}$.

Thus we see that Maclaurin is wrong in his denominator.

There is, however, a very serious objection to the process just given. If Maclaurin retained the term in $j\lambda$ in the denominator, he ought to have carried on the approximations in the numerator to a higher order; for instance, e^4 ought to have been retained: and then when the value of e^2 in terms of j is substituted in the numerator the square of j must be retained. But, in order to determine in a satisfactory manner how far the approximations are to be carried, we must make some hypothesis as to the value of λ. Suppose, for instance, that λ is $\frac{1}{3}$ or $\frac{1}{4}$; then $j\lambda$ will be of the same order as j; and in the numerator of the ratio we shall have to retain the squares of e^2 and j, and the product e^2j. But if we suppose that λ is of the same order as j, and retain the term $j\lambda$ in the denominator, then we must make our numerator accurate to the *third* order of small quantities, and our denominator accurate to the *second* order, considering e^2 or j as of the first order.

I have taken $G=\frac{1}{2}(Y+X)$ as Maclaurin's words certainly imply. I do not retain his notation nor his language; but use what I find most convenient. Maclaurin himself, in his Art. 666, explains that in what follows he uses *gravitation* for the excess of gravity above centrifugal force: so that his gravity corresponds to my attraction, and his gravitation to my gravity.

It is possible however that, with Maclaurin, $G=\frac{1}{2}(Y+X-jX)$.

This meaning of G makes $Y-X(1-j)=2G\frac{1-\sqrt{(1-e^2)}}{1+\sqrt{(1-e^2)}}$;

and the ratio of the difference of the polar gravity and the equatorial gravity to the half-sum becomes accurately

$$\frac{2\frac{1-\sqrt{(1-e^2)}}{1+\sqrt{(1-e^2)}}-\lambda\left\{\frac{1}{1-e^2}-(1-j)\right\}}{1-\frac{1}{2}\lambda\left\{\frac{1}{1-e^2}+1-j\right\}}.$$

Instead of the expression $4+\frac{5j}{2}-4\lambda-\frac{15}{2}j\lambda$, which we obtained before for the denominator by approximation we should now have $4-4\lambda-3j\lambda$, which is still different from Maclaurin's result.

However, though Maclaurin's process is very unsatisfactory, his conclusion is true that the ratio of the difference to the half-sum of the gravities is diminished by removing matter from the centre. The best way of shewing this, is to start from the algebraical fact that $\frac{p-p'}{q-q'}$ is less than $\frac{p}{q}$ if $\frac{p'}{q'}$ is greater than $\frac{p}{q}$. Accordingly we have only to shew that $\frac{\frac{1}{1-e^2}-(1-j)}{\frac{1}{1-e^2}+1-j}$ is greater than $\frac{1-\sqrt{(1-e^2)}}{1+\sqrt{(1-e^2)}}$; this reduces to shewing that $1-j$ is less than $\frac{1}{\sqrt{(1-e^2)}}$, which is obviously true.

There would, however, be little interest in ascertaining that the ratio is diminished without any estimate of the amount of diminution; but, in order to form such an estimate, it would be necessary to make an hypothesis as to the value of λ, and then to approximate to a suitable degree of accuracy.

Hitherto in this Article we have not paid any regard to the supposition that the oblatum is *fluid*; but let us now adopt that supposition. Maclaurin finds by Newton's method of balancing columns that when matter is removed from the centre, the polar diameter will be diminished, and the equatorial diameter increased, and so the excentricity increased. The process is not satisfactory; for Maclaurin does not shew that the fluid can remain in equilibrium when matter is removed from the centre: and in fact we now know that it will be necessary to make some fresh hypothesis. We may suppose that there is a solid spherical nucleus, surrounded by a fluid of greater density. In this case it will be found that relative equilibrium will subsist, when the bounding surface is an oblatum of certain excentricity; and this excentricity is greater than when the body is entirely fluid and homogeneous. But the value of λ cannot be taken quite arbitrarily: it must fall below $\frac{2}{5}$. The problem in fact was solved by Clairaut in the more general form of a central nucleus which is not a sphere but an ellipsoid of revolution, having for its axis of revolution the axis of rotation. See his *Figure de la Terre*, page 219.

We will briefly solve the problem, when the nucleus is spherical, in the modern way. Let M denote the mass of the body, supposed entirely fluid and homogeneous; then λM is the mass which is supposed to be removed, so as to make the central nucleus less dense than the fluid. We may consider that the attraction at any point of the fluid is produced by the action of the whole oblatum of fluid, diminished by the action of the sphere of mass λM.

Take the axis of z for that of revolution. Let ω be the angular velocity. The attraction of the oblatum at the point (x, y, z) parallel to the axes will be Ax, Ay, Cz, respectively, where A and C are constants. The attraction of the sphere will be $\frac{\lambda x M}{r^3}$, $\frac{\lambda y M}{r^3}$, and $\frac{\lambda z M}{r^3}$ respectively, where $r^2 = x^2 + y^2 + z^2$.

Hence the equation to the surface of the fluid must be

$$\frac{A(x^2+y^2)}{2} - \frac{\omega^2(x^2+y^2)}{2} + \frac{Cz^2}{2} + \frac{\lambda M}{r} = \text{constant}.$$

Suppose $2a$ and $2c$, the equatorial and polar diameters; then we get

$$\frac{Aa^2}{2} - \frac{\omega^2 a^2}{2} + \frac{\lambda M}{a} = \text{constant},$$

and $$\frac{Cc^2}{2} + \frac{\lambda M}{c} = \text{constant};$$

therefore by subtraction

$$Aa^2 - Cc^2 - \omega^2 a^2 + 2\lambda M\left(\frac{1}{a} - \frac{1}{c}\right) = 0.$$

Now by hypothesis we have

$$a\omega^2 = j\left(Aa - \frac{\lambda M}{a^2}\right),$$

so that $$Aa^2 - Cc^2 - j\left(Aa^2 - \frac{\lambda M}{a}\right) + 2\lambda M\left(\frac{1}{a} - \frac{1}{c}\right) = 0.$$

If we suppose that e is very small, we find by Article 262 that approximately

$$Aa^2 = \frac{M}{a}\left(1 + \frac{3}{10}e^2\right), \quad Cc^2 = \frac{M}{a}\left(1 - \frac{1}{10}e^2\right);$$

and $$\frac{1}{a} - \frac{1}{c} = -\frac{e^2}{2a};$$

so that $$e^2 = \frac{j(1-\lambda)}{\frac{2}{5} - \lambda} = \frac{\frac{5j}{2}(1-\lambda)}{1 - \frac{5\lambda}{2}}.$$

It is obvious that if we suppose e^2 and j to be of the same order of magnitude, this process is not satisfactory for every value of λ: for instance, λ must not be nearly equal to $\frac{2}{5}$. And if λ is itself of the same order as e^2 and j, the result is not admissible, for then we ought to have retained e^4 and $e^2 j$ as well as $j\lambda$ and $e^2\lambda$.

We may accept the investigation as sufficiently accurate for such cases as $\lambda = \frac{1}{5}$, or $\lambda = \frac{1}{10}$; and we see that the excentricity is greater than for the case of the oblatum entirely fluid and homogeneous: so far then we agree with Maclaurin.

Maclaurin, however, asserts, that in consequence of this increase of the excentricity, the ratio of the difference of the gravities to their half-sum is rendered still less than it was before we adopted the supposition of fluidity. This is a mere assertion unsupported by evidence. So far as the influence of the removal of central matter is concerned, we may admit that the increase of the excentricity tends to bring the polar gravity and the equatorial *nearer* to equality; but, on the other hand, considering all the other matter as forming a homogeneous oblatum, we see that the increase of the excentricity tends to bring the polar gravity and the equatorial *further* from equality. Thus, to obtain the actual result, we must strike a balance between opposing influences; and this Maclaurin has not done.

We can easily submit the question to calculation. Before the hypothesis of fluidity was adopted, taking λ less than $\frac{2}{5}$ but not so small as j, we have for the approximate value of this ratio $\dfrac{\frac{5j}{4}\left(1-\frac{14\lambda}{5}\right)}{1-\lambda}$ to the order of accuracy necessary: to this order, in fact, Maclaurin's result agrees with that which we obtained.

Now with the hypothesis of fluidity we may find the ratio by the aid of Clairaut's theorem; for the ratio of the difference of the gravities to the half-sum is the same as the ratio of the difference of the gravities to the equatorial gravity, to our order of accuracy. Thus, by Art. 171, the ratio is

$$\frac{5j}{2}-\frac{e^2}{2},\ \text{that is}\ \frac{5j}{2}-\frac{\frac{5j}{4}(1-\lambda)}{1-\frac{5\lambda}{2}},\ \text{that is}\ \frac{\frac{5j}{4}(1-4\lambda)}{1-\frac{5}{2}\lambda}.$$

Now this is not necessarily *less* than the former value; it is in fact *greater* if λ is less than $\frac{1}{10}$.

265. Maclaurin considers in his Articles 668...671 the attraction of an ellipsoid of revolution made up of similar and concentric shells of varying density. He shews theoretically how to determine the attraction on a particle on the axis, or in the plane of the equator, either external or internal. In modern language we should say that he reduces the general problem to depend on a single integration: see Art. 261. Maclaurin then takes special

cases; he treats briefly the case in which the density varies inversely as the diameter of the shell, and the case in which it varies inversely as the square of the diameter; and more fully the case in which it varies as the diameter.

Jacobi has made an important remark on the subject of the similar concentric shells when the ellipsoid is not of revolution: see Poggendorff's *Annalen*, Vol. XXXIII. 1834, page 233. Pontécoulant, *Théorie Analytique, Supplément au Livre* V. page 22.

266. It will be interesting to discuss analytically some cases of similar concentric shells with varying density.

I. Suppose the density to vary inversely as the diameter. Put x for ea; then the density varies inversely as x; say that the density $= \frac{\mu}{x}$. Take the formulæ of Art. 261, omitting the common factor $\frac{4\pi\mu\sqrt{(1-e^2)}}{e^3}$; thus we find that the attractions for an external particle in the plane of the equator and on the axis respectively are

$$\int \frac{xdx}{r\sqrt{(r^2-x^2)}} \text{ and } \int \frac{xdx}{r^2+x^2},$$

r being the distance of the particle from the common centre of the shells. The limits of integration are 0 and ce, where c is the semi-axis major of the bounding shell of the solid. Thus we obtain respectively

$$\frac{r-\sqrt{(r^2-c^2e^2)}}{r}, \text{ and } \tfrac{1}{2}\log\frac{r^2+c^2e^2}{r^2}.$$

Suppose now that we take the external particle to be *on* the surface of the oblatum; then in the former expression we put $r=c$, and in the latter we put $r^2=c^2(1-e^2)$. In both cases we obtain a result independent of c. Thus the attractions at the equator and at the pole are independent of the size of the oblatum. Maclaurin gives this result so far as relates to the attraction at the equator.

It is also true that for a particle situated at *any* point of the surface, the attraction will be independent of c; this may be shewn by reasoning of the kind given in Art. 242.

II. Suppose the density to vary inversely as the square of the diameter.

In this case we find, omitting the same common factor as before, that the attractions for a particle in the plane of the equator and on the axis are respectively

$$\int \frac{dx}{r\sqrt{(r^2 - x^2)}} \text{ and } \int \frac{dx}{r^2 + x^2},$$

that is, $$\frac{1}{r}\sin^{-1}\frac{ce}{r} \text{ and } \frac{1}{r}\tan^{-1}\frac{ce}{r}.$$

Suppose now we take the external particle to be *on* the surface of the oblatum; then in the former expression we put $r = c$, and in the latter $r = c\sqrt{(1-e^2)}$. Hence we see, that for oblata similar in form but different in size, each result varies inversely as c. Maclaurin gives this result so far as relates to the attraction at the equator.

It is also true that for a particle situated at any point of the surface the attraction will vary inversely as c; this may be shewn by reasoning of the kind given in Art. 242.

Also, since $\sin^{-1} e = \tan^{-1} \frac{e}{\sqrt{(1-e^2)}}$, we see that for the *same* ellipsoid, the equatorial and polar attractions for a particle on the surface are inversely as the equatorial and polar diameters. Maclaurin does not mention this. I add, that the law of density under consideration is the only law which gives the result just obtained; the density being assumed to be a function of the diameter of the shell. To prove this: assume the law of density to be represented by $\frac{\phi(x)}{x^2}$. Then we require that $\int_0^{ce} \frac{\phi(x)\, dx}{c\sqrt{(c^2-x^2)}}$ should be to $\int_0^{ce} \frac{\phi(x)\, dx}{c^2 - c^2e^2 + x^2}$ as $c\sqrt{(1-e^2)}$ is to c.

Assume in the first integral $x = c\sin\theta$, and in the second $x = c\sqrt{(1-e^2)}\tan\theta$: then we arrive at

$$\int_0^{\sin^{-1} e} [\phi(c\sin\theta) - \phi\{c\sqrt{(1-e^2)}\tan\theta\}]\, d\theta = 0.$$

Differentiate with respect to c and to e; thus

$$0 = \int_0^{\sin^{-1} e} [\sin\theta\phi'(c\sin\theta) - \sqrt{(1-e^2)}\tan\theta\phi'\{c\sqrt{(1-e^2)}\tan\theta\}]\,d\theta,$$

and $$0 = \int_0^{\sin^{-1} e} \frac{ce}{\sqrt{(1-e^2)}}\tan\theta\phi'\{c\sqrt{(1-e^2)}\tan\theta\}\,d\theta.$$

Multiply the latter by $\frac{1-e^2}{ce}$, and add to the former; thus we obtain

$$\int_0^{\sin^{-1} e} \sin\theta\phi'(c\sin\theta)\,d\theta = 0;$$

and by differentiating with respect to e we see that $\phi'(ce) = 0$. This shews that $\phi(x)$ must be a constant.

III. Suppose the density to vary as the diameter.

In this case, omitting the same common factor as before, the attractions for a particle in the plane of the equator and on the axis are respectively

$$\int \frac{x^3dx}{r\sqrt{(r^2-x^2)}} \text{ and } \int \frac{x^3dx}{r^2+x^2}.$$

Thus we shall obtain when the external particle is *on* the surface

$$\frac{c^2}{3}\{2 - 3\sqrt{(1-e^2)} + (1-e^2)^{\frac{3}{2}}\} \text{ and } \frac{c^2}{2}\{e^2 + (1-e^2)\log(1-e^2)\}.$$

Each varies directly as the square of c. And, as before, for a particle situated at any point of the surface the attraction will vary as the square of c. In this case the ratio of the equatorial attraction to the polar is

$$\frac{2}{3}\,\frac{2-3\sqrt{(1-e^2)}+(1-e^2)^{\frac{3}{2}}}{e^2+(1-e^2)\log(1-e^2)}.$$

Expanding in powers of e^2 we shall find that this becomes

$$\frac{1+\frac{e^2}{3}+\frac{3e^4}{16}+\dots}{1+\frac{e^2}{3}+\frac{e^4}{6}+\dots},$$

thus if we neglect the square and higher powers of e^2, the two attractions are equal. This agrees with a statement in Maclaurin's Article 673.

IV. Suppose the density to vary as the cube of the diameter.

In this case, omitting the same common factor as before, the attractions for a particle in the plane of the equator and on the axis are respectively

$$\int \frac{x^5 dx}{r\sqrt{(r^2 - x^2)}} \text{ and } \int \frac{x^5 dx}{r^2 + x^2}.$$

Thus we obtain when the external particle is *on* the surface

$$\frac{c^4}{15}\{8 - 15(1-e^2)^{\frac{1}{2}} + 10(1-e^2)^{\frac{3}{2}} - 3(1-e^2)^{\frac{5}{2}}\},$$

and $$\frac{c^4}{4}\{e^4 - 2e^2(1-e^2) - 2(1-e^2)^2 \log(1-e^2)\}.$$

Each varies as the fourth power of c. And, as before, the same result will hold for a particle situated at any point of the surface.

The ratio of the former to the latter when we neglect the square and higher powers of e^2 is $\dfrac{1+\frac{3e^2}{8}}{1+\frac{e^2}{4}}$.

267. Maclaurin in his Articles 672 and 673 supposes that his shells are fluid, and that the density varies as the diameter. He comes to the conclusion that the ellipticity is rather greater than it is for the case of uniform density; but that the increase of gravity in passing from the equator to the pole is less than for the case of uniform density. He also briefly states the results for the case in which the density varies as the cube of the diameter.

The results are of no value, for Maclaurin merely assumes Newton's principle of columns of fluid balancing at the centre, and does not shew that the whole fluid will be in equilibrium. In fact it is known that the whole fluid will not be in equilibrium. If the density of the shells varies the excentricity can not be constant. The objection to Maclaurin's investigations was noticed by Clairaut: see his *Figure de la Terre*, pages 229...232.

For an example we will give the investigation, on Maclaurin's principles, of the case in which the density varies as the cube of the diameter.

Denote the attraction for a particle *on* the surface at the equator by E, and at the pole by P, the density at the surface by ρ, and the centrifugal force at the equator by V: let $2a$ and $2b$ be the equatorial and polar diameters.

For the equatorial column, at a distance x from the centre, the attraction is $E\frac{x^4}{a^4}$, the centrifugal force is $V\frac{x}{a}$, and the density is $\rho\frac{x^3}{a^3}$: hence the weight of the column

$$=\int_0^a \left(E\frac{x^4}{a^4} - V\frac{x}{a}\right)\rho\frac{x^3}{a^3}\,dx = \left(\frac{E}{8} - \frac{V}{5}\right)\rho a.$$

Similarly the weight of the polar column $=\frac{P}{8}\rho b$.

Therefore
$$\left(\frac{E}{8} - \frac{V}{5}\right)\rho a = \frac{P}{8}\rho b.$$

We take from observation

$$\frac{V}{E} = \frac{1}{289}, \text{ so that } Pb = E\left(1 - \frac{8}{5\times 289}\right)a.$$

Therefore
$$\frac{E}{P} = \frac{\sqrt{(1-e^2)}}{1 - \frac{8}{5\times 289}} = \frac{1 - \frac{e^2}{2}}{1 - \frac{8}{5\times 289}} \text{ approximately.}$$

But we saw in Art. 266 that $\frac{E}{P} = 1 + \frac{e^2}{8}$ nearly;

therefore
$$1 - \frac{e^2}{2} = \left(1 - \frac{8}{5\times 289}\right)\left(1 + \frac{e^2}{8}\right);$$

so that
$$\frac{e^2}{2} = \frac{4}{5}\times\frac{8}{5}\times\frac{1}{289} = \frac{1}{226}.$$

The ratio of the polar gravity to the equatorial

$$= \frac{P}{E-V} = \frac{1}{\left(1+\frac{e^2}{8}\right)\left(1 - \frac{1}{289}\right)} = 1 + \frac{1}{289} - \frac{e^2}{8} = 1 + \frac{13}{25}\times\frac{1}{289} \text{ nearly.}$$

Thus we obtain an excentricity slightly greater than for the case of uniform density, where $\frac{e^2}{2} = \frac{1}{230}$; but the increase of gravity in passing from the equator to the pole is much less than for the case of uniform density, where it is $\frac{1}{230}$ of the whole.

268. Maclaurin devotes his Articles 674...678 to the discussion of the case in which the density involves two terms, one constant, and the other varying as the diameter of the shells. Let x represent the diameter of any shell, a the diameter of the outside shell; then he takes the density to vary as $\frac{na}{n-1} - x$. This obviously amounts to supposing the density to vary as the distance from some point *beyond* the outside shell. Maclaurin's discussion of the attractions at the equator and at the pole is very clear and satisfactory.

Assuming as before that the body is fluid, and using Newton's principle of columns balancing at the centre, Maclaurin arrives at the following results:

If e and j have their usual meanings

$$\frac{e^2}{4} = \frac{5j(n+2)(n+3)}{17n^2+34n+45} = \frac{5j}{8}\left\{1 - \frac{3(3n+1)(n-1)}{17n^2+34n+45}\right\}$$

The ratio of the difference of polar and equatorial gravities to their half-sum is

$$\frac{5j}{4}\left\{1 + \frac{3(n+3)(n-1)}{17n^2+34n+45}\right\}.$$

Maclaurin says in his Art. 678,

...no supposition of this kind can account for a greater variation from the spherical figure, and at the same time for a greater increase of gravitation from the equator to the poles....

If we put $n=0$ in the above value of e^2 we get $e^2 = \frac{8j}{3}$; the density now varies as the diameter: the result coincides with that obtained by Maclaurin in his Art. 673.

Maclaurin in his Article 679 states the results obtained by substituting for n in the above general formulæ the values 2, 3 and infinity.

269. Problems of the kind considered by Maclaurin in his Articles 672 ... 679 had previously engaged the attention of Clairaut: see Chapter VI. Both Clairaut and Maclaurin however failed, from not knowing that the equilibrium of the whole fluid was impossible on their hypotheses. Considered merely with respect to *attractions* both supplied interesting results: Clairaut gave *approximate* values of the attraction at any point of the surface, and Maclaurin gave *exact* values of the polar and equatorial attractions. The failure as regards the hydrostatical part of the problems was recognised by Clairaut himself: see his *Figure de la Terre*, pages 155 and 259.

270. Maclaurin in his Article 680 takes the case of an oblatum which is composed of shells of finite thickness; each shell is of uniform density, but the density varies from shell to shell, increasing towards the centre: the bounding surfaces of the shells are supposed to be all similar and concentric. He gives, in fact, an approximate expression for the excentricity in the case of one shell surrounding a central portion, from which it appears that the excentricity is less than for the case of a homogeneous fluid; and he states that a similar result will hold when there are more shells.

Let us investigate the general result which is briefly indicated in Maclaurin's Article 680.

First, let there be one shell surrounding a central part. Denote the density of the shell by 1, and that of the central part by $1+\sigma$. Let the equatorial diameter of the central part be $\frac{2a}{n}$, where $2a$ is the outer equatorial diameter of the shell.

We proceed with Maclaurin to equate the weights of the equatorial and polar columns.

We begin with finding the weight of the equatorial column. Let x denote a distance from the centre, γ the density at this point, $\phi(x)$ the attraction at this point. Then the weight of the column will be denoted by $\int_0^a \gamma\phi(x)\,dx$; and we must observe that γ and $\phi(x)$ have different forms at different points.

Put k for $\frac{4\pi}{3}\sqrt{(1-e^2)}$. Then, from $x=0$ to $x=\frac{a}{n}$ we have $\gamma=1+\sigma$, and $\phi(x)=k(1+\sigma)\left(1+\frac{3e^2}{10}\right)x$; and from $x=\frac{a}{n}$ to $x=a$ we have $\gamma=1$, and $\phi(x)=k\left(1+\frac{3e^2}{10}\right)x+k\sigma\left(\frac{a^3}{n^3x^2}+\frac{3a^5e^2}{10n^5x^4}\right)$.

Here we only retain the first power of e^2; and this we shall do throughout the investigation. See Art. 261.

Hence we shall find that $\int_0^a \gamma\phi(x)\,dx$ becomes

$$k(1+\sigma)^2\left(1+\frac{3e^2}{10}\right)\frac{a^2}{2n^2}+k\left(1+\frac{3e^2}{10}\right)\left(1-\frac{1}{n^2}\right)\frac{a^2}{2}+k\sigma\left\{\frac{n-1}{n^3}+\frac{e^2(n^3-1)}{10n^5}\right\}a^2.$$

If V denote the centrifugal force at the equator, the effect of the centrifugal force on the column is $V\left(1+\frac{\sigma}{n^2}\right)\frac{a}{2}$. We put as usual $\frac{V}{\left(k+\frac{k\sigma}{n^3}\right)a}=j$; for the denominator on the left-hand side expresses the attraction at the equator to the order which we are here considering. Thus the effect of the centrifugal force on the column is $jk\left(1+\frac{\sigma}{n^2}\right)\left(1+\frac{\sigma}{n^3}\right)\frac{a^2}{2}$.

In a similar manner we find that if $2b$ be the outer polar diameter of the shell the weight of the polar column is denoted by

$$k(1+\sigma)^2\left(1-\frac{e^2}{10}\right)\frac{a^2}{2n^2}+k\left(1-\frac{e^2}{10}\right)\left(1-\frac{1}{n^2}\right)\frac{a^2}{2}$$
$$+k\sigma\left\{\frac{n-1}{n^3}\;\frac{a}{b}-\frac{e^2(n^3-1)}{5n^5}\cdot\frac{a^3}{b^3}\right\}a^2.$$

The factor $\left(1-\frac{e^2}{10}\right)\frac{a^2}{2}$ may be obtained thus: the attraction at the pole of an oblatum of density unity is $k\left(\frac{a^3}{b^2}-\frac{3e^2}{5}\cdot\frac{a^5}{b^4}\right)$, that is,

$kb\left(1+\frac{9e^2}{10}\right)$ nearly; thus the weight of the polar column, if the density were unity throughout, would be $k\frac{b^2}{2}\left(1+\frac{9e^2}{10}\right)$, that is, $k\left(1-\frac{e^2}{10}\right)\frac{a^2}{2}$.

Equate the weights of the columns; thus we get

$$(1+\sigma)^2\frac{2e^2}{5n^2}+\frac{2e^2}{5}\left(1-\frac{1}{n^2}\right)-\sigma e^2\frac{n-1}{n^3}+\frac{3\sigma e^2}{5}\frac{n^3-1}{n^5}$$

$$=j\left(1+\frac{\sigma}{n^2}\right)\left(1+\frac{\sigma}{n^3}\right);$$

therefore

$$e^2=\frac{\frac{5j}{2}(n^2+\sigma)(n^3+\sigma)}{n^5+n^3\sigma^2+n^3\sigma+n^2\sigma+\frac{3}{2}(n^2-1)\sigma}:$$

this is less than $\frac{5j}{2}$, since σ is positive and n greater than unity. Maclaurin gives this result.

Let us now suppose that there are three portions of fluid, an outer shell of density 1, a second shell of density $1+\rho$, and an inner part of density $1+\rho+\sigma$. Let the equatorial diameter of the inner part be $\frac{2a}{n}$; and let the outer equatorial diameter of the second shell be $\frac{2a}{m}$. It is easy to see that the value of e^2 will now be determined by an equation of the form

$$e^2=\frac{\frac{5j}{2}\left(1+\frac{\rho}{m^2}+\frac{\sigma}{n^2}\right)\left(1+\frac{\rho}{m^3}+\frac{\sigma}{n^3}\right)}{1+\text{terms of the first and second degree in }\rho\text{ and }\sigma}.$$

Now with respect to the denominator on the right-hand side, we know that if $\rho=0$ it reduces to

$$1+\frac{\sigma^2}{n^2}+\left(\frac{1}{n^2}+\frac{1}{n^3}\right)\sigma+\frac{3}{2}\frac{n^2-1}{n^5}\sigma;$$

and if $\sigma=0$ it will reduce to a similar expression in ρ and m.

Hence, in fact, we have only the term in $\rho\sigma$ to find. Proceed as before: we see that in estimating the weight of the equatorial column we have a term

$$2k\rho\sigma\left(1+\frac{3e^2}{10}\right)\frac{a^2}{2n^2}+k\rho\sigma\left\{\frac{n-m}{n^3}+\frac{e^2(n^3-m^3)}{10n^5}\right\}a^2,$$

and in estimating the weight of the polar column we have a term

$$2k\rho\sigma\left(1-\frac{e^2}{10}\right)\frac{a^2}{2n^2}+k\rho\sigma\left\{\frac{n-m}{n^3}\cdot\frac{a}{b}-\frac{e^2(n^3-m^3)}{5n^5}\cdot\frac{a^3}{b^3}\right\}a^2.$$

This shews that the term we are seeking is

$$\rho\sigma\left(\frac{2}{n^2}-\frac{5}{2}\cdot\frac{n-m}{n^3}+\frac{3}{2}\cdot\frac{n^3-m^3}{n^5}\right),\text{ that is }\rho\sigma\left\{\frac{1}{n^2}+\frac{m}{n^3}+\frac{3m(n^2-m^2)}{2n^5}\right\}.$$

The term which involves $\rho\sigma$ in the numerator is $\dfrac{1}{m^3n^2}+\dfrac{1}{m^2n^3}$, which is certainly *less* than the term which involves $\rho\sigma$ in the denominator.

There will be no difficulty in extending this. Suppose that there are four portions of fluid, and that their densities are 1, $1+\varpi$, $1+\varpi+\rho$, $1+\varpi+\rho+\sigma$; and the corresponding equatorial semi-diameters a, $\dfrac{a}{l}$, $\dfrac{a}{m}$, $\dfrac{a}{n}$. Then the numerator of e^2 will now be

$$\frac{5j}{2}\left(1+\frac{\varpi}{l^2}+\frac{\rho}{m^2}+\frac{\sigma}{n^2}\right)\left(1+\frac{\varpi}{l^3}+\frac{\rho}{m^3}+\frac{\sigma}{n^3}\right).$$

The terms in the denominator can easily be written down; that in $\rho\sigma$ is the same as before; that in $\varpi\rho$ will in like manner be $\varpi\rho\left\{\dfrac{1}{m^2}+\dfrac{l}{m^3}+\dfrac{3l(m^2-l^2)}{2m^5}\right\}$; and that in $\varpi\sigma$ will be $\varpi\sigma\left\{\dfrac{1}{n^2}+\dfrac{l}{n^3}+\dfrac{3l(n^2-l^2)}{2n^5}\right\}$.

The problem is of no importance; for, as we have said, the whole fluid mass will not be in equilibrium: but still there is something curious in the simplicity of the solution when considered with regard to the complexity of the hypothesis.

271. Maclaurin in his Article 681 takes the following hypothesis: let there be a shell of fluid, the bounding surfaces of which are concentric and similar oblata; and within the inner surface let there be a solid concentric sphere. He again equates the

weights of the equatorial and polar columns of fluid. It is obvious that the hypothesis is not consistent with the known conditions for fluid equilibrium, unless he supposes the inner surface of the fluid to become rigid; and if this is supposed, the weights of the columns will not be equal. Clairaut pointed out that the hypothesis is untenable: see his *Figure de la Terre*, page 256.

We will state the results which will be obtained on Maclaurin's principles. Take the density of the solid and of the fluid to be the same, and uniform; let $2a$ and $\frac{2a}{n}$ be the external and internal equatorial diameters of the shell. Suppose the volume of the sphere to be $\frac{1}{N}$ of the volume of the oblatum if complete; then we shall obtain

$$e^2 = \frac{\frac{5j}{2}(n+1)\left(n^3 - 1 + \frac{n^3}{N}\right)}{n^4 + n^3 + n^2 - \frac{3n+3}{2} - \frac{5n^5}{2N}}.$$

Maclaurin's result agrees with this; but he uses the word *area* for *volume*.

The ratio of the difference of the polar and equatorial gravity to the semi-sum will be found to be

$$j + \frac{j(n+1)(n^5 - 10n^2 + 9) + \frac{10jn^3}{N}}{2n^2\left(2n^4 + 2n^3 + 2n^2 - 3n - 3 - \frac{5n^5}{N}\right)}.$$

Maclaurin has n^5N where we have $\frac{n^5}{N}$, and he has $-30n^2$ where we have $-10n^2$. We may verify by putting N infinite and $n=1$; then we have only an indefinitely thin shell, and we get $e^2 = 2j$: and the excess of polar over equatorial gravity becomes zero by our formula, as it should. If we put $n=2$, we find that Maclaurin's result would in general be negative, supposing we make the correction for N.

Maclaurin next supposes that the central part instead of being a sphere is an ellipsoid of revolution; he gives the correct result on his principles, supposing the ellipticity of the central part *to be small:* he has not formally stated this condition, though he has certainly used it. The following is his result: let the distance from the centre to a focus of the inner part be $\frac{a}{r}$; then the rest of of the notation being as before,

$$e^2 = \frac{\frac{5j}{2}(n+1)\left(n^3 - 1 + \frac{n^3}{N}\right) - \frac{3n^5}{2r^2N}(n^2+n+1)}{n^4 + n^3 + n^2 - \frac{3n+3}{2} - \frac{5n^5}{2N}}.$$

Suppose, for example, that the surface of the solid part coincides with the inner surface of the fluid, so that $\frac{1}{r} = \frac{e}{n}$, and $N = n^3$: then we obtain $e^2 = \frac{5j}{2}$, as it should be.

Maclaurin goes on to say that other suppositions might be made, but implies that it is not desirable to dwell on them. He makes the following very judicious remark:

When more degrees shall be measured accurately on the meridian, and the increase of gravitation from the equator towards the poles determined by a series of many exact observations, the various *hypotheses*, that may be imagined concerning the internal constitution of the earth, may be examined with more certainty.

272. Maclaurin gives in his Articles 682...685 some remarks on the shape of the planet Jupiter.

Suppose a satellite to describe round its primary in the plane of the primary's equator, a circle of radius r in time T; let the primary revolve on its axis in time t; let a and $a\sqrt{(1-e^2)}$ be the semi-axes of the primary. Maclaurin puts N for $\frac{r^3}{a^3} \times \frac{t^2}{T^2}$.

To connect N with j and e we have the following equations: see Art. 261:

$$j = \frac{\left(\frac{2\pi}{t}\right)^2}{\frac{4\pi\rho}{3}\sqrt{(1-e^2)}\left\{1 + \frac{3e^2}{10} + \frac{9e^4}{56} + \dots\right\}},$$

$$r\left(\frac{2\pi}{T}\right)^2 = \frac{4\pi\rho\sqrt{(1-e^2)}}{3}\left\{\frac{a^3}{r^2} + \frac{3e^2}{10}\frac{a^5}{r^4} + \frac{9e^4}{56}\frac{a^7}{r^6} + \dots\right\};$$

therefore $$Nj = \frac{1 + \frac{3e^2}{10M^2} + \frac{9e^4}{56M^4} + \dots}{1 + \frac{3e^2}{10} + \frac{9e^4}{56} + \dots},$$

where M stands for $\frac{r}{a}$.

Put for j its value from Art. 262; thus

$$N\left(\frac{2}{5}e^2 + \frac{9}{35}e^4 + \frac{5}{28}e^6 + \dots\right) = 1 + \frac{3e^2}{10M^2} + \frac{9e^4}{56M^4} + \dots$$

Now Maclaurin says in his Article 660, that "the excess of the semidiameter of the equator above the semiaxis is to the mean semidiameter as 5 to $4N + \frac{10}{7} - \frac{3}{MM}$ nearly;" and in his Article 682 he says, "By continuing the series in art. 660 one step further, the excess of the semidiameter of the equator above the semiaxis is to the mean semidiameter as 5 is to $4N + \frac{10}{7} - \frac{3}{MM} + \frac{4825}{336N}$, ..." Let us examine the last statement.

We have just seen that

$$e^2 = \frac{\frac{5}{2N}\left(1 + \frac{3e^2}{10M^2} + \frac{9e^4}{56M^4} + \dots\right)}{1 + \frac{9e^2}{14} + \frac{25}{56}e^4 + \dots} \dots\dots\dots\dots\dots (1).$$

We can infer from Maclaurin's result that he rejects the squares of $\frac{e^2}{M^2}$; and, indeed, if we look at his numerical values, it will appear that to the order he considers, he might have rejected $\frac{e^2}{M^2}$ also. However, retaining $\frac{e^2}{M^2}$, we have from (1),

$$e^2 = \frac{\frac{5}{2N}}{1 + \left(\frac{9}{14} - \frac{3}{10M^2}\right)e^2 + \frac{25}{56}e^4 + \dots} \dots\dots\dots\dots\dots (2).$$

For a first approximation we have from (2)

$$e^2 = \frac{5}{2N}.$$

Substitute this value in the denominator of (2), neglecting e^4, then for a second approximation

$$e^2 = \frac{\frac{5}{2N}}{1 + \left(\frac{9}{14} - \frac{3}{10M^2}\right)\frac{5}{2N}} = \frac{\frac{5}{2N}}{1 + \frac{45}{28N} - \frac{3}{4NM^2}}:$$

this agrees with what Maclaurin gives at the beginning of his Article 660.

For a third approximation we substitute for e^2 in the denominator of (2) the value $\frac{5}{2N}\left(1 - \frac{45}{28N}\right)$; and so we get

$$e^2 = \frac{\frac{5}{2N}}{1 + \frac{45}{28N} - \frac{3}{4NM^2} + \left(\frac{5}{2N}\right)^2\left(\frac{25}{56} - \frac{81}{196}\right)}$$

$$= \frac{\frac{5}{2N}}{1 + \frac{45}{28N} - \frac{3}{4NM^2} + \frac{25 \times 13}{8 \times 196N^2}}.$$

Now we require the value of $\frac{\frac{e^2}{2}}{1 - \frac{e^2}{2}}$; and this is

$$\frac{\frac{5}{4N}}{1 + \frac{45}{28N} - \frac{3}{4NM^2} + \frac{25 \times 13}{8 \times 196N^2} - \frac{5}{4N}},$$

that is,

$$\frac{5}{4N + \frac{10}{7} - \frac{3}{M^2} + \frac{25 \times 13}{2 \times 196N}}.$$

Thus instead of Maclaurin's large coefficient $\frac{4825}{336}$ we get only $\frac{325}{392}$.

273. Maclaurin finds that his calculation brings out too great an ellipticity for Jupiter, making the longer diameter to be the shorter, about as 10·3 to 9·3; whereas, according to Cassini, the difference of the diameters was about $\frac{1}{15}$ of the longer diameter, and according to Pound between $\frac{1}{12}$ and $\frac{1}{15}$. Maclaurin then makes the supposition which we have noticed in Art. 268, that the density varies as $\frac{na}{n-1} - x$; he gives the general result, and putting $n = 4$ in this, he finds a tolerable agreement with observation.

But I am unable to verify his general result. By the aid of the expression given in Art. 261 for the attraction of a shell on a particle in the plane of the equator, I obtain with the notation of Art. 272,

$$\frac{1}{Nj} = \frac{n + 3 + \frac{e^2}{5}(n+5)}{n + 3 + \frac{e^4(n+5)}{5M^2}}.$$

Maclaurin's result in this notation is

$$\frac{1}{Nj} = \frac{n + 3 + \frac{7ne^2}{10} + \frac{5e^2}{2}}{\left(1 + \frac{e^2}{2}\right)(n+3) + \frac{6ne^4}{5M^2}};$$

if we multiply both numerator and denominator of the last fraction by $1 - \frac{e^2}{2}$, and neglect e^4, we get

$$\frac{n + 3 + \frac{e^2}{5}(n+5)}{n + 3 + \frac{6ne^4}{5M^2}}.$$

Maclaurin cannot be correct; for it is certain that if $M = 1$ we ought to have $Nj = 1$.

274. Some other investigations respecting attractions are contained in Articles 900...905 of Maclaurin's Fluxions.

Here he supposes the law of attraction to be that of the nth power of the distance; he says that n is to be less than 3: it will be found on examination that he means n to be *algebraically* less than 3, and does not assume n to be necessarily an integer, so that in fact $3-n$ must be positive. Maclaurin considers the attraction of an ellipsoid of revolution on a particle at the equator or at the pole; as we should say in modern language he reduces the problem to a single integration. He says in his Article 904 as his general conclusion, "Hence, therefore, the gravity at the equator, as well as the gravity at the poles, is measured by circular arks or logarithms when n is any integer number less than $+3$."

Maclaurin refers in his Article 905 to "a late ingenious essay, *Phil. Trans.* N. 449. by Mr Clairaut:" see Art. 167.

275. We will now notice the bearing on our subject of Maclaurin's Prize Essay on the Tides, which was mentioned in Art. 232.

Maclaurin in Lemma III. of his *Essay* gives matter equivalent to Articles 628...630 of the *Fluxions:* see Arts. 242 and 243. In Lemma IV. he gives matter equivalent to Articles 631...634 of the *Fluxions:* see Art. 244. The Propositio I. Theorema Fundamentale of the *Essay* contains the important results enunciated in Article 636 and demonstrated in the following three Articles of the *Fluxions;* see Art. 245. Maclaurin briefly indicates the application of this fundamental theorem to the Figure of the Earth, supposing that the Earth is a fluid of uniform density; the theorem gives the ratio of the axes, and the direction of gravity at any point. He says: "Hæc omnia accuratè demonstrantur ex hac Propositione; quæ quamvis in disquisitione de figura Terræ eximii usûs sint, hic obiter tantum monere convenit."

Lemma V. of the *Essay* corresponds to Article 642 of the *Fluxions;* though it is rather less general: see Art. 250. By means of this Lemma the calculation of the attraction of a solid of revolution on a particle at its pole is made to depend on finding the area of a certain curve.

Propositio II. of the Essay determines the attraction of an oblongum on a particle at its pole; the method is substantially the same as that in Article 647 of the Fluxions, but in the Essay the notation is that of the Differential and Integral Calculus, not that of Fluxions and Fluents: see Art. 252. At the end of the proposition Maclaurin briefly indicates the result for the case of an oblatum; this case is worked out in Article 646 of the *Fluxions.* For the subject of the Tides the *oblongum* is the important figure, while for the subject of the Figure of the Earth the *oblatum* is the important figure

In Lemma VI. and Proposition III. of the Essay, Maclaurin estimates the attraction of an oblongum on a particle at the equator, and briefly indicates the result for an oblatúm; the method is substantially the same as in Articles 646 and 647 of the Fluxions.

Thus we see that at the date of the *Essay on the Tides* Maclaurin had completely solved the problem of the attraction of a homogeneous ellipsoid of revolution on an *internal* particle. The *Treatise of Fluxions* contains in addition the theorem respecting the attraction on an *external* particle which we have noticed in Art. 259; and also the propositions respecting ellipsoids of revolution, not homogeneous, which we have noticed in Arts. 256, 264 and 265.

276. Maclaurin died in 1746, so that he survived the publication of Clairaut's *Figure de la Terre.* It does not however appear that he published anything on our subject after his *Fluxions.* In the last year of his life he was obliged to leave his home in consequence of the rebellion in favour of the Stuarts; and the hardships he thus encountered seem to have laid the foundation of his mortal illness: in the premature death of the most famous of her sons Scotland paid a heavy price for the temporary success of the Pretender's enterprise.

The importance of Maclaurin's investigations may be seen by observing how great has been his influence on succeeding writers. Clairaut, D'Alembert, Lagrange, Legendre, Laplace, Gauss, Ivory and Chasles shew by reference explicit or implicit their obligations to the creator of the theory of the attraction of ellipsoids. Maclaurin well deserves the memorable association of his name with that of the great master in the inscription which records that he was appointed professor of mathematics at Edinburgh, *ipso Newtono suadente.*

In the application of the theory of Attraction to the Figure of the Earth Maclaurin was impeded by the imperfect state at that time of the knowledge of the conditions of fluid equilibrium, and also by the want of accurate measurements; the latter circumstance led him to suppose that the ellipticity was greater than it really is. Nevertheless he was the first to demonstrate exactly the possibility of the relative equilibrium of an oblatum of rotating fluid. See Art. 249.

CHAPTER X.

THOMAS SIMPSON.

277. THOMAS SIMPSON published in 1743 a volume entitled *Mathematical Dissertations on a variety of Physical and Analytical Subjects*. The volume is in quarto; the Title, Dedication, and Preface occupy viii pages, and the text occupies 168 pages.

278. The first essay extends over 30 pages; it is entitled *A Mathematical Dissertation on the Figure of the Earth*. In the preface Simpson speaks of this as "one of the most considerable Papers in the whole Work,..."; and after referring to the contents of the essay he says:

...I must own that, since my first drawing up this Paper, the World has been obliged with something very curious on this Head, by that celebrated Mathematician Mr. Mac-Laurin, in which many of the same Things are demonstrated. But what I here offer was read before the Royal Society, and the greater Part of this Work printed off, many Months before the Publication of that Gentleman's Book; for which Reason I shall think myself secure from any Imputations of Plagiarism, especially as there is not the least Likeness between our two Methods.

In a foot-note he says

It was read before the Royal-Society in *March* or *April*, 1741, and had been printed in the Philosophical Transactions, had not I desired the contrary.

The preceding extract might seem to establish for Simpson the priority over Maclaurin in the first enunciation of some of the most important results in our subjects; but Simpson makes no reference to Maclaurin's prize Essay on the Tides which belongs to an earlier date than March, 1741, and contains the essence of much that was expanded in the *Treatise of Fluxions:* see Art. 275. Thus Maclaurin's claims remain indisputable; but as we shall shew there are some very important points in which Simpson had no predecessor.

Simpson's essay is very remarkable, as we shall see by an analysis of its contents.

279. The first fourteen pages bring out exact expressions for the attraction of an oblatum on a particle at the surface; Maclaurin as we have seen had previously effected as much. The following is the essential part of Simpson's method: suppose an ellipse to revolve round a tangent at one end of an axis, through an indefinitely small angle; a wedge-shaped element is thus produced, and Simpson calculates the attraction which this element exerts on a particle placed at the point of tangency. The whole oblatum is cut up into such wedge-shaped elements, and so the resultant attraction is determined. Instead of the elegant geometry of Maclaurin, Simpson employs analysis, the style of which for its rude strength reminds the reader of that of Laplace.

280. In the course of his investigation on his page 3, Simpson has in effect to determine the value of $\frac{1}{\sqrt{(1+g)}}\int_0^1 \frac{\sqrt{x}\,dx}{\sqrt{(1+gx)}}$, in the form of a series proceeding according to ascending powers of g. He expands the expression under the integral sign in powers of g, and effects the integration; then he multiplies this by the expansion of $\frac{1}{\sqrt{(1+g)}}$, and arranges the product. He does not however *demonstrate* the form of the general term, but seems to assume it from observation of a few simple cases. As all his subsequent investigations rest on this, it seems strange that he did not proceed here with rigid exactness.

We may of course obtain the required result easily in another way. Assume $x = \frac{\sin^2\theta}{1+g\cos^2\theta}$; thus we find that the integral is transformed into $2\int_0^{\frac{1}{2}\pi} \frac{\sin^2\theta\cos\theta\,d\theta}{(1+g\cos^2\theta)^2}$.

Then, expanding in powers of g, we see that the general term

$$= 2(n+1)(-1)^n g^n \int_0^{\frac{1}{2}\pi} (1-\cos^2\theta)\cos^{2n+1}\theta\,d\theta$$

$$= 2(n+1)(-1)^n g^n \left(1-\frac{2n+2}{2n+3}\right)\int_0^{\frac{1}{2}\pi} \cos^{2n+1}\theta\,d\theta$$

$$= (-1)^n g^n \frac{2.4.6\ldots(2n+2)}{3.5.7\ldots(2n+3)}.$$

This agrees with Simpson's result.

281. The preceding Article furnishes the only instance of an imperfect investigation which I have noticed in Simpson's essay: there are however, as might be expected, cases in which his processes may be simplified. Perhaps the most important part of his analysis consists of the evaluation, on his page 10, of the following definite integrals:

$$\int_0^{\frac{1}{2}\pi} \{(a\cos\theta + A\sin\theta)^{2n} - (a\cos\theta - A\sin\theta)^{2n}\}\sin\theta\cos\theta\,d\theta,$$

and $$\int_0^{\frac{1}{2}\pi} \{(a\cos\theta + A\sin\theta)^{2n} + (a\cos\theta - A\sin\theta)^{2n}\}\sin^2\theta\,d\theta.$$

We will consider the second of these; our remarks will be easily applicable to the first.

Simpson expands $(a\cos\theta + A\sin\theta)^{2n}$ and $(a\cos\theta - A\sin\theta)^{2n}$, and then integrates each term separately; the following is a simpler method.

It is obvious that if we were to expand, our final expression would involve only *even* powers of $\sin\theta$ and $\cos\theta$; and thus we may use 0 and 2π as the limits of integration, and take one fourth of the result.

Assume $a = k\cos\beta$, and $A = k\sin\beta$; so that $k^2 = a^2 + A^2$; then the definite integral becomes

$$\frac{1}{4}(a^2+A^2)^n \int_0^{2\pi} \{\cos^{2n}(\theta-\beta) + \cos^{2n}(\theta+\beta)\}\sin^2\theta\,d\theta.$$

Consider $\int_0^{2\pi} \cos^{2n}(\theta-\beta)\sin^2\theta\,d\theta.$

Put $\sin\theta = \sin(\beta + \theta - \beta) = \sin\beta\cos(\theta - \beta) + \cos\beta\sin(\theta - \beta)$.

Thus we get

$$\int_0^{2\pi} \cos^{2n}(\theta - \beta)\,\{\sin\beta\cos(\theta - \beta) + \cos\beta\sin(\theta - \beta)\}^2\,d\theta.$$

Put ϕ for $\theta - \beta$; then the limits of the integration for ϕ are $-\beta$ and $2\pi - \beta$. The integral $\int \cos^{2n+1}\phi\sin\phi\,d\phi$ is zero between these limits; so that we are left with

$$\int_{-\beta}^{2\pi-\beta} (\sin^2\beta\cos^{2n+2}\phi + \cos^2\beta\sin^2\phi\cos^{2n}\phi)\,d\phi.$$

The limits may be changed to 0 and 2π, because the expression to be integrated has the same value when $\phi = 2\pi - \alpha$ as when $\phi = -\alpha$.

Transform $\int_0^{2\pi} \cos^{2n}(\theta + \beta)\sin^2\theta\,d\theta$ in a similar manner.

Thus finally we obtain

$$\int_0^{2\pi} \{\cos^{2n}(\theta - \beta) + \cos^{2n}(\theta + \beta)\}\sin^2\theta\,d\theta$$

$$= 2\cos^2\beta\int_0^{2\pi} \cos^{2n}\phi\,d\phi + 2(\sin^2\beta - \cos^2\beta)\int_0^{2\pi} \cos^{2n+2}\phi\,d\phi;$$

we have now a well-known definite integral form.

282. It should however be observed that Simpson's series are not always convergent. For example, on his page 13 he has the series which results from expanding $\tan^{-1}\sqrt{B}$ in powers of $\sqrt{B}$, and B is not necessarily less than unity.

283. Having obtained accurate expressions for the attraction of an oblatum on a particle at the surface, Simpson considers the relative equilibrium of a mass of rotating fluid. He says on his page 16, "the Form which that Fluid must be under, to preserve this Equilibrium of its Parts, is that of an oblate Spheroid." It is almost needless to remark that Simpson does not demonstrate this; he demonstrates that the figure which he assigns is a possible

figure of relative equilibrium, and not that it is the *only* figure: see Art. 168.

Simpson contents himself with shewing that Huygens's condition for fluid equilibrium is satisfied.

Laplace gives, in the *Mécanique Céleste*, Livre III. § 20, the following equation which connects the excentricity of the oblatum, supposed small, with the angular velocity

$$\lambda^2 = \frac{5}{2}q + \frac{75}{14}q^2 + \dots$$

Simpson gives this on his page 19 in his own notation, and supplies the third term of the series, namely $\frac{125 \times 37}{8 \times 49} q^3$ in Laplace's notation: Simpson remarks that this is very nearly the same as

$$\lambda^2 = \frac{35q}{14 - 30q};$$

and the approximation will be found extremely close as far as q^3.

Simpson on his pages 15 and 20 demonstrates the truth of some approximations given by Stirling; see Chapter V.

284. We now arrive at the most important part of the Essay. Simpson shews, to use modern language, that if the angular velocity of rotation exceeds a certain limit, the oblatum is no longer a possible form of equilibrium. This proposition has since been incorporated in the *Mécanique Céleste*, without any reference to Simpson: see Livre III. § 20.

Laplace uses $\sqrt{(1 + \lambda^2)}$ to express the ratio of the major axis to the minor axis in the oblatum, and Simpson uses $\sqrt{(1 + x^2)}$; for the extreme case in which equilibrium is possible, Simpson gives $x = 2{\cdot}5293$, while Laplace gives $\lambda = 2{\cdot}5292$.

Pontécoulant agrees with Laplace; see his *Théorie Analytique...*, Vol. II. page 399. Poisson agrees with Simpson; see his *Mécanique*, Vol. II. page 542: so also does Résal; see his *Traité Elémentaire de Mécanique Céleste*, page 196.

Simpson's investigation, though less elaborate than Laplace's, is adequate and satisfactory.

285. For any angular velocity less than the limit to which we have alluded in the preceding Article, there are two and only two possible oblata; this has been shewn by Laplace in the section already cited. According to Laplace, D'Alembert first observed that more than one figure of equilibrium might correspond to the same angular velocity without however determining the number of such figures: see Laplace's *Figure des Planetes*, page 124, and the *Mécanique Céleste*, Livre XI. § 1. Ivory makes a similar remark in the *Philosophical Transactions*, 1834, page 513. But it should be observed that although D'Alembert may have first explicitly published the statement, yet Simpson gives a Table which distinctly implies the fact.

The Table in substance is the following:

1 to 1·01	11·236	·08925
1 to 1·05	5·137	·1978
1 to 1·5	2·056	·5568
1 to 2	1·814	·6944
1 to 2·7198	1·7226	·8105
1 to 4	1·810	·8774
1 to 7·57	2·118	·92705
1 to 10	2·338	·9216
1 to 20	3·110	·8728
1 to 40	4·275	·8000
1 to 100	6·600	·7033
1 to 1000	20·640	·4845

This Table is given on Simpson's page 24, with the exception of two lines which I have supplied from other parts of the essay. The first column expresses the ratio of the minor axis to the major axis of the revolving oblatum; in Laplace's notation it is $\frac{1}{\sqrt{(1+\lambda^2)}}$. The second column is Laplace's $\frac{1}{\sqrt{q}}$; thus it is inversely proportional to the angular velocity, and so directly proportional to the time of rotation; it may be considered to express the time of rotation if we take a certain unit of time, the unit being the time in which a satellite would revolve round a sphere equal in volume and density to the oblatum, moving close to the surface: this is Simpson's own interpretation. The third column we will speak of presently.

An inspection of this Table shews that in the second column the figures decrease down to some minimum, and then increase again: thus it is obvious that corresponding to an assigned angular velocity there are in general two values of $\sqrt{(1+\lambda^2)}$.

286. Let us now explain the third column of the Table.

Simpson uses the term *momentum of rotation* for the sum of the products of the mass of every particle into its velocity. Let ω be the angular velocity, $2a$ the major axis, $2b$ the minor axis, ρ the density; then it is easy to shew that the *momentum of rotation* of the oblatum is $\frac{\pi^2}{4}\rho\omega a^3 b$. Now suppose a sphere, equal in density and volume to the oblatum, rotating in the unit of time specified in Art. 285. The *momentum of rotation* for the sphere would be $\frac{\pi^2}{4}\rho\omega_1 R^4$, where $R^3 = a^2 b$; so that it would be $\frac{\pi^2}{4}\rho\omega_1 (a^2 b)^{\frac{4}{3}}$. The ratio of the former value to the latter is therefore $\frac{\omega}{\omega_1}\left(\frac{a}{b}\right)^{\frac{1}{3}}$, that is $\left(\frac{q}{q_1}\right)^{\frac{1}{2}}(1+\lambda^2)^{\frac{1}{6}}$. But Simpson has taken the unit of time so that $q_1 = 1$; hence the ratio becomes $q^{\frac{1}{2}}(1+\lambda^2)^{\frac{1}{6}}$. Thus the third column can be obtained from the first and the second; we must divide the cube root of $(1+\lambda^2)^{\frac{1}{2}}$ which is given in the first column by $\frac{1}{\sqrt{q}}$ which is given in the second column.

Simpson's third column has not any physical interpretation, though he himself by mistake supposed that it had. For he uses the term *quantity of motion* on his page 21 in the same sense as angular momentum; and he erroneously says that it "will be no ways affected by the Action of the Particles upon one another while the Figure of the Fluid is changing." Then on his page 22 he gives a discussion as to the greatest possible value of the quantity of motion for a given mass.

What he must have intended to employ is the principle which in modern language we call the *Conservation of Areas.* This is plain from what he says in a note on page 135 of his *Miscellaneous Tracts*, 1757; here he admits the mistake in the present work. Instead of the sum of the products of the mass of every particle into its velocity, he should have considered the sum of the products of the mass of every particle into what we may call its *areal* velocity. Laplace uses this sum in the *Mécanique Céleste*, Livre III. § 21. He there has an equation $\phi = 0$, which agrees substantially with one given by Simpson on page 136 of his *Miscellaneous Tracts*. Simpson however does not discuss the equation; Laplace shews that it has only one solution.

287. In the Table of Art. 285, the fifth line and the seventh line are not inserted by Simpson, though he has supplied the materials for them in the course of his essay.

In the fifth line the entry in the second column gives the minimum value of that column; it really occurs in page 20 of Simpson's essay in the form $\frac{1}{\cdot 58053}$, so that $\cdot 58053$ is the value of $\sqrt{q}$ which corresponds to Laplace's value of $\cdot 337007$ for q. The corresponding number in the third column by Art. 286 is therefore $(2\cdot 7198)^{\frac{1}{3}} \times \cdot 58053$.

In the seventh line the entry in the third column gives the maximum value of that column; Simpson finds on his page 22 that for this case $\lambda = 7\cdot 5$ nearly. The corresponding number in the second column by Art. 286 is therefore $(7\cdot 57)^{\frac{1}{3}} \div \cdot 92705$.

288. Simpson shews on his page 22 that the gravity at any point of the surface of the oblatum varies as the length of the normal between the point and the axis of revolution. See Arts. 153 and 247.

289. Simpson having thus discussed the case of a homogeneous oblatum, proceeds to the case in which the oblatum is not homogeneous. He supposes that the oblatum consists of a central portion which is spherical and denser than the rest, and of an outer portion; each portion is supposed homogeneous.

If we change the sign of λ in a result which was obtained in Art. 264, page 156, we have

$$e^2 = \frac{\frac{5j''}{2}(1+\lambda)}{1+\frac{5\lambda}{2}};$$

and Simpson's result agrees with this.

Simpson does not shew that his fluid mass will remain in equilibrium; he contents himself with making the resultant force at the surface normal to the surface: if we suppose his central portion to be solid, the conditions of equilibrium will be satisfied. With the exception of this defect, Simpson's investigation of the value of the ellipticity and of the variation of gravity along the surface is quite satisfactory. In finding a definite value for the ellipticity, Simpson gives a better treatment of the problem than Maclaurin did in his Articles 666 and 667.

Simpson briefly applies his result to the case of the planet Jupiter. He concludes thus:

... but as no Hypothesis, for the Law of Variation of Density, can (from the Nature of the Thing) be verified either by Experiments, made on Pendulums in different Latitudes, or an actual Mensuration of the Degrees of the Meridian, I shall insist no further on this Matter, but content myself with having proved in general, that the greater the Density is towards the Centre, the less will the Planet differ from a Sphere, and the greater will be the Variation of Gravitation at its Surface.

290. The second essay in Simpson's *Mathematical Dissertations* is contained in pages 31...37; it is entitled *A General Investigation of the Attraction at the Surfaces of Bodies nearly spherical.*

The essay begins with investigating the attraction of a wedge-shaped element like that in Art. 279 on a particle in a certain position; the boundary however is now not an ellipse but any curve which is nearly circular. Take for the equation to this boundary

$$y^2 = cx - x^2 + b_2x^2 + b_3x^3 + b_4x^4 + \ldots\ldots\ldots\ldots\ldots (1),$$

where $b_2, b_3, b_4, \ldots$ are supposed to be so small that their squares and products may be neglected; the boundary passes through the origin: suppose that it cuts the axis of x again at the point for which $x = a$. Let the figure revolve round the axis of y through an infinitesimal angle $\delta\phi$; then the attraction of the element generated by the revolution of the area $2y\delta x$ on a particle at the origin, resolved along the axis of x is $x\delta\phi \dfrac{2y\delta x}{x\sqrt{(x^2+y^2)}}$. Hence the attraction of the whole wedge-shaped element is $2\delta\phi \displaystyle\int_0^a \frac{y\,dx}{\sqrt{(x^2+y^2)}}$.

As in Art. 280, Simpson gives the correct value of this integral; but he does not strictly *demonstrate* his result.

We will supply the demonstration

$$\int_0^a \frac{y\,dx}{\sqrt{(x^2+y^2)}} = \int_0^a \frac{\sqrt{(cx - x^2 + b_2x^2 + b_3x^3 + b_4x^4 + \ldots)}}{\sqrt{(cx + b_2x^2 + b_3x^3 + b_4x^4 + \ldots)}}\,dx.$$

Now by supposition $c = a - b_2a - b_3a^2 - b_4a^3 - \ldots$; substitute this value of c in the expression under the integral sign, divide both numerator and denominator by $\sqrt{x}$, and expand. Hence we find that the above integral becomes

$$\int_0^a \sqrt{\frac{a-x}{a}}\left\{1 - \ldots - \frac{b_n}{2(a-x)}(a^{n-1} - x^{n-1}) + \frac{b_n}{2a}(a^{n-1} - x^{n-1}) - \ldots\right\} dx,$$

where n is to have all positive integral values beginning with 2.

To effect the integration put $x = a\sin^2\theta$; thus we get

$$2a\int_0^{\frac{1}{2}\pi} \sin\theta\cos^2\theta \left\{1 - \ldots \right.$$
$$\left. - \frac{b_na^{n-2}}{2\cos^2\theta}(1 - \sin^{2n-2}\theta) + \frac{b_na^{n-2}}{2}(1 - \sin^{2n-2}\theta) - \ldots\right\} d\theta,$$

that is,

$$\int_0^{\frac{1}{2}\pi} \sin\theta \left\{2a\cos^2\theta - \ldots - b_na^{n-1}(1 - \sin^{2n-2}\theta)\sin^2\theta - \ldots\right\} d\theta.$$

Thus finally we obtain for the attraction required

$$\int_0^{\frac{1}{2}\pi} \sin\theta \left\{2a\cos^2\theta - (b_2a + b_3a^2 + b_4a^3 + \ldots)\sin^2\theta\right\} d\theta$$

$$+ \text{ a series whose general term is } \int_0^{\frac{1}{2}\pi} b_n a^{n-1}\sin^{2n+1}\theta\,d\theta.$$

Then making use of the value of c, we find that this becomes $\frac{2}{3}c +$ a series whose general term is $b_n a^{n-1} \frac{2.4 \ldots 2n}{3.5 \ldots (2n+1)}$.

In the small terms we may put c for a, so that our result is

$$\frac{2}{3}c + \frac{2.4}{3.5}b_2 c + \frac{2.4.6}{3.5.7}b_3 c^2 + \frac{2.4.6.8}{3.5.7.9}b_4 c^3 + \ldots$$

This agrees with Simpson's result.

291. Having thus obtained the attraction of the wedge-shaped element, Simpson proceeds to the attraction of any solid of revolution which is nearly spherical: his final result on his page 37 gives the expressions for the resolved attractions, along the normal and along the meridian tangent, which such a body produces on a particle at its surface.

292. The pages 41...45 of Simpson's *Mathematical Dissertations* contain an essay entitled *To determine the Length of a Degree of the Meridian, and the meridional Parts answering to any given Latitude, according to the true spherodical Figure of the Earth.*

This essay gives an approximate expression for the length of a degree of the meridian, on the hypothesis that the earth is an oblatum; a small Table is supplied of the length of a degree of the meridian in various latitudes, calculated on the hypothesis that the ratio of the axes of the earth is that of 231 to 230.

293. The subject of attraction is discussed by Simpson in his work, entitled, *The Doctrine and Application of Fluxions.* I have not seen the first edition of this work, which appears to have been published in 1750. The second edition is dated 1776, which is subsequent to the author's death: I presume that this is a reprint of the first edition. This contains 576 octavo pages, besides the Title, Dedication, and Preface on xi pages in the first volume, and the Title of the second volume.

Section IX. on pages 445...479 is entitled, *The Use of Fluxions in determining the Attraction of Bodies under different Forms.*

We have investigations, on the ordinary law, of the attractions of a straight line, of a circular lamina on an external particle which is perpendicularly over the centre, of a cone on a particle at the vertex, of a cylinder on a particle on the axis, and of a sphere on an external particle. With respect to the circular lamina and the sphere, the investigation is also given for the case in which the attraction varies as the nth power of the distance. The processes are all satisfactory, though some of them are rather artificial.

The attraction of an oblatum on a particle at the surface is determined in essentially the same manner as in the *Mathematical Dissertations;* but the analysis is a little simplified in some parts. In the *Dissertations* Simpson resolves the attraction in the directions of the tangent and the normal; in the *Fluxions* he resolves it parallel to the axes of the generating ellipse.

Simpson remarks on his page 455 that the integral which we have considered in Art. 280 might be expressed in finite terms instead of an infinite series; and this is obviously true.

On his page 463 Simpson demonstrates exact results corresponding to the approximate results enunciated by Stirling: see the diagram to Art. 153. Simpson shews that if PH be the direction of the attraction at P, then H divides CG in a constant ratio, and the attraction varies as PH. These results may be established immediately by the aid of the modern formulæ which are given in Art. 261.

On his page 466 Simpson determines the attraction of an oblatum on any internal particle. This enables him to give a more elaborate investigation than that in his *Dissertations* of Newton's postulate.

On his page 474 Simpson gives 2 hours 26 minutes as the least time in which the Earth, supposed a homogeneous fluid, could rotate: this however might have been stated in the *Dissertations,* as the necessary elements for the result are there supplied. It corresponds to Laplace's 1009 of a day: see the *Mécanique Céleste,* Livre III. § 20.

The Table which we have given, from the *Dissertations,* in Art. 285 is not reproduced in the *Fluxions.*

294. Thus we see that the contributions of Thomas Simpson to our subject are of eminent importance. In the homogeneous

Figure of the Earth he first determined the existence of a limiting angular velocity, for which the relative equilibrium is possible; and he implicitly shewed that different oblata might correspond to the same angular velocity. In Attraction he gave an accurate investigation for the case of an oblatum when the attracted particle is at the surface; and also an approximate investigation for the case of any nearly spherical body of revolution, and the analysis which he employed would not have been unworthy of Laplace himself.

Thomas Simpson was a mathematician of the highest order; and his merit is increased by reason of the great difficulties which impeded him. He has been pronounced "an analyst of first-rate genius," by one who like himself had risen to distinction in spite of adverse circumstances, and whose life like his closed prematurely in gloom and trouble. He has been placed at the head of the non-academical body of English mathematicians by a member of that body, whose ability and learning well qualified him for forming an opinion. It may be doubted whether the eighteenth century, after the death of Newton, supplies any mathematician in England more illustrious than the weaver whose genius raised him to the professorship of mathematics at Woolwich.

See the life prefixed to Hutton's edition of Simpson's *Select Exercises*; Murphy's *Theory of Equations*, page 54; *Philosophical Magazine*, September, 1850, page 209.

CHAPTER XI.

CLAIRAUT.

295. We now arrive at the great work of Clairaut, which is entitled *Théorie de la Figure de la Terre, tirée des Principes de l'Hydrostatique; par Clairaut, de l'Académie royale des Sciences, et de la Société royale de Londres.*

The work was published in 1743, and was reprinted in 1808. A note to the reprint states that the subject has been much considered by mathematicians, and that the actual state of the theory will be found in the third book of the *Mécanique Céleste;* but on account of its historical interest the treatise of Clairaut may be studied with advantage, and so it has been reproduced without change or addition: the reprint in fact corresponds nearly page for page with the original. It is stated that nothing has been neglected in order to remove the old errors of the press, and to avoid fresh errors: there is however an adequate supply of errors in the reprint.

A reason for adding no notes is assigned in these words: "Elles auraient dénaturé un ouvrage original, sans le rendre plus utile au public." The principle involved in these words is known to have been held by Laplace; and the conjecture has occurred to me that the reprint of Clairaut's work might have been suggested or encouraged by Laplace. The reprint is said to have been edited by Poisson: see the *Catalogue des ouvrages...de Siméon-Denis Poisson*, 1851.

I proceed to give an account of Clairaut's work; I use the edition of 1808: both editions are in octavo. The preliminary note to which I have just referred is of course peculiar to the edition of 1808; it occupies two pages; a Dedication to the Comte de Maurepas occupies two pages; then an Introduction follows on pages vii...xl; and the text, including a Table of Chapters, occupies pages 1...308.

296. The introduction gives a general notion of the subject of the work. Let us briefly consider what was the state of knowledge in 1743. With respect to fluid equilibrium Newton's principle of columns balancing at the centre, and Huygens's principle of the plumb-line were allowed to be *necessary*, but it was not known what principles were *sufficient*. Maclaurin had advanced far in the theory of the attractions of ellipsoids of revolution; and had well discussed the homogeneous figure of the Earth; and from the fact that his researches appeared originally in Latin they obtained a currency which the important additions made to the theory by Thomas Simpson, published only in English, probably never enjoyed. The measurement of a degree of the meridian in Lapland had been made, and from a comparison of this with the measurements made in France, it had been inferred that the ratio of the axes of the earth was that of 177 to 178; but the return of the expedition which had been sent to Peru was anxiously expected, in order to obtain more information on this point: see Clairaut's pages 299, 304. The diminution of gravity in proceeding from the equator to the pole was well established; and it was plain that the whole diminution of gravity must be greater than $\frac{1}{230}$ of the gravity at the pole: see Clairaut's page 297.

297. The Cartesians, according to Clairaut, enlightened by Newton held that all bodies were attracted to the *centre* of the Earth by a force which varied inversely as the square of the distance; from this Clairaut infers that the ratio of the axes of the Earth would be that of 576 to 577; see Clairaut's pages xiv, xvii, 143: in fact Clairaut shews on his page 143 that this is true whatever be the law of attraction provided the direction always passes through the centre: see also Art. 56.

But if we admit with Newton that every particle of matter attracts every other particle with a force varying inversely as the square of the distance, bodies will no longer necessarily be attracted exactly towards the centre of the earth; the direction of the resultant attraction on any particle will depend on the form of the earth, and on the position of the particle. Clairaut states the result which is demonstrated in the work, that considering the earth a homogeneous fluid in relative equilibrium the ratio of the axes will be that of 230 to 231: see Clairaut's pages xxiii and 195.

Clairaut remarks that the Newtonians may consistently with their fundamental principle obtain other results besides that just given; for they have only to suppose that the earth is not homogeneous. Clairaut considers that the result already given is that which the Cartesians ought to hold as following from their principles; but he suggests for them various expedients by which they might escape from the conclusion: see Clairaut's pages xxiv, xxv.

298. Clairaut draws attention to his own methods for discussing the equilibrium of fluids. He says Bouguer first remarked that there are hypotheses as to the nature of attraction under which a fluid could not be in equilibrium: see Clairaut's page xxxi, and our Art. 219. Clairaut says on his page xxxiii:

J'ai bientôt reconnu qu'il était vrai, ainsi que je l'avais soupçonné, que l'accord des deux principes ordinaires, c'est-à-dire l'équilibre des colonnes et de la tendance perpendiculaire à la surface, n'assurait pas l'équilibre d'une masse fluide; car j'ai trouvé qu'il y avait une infinité d'hypothèses de pesanteur où ces deux principes donneraient la même courbe, sans que pour cela les efforts de toutes les parties du fluide se contrebalançassent mutuellement. J'ai trouvé ensuite deux méthodes générales et sûres, pour reconnaître les hypothèses de pesanteur dans lesquelles les fluides peuvent être en équilibre, et pour déterminer la figure que les planètes doivent avoir dans ces hypothèses.

The two general and sure methods to which Clairaut alludes in the preceding extract may be called the *Principle of Canals*, and the *Principle of Level Surfaces:* we shall give an account of them in our analysis of the work. It would appear from Clairaut's words on his page xxxiv, that he intended to furnish some explanation of these methods in his Introduction; but the intention is not carried out, and the Introduction terminates somewhat abruptly.

299. The following points of interest in the Introduction may be noticed.

On page xiii. Clairaut says in a note:

Je fais ici la même distinction que M. de Maupertuis (la Figure de la Terre déterminée, etc.) entre la pesanteur et la gravité; j'entends par pesanteur, la force naturelle avec laquelle tout corps tombe, et j'appelle gravité la force avec laquelle ce corps tomberait, si la rotation de la Terre n'altérait pas son effort et sa direction.

I have already drawn attention to the distinction here explained: see Art. 25. It must however be observed that Clairaut does not adhere strictly to the language which he here professes to adopt. Thus on his page 28 he uses *pesanteur*, and on his page 30 he uses *gravité*, meaning the same thing in both cases, namely my *attraction;* and on his page 144 he uses *gravité* where he ought to use *pesanteur*.

On his page xxix. he enunciates the theorem which we call *Clairaut's Theorem:* see Art. 171.

On his page xxxviii. Clairaut is treating of rotation. He has supposed that an atom has described in an infinitesimal time a straight line Mm, so that if left to itself it would describe in the next equal infinitesimal time a straight line mn in the prolongation of Mm and equal to Mm. Then he says:...au lieu de la force qu'il aurait pour parcourir mn, on peut lui en substituer deux autres... Thus he uses the word *force* where we should now use *velocity*. In reading Clairaut's work, we are struck with the fact that although his conclusions are correct, his language is sometimes extremely inaccurate according to our modern notions.

300. Clairaut's work is divided into two parts. The first part treats of the general principles of fluid equilibrium; the second part treats of the Figure of the Earth and the other planets, assuming the ordinary law of attraction. The first part consists of twelve Chapters, and occupies pages 1...151; the second part consists of five Chapters, and occupies pages 152...304.

301. Clairaut's treatment of the theory of fluid equilibrium is a great advance beyond what his predecessors had given; but it is not free from obscurity. Clairaut never uses, as we now do, a symbol p to denote the pressure at any point of the fluid; this important step was first taken by Euler in the Berlin *Mémoires* for 1755. I am little likely to undervalue any improvement in the *Calculus of Variations*, but I attach less importance to the well-known introduction of the symbol δ into that subject by Lagrange, than to the introduction of the symbol p into Hydrostatics by Euler. Before Euler thus illustrated the subject, there had been *demonstrations* in Hydrostatics, but I cannot consider that these demonstrations were altogether intelligible.

302. Clairaut's first Chapter occupies pages 1...16; it expounds what may be called the *Principle of Canals.* Let there be a mass of fluid in equilibrium; we may imagine any portion of it to become solid, and the remainder will still be in equilibrium. Thus we may solidify all the fluid except that contained in an infinitesimal canal; and so the fluid in such a canal will remain in equilibrium. This canal may be of any form, straight or curved; it may pass completely through the mass, or it may be altogether within the mass returning to itself.

The principle of canals had already in effect been used by Newton, Huygens, and Maclaurin; though in general straight canals, which for distinction I call *columns*, had sufficed for their purposes; see Arts. 24, 55, and 245.

Although the *Principle of Canals* as stated in the preceding Article will be admitted to be obvious, yet Clairaut's method in applying the principle is not always clear. Thus, for example, on his page 2, he has a canal ORS passing entirely through a mass of fluid, which is in equilibrium; he says: "or cela ne peut arriver que les efforts de OR pour sortir vers S, ne soient égaux à ceux de SR pour sortir vers O." But how are we to measure the efforts which OR makes to escape towards S; or in fact what distinct idea can we form of these efforts?

Again take an example from his page 12. He has two canals of fluid HI and KL under certain circumstances; and he says that the weights of these two canals will be the same. But it is not immediately obvious how these *weights* are to be measured: the fact in modern language is that the pressure at H is equal to the pressure at K, and the pressure at I is equal to the pressure at L.

303. Clairaut's second Chapter occupies pages 16...28; it consists of general reasoning to shew that under certain attractive forces a fluid mass will remain in equilibrium. The Chapter seems superfluous, for in the sixth Chapter we have substantially a more satisfactory treatment of the subject. In reading the second Chapter it may assist the understanding if we conceive the fluid to be all enclosed within a rigid envelope; and then the sixth Chapter will in fact shew that we may dispense with this envelope.

304. Clairaut's third Chapter occupies pages 28...33; it considers a law of attraction under which a fluid mass could not be in equilibrium. The law is that in which the attraction towards a fixed centre is not a function of the *length* of the radius vector alone, but also of the *position* of the radius vector. The following is the demonstration, translated into modern language, of the impossibility of fluid equilibrium under such a law of force. Let MN be an arc of a circle having the centre of force C for centre; let PQ be an arc of a concentric circle, such that MPC is a straight line, and also NQC a straight line. Conceive the fluid in an infinitesimal canal MN to become solid; take the moments round C of the forces which act on it: thus we see that for equilibrium the pressure at M must be equal to the pressure at N. Similarly the pressure at P must be equal to the pressure at Q. But since the attraction along PM is not the same at equal distances from C as the attraction along QN, the change of pressure in passing from P to M is *not* equal to the change of pressure in passing from Q to N. This contradicts the former result.

305. Clairaut infers that there are innumerable cases in which a fluid mass will not be in equilibrium even although the conditions of Newton and Huygens are both satisfied. Clairaut is brief; we may expand his remarks. Let there be a curve $r = \phi(\theta)$ which revolves round the initial line; suppose we want to have a

mass of fluid in relative equilibrium when rotating with a given angular velocity round the initial line under an attractive force to the pole, and taking the form of the solid of revolution just generated. Since the angular velocity is given, the centrifugal force is known at every point of the boundary; hence the amount of the attractive force can be determined which must act at any point of the boundary, along the radius vector, so as to satisfy Huygens's principle of the plumb line: let $\psi(\theta)$ denote the amount of this attractive force at the point for which θ is the angular coordinate. Assume for the formula of attractive force $f(\theta)\{\phi(\theta)-r\}^n+\psi(\theta)$, a function of r and θ, in which $f(\theta)$ is at present undetermined; then it is obvious that Huygens's principle is satisfied. To satisfy Newton's principle we require that the expression $\int_0^{\phi(\theta)}[f(\theta)\{\phi(\theta)-r\}^n+\psi(\theta)]\,dr$, which measures the weight of a column, should be constant, the integration being taken with respect to r. This gives $\frac{f(\theta)\{\phi(\theta)\}^{n+1}}{n+1}+\phi(\theta)\psi(\theta)$ equal to a constant; and so $f(\theta)$ is determined. Thus Newton's principle is also satisfied. But by Art. 304 the fluid cannot be in equilibrium under the law of force which we have assigned.

306. Clairaut's fourth Chapter occupies pages 33...39; it determines the form of a mass of fluid in relative equilibrium acted on by certain forces. Suppose fluid to rotate round the axis of x, with angular velocity ω, under forces of which the acceleration parallel to the axis of x is X, that parallel to the axis of y is $\frac{Ry}{r}$, and that parallel to the axis of z is $\frac{Rz}{r}$; where $r^2=y^2+z^2$: then the equation to the free surface when there is relative equilibrium is

$$\int(Xdx+Rdr)+\frac{\omega^2r^2}{2}=\text{constant},$$

and the condition $\frac{dX}{dr}=\frac{dR}{dx}$ must hold.

This is not quite Clairaut's notation, but the difference is unimportant.

The demonstration of these results will be found in our ordinary treatises on Hydrostatics. I do not regard Clairaut's process as quite satisfactory until it is translated into our modern language.

Clairaut, after giving the equation of condition which we express as $\frac{dX}{dr} = \frac{dR}{dx}$, says briefly and authoritatively: "Toutes les fois que cette équation aura lieu, on sera sûr qu'il y aura équilibre dans le fluide." To me there appears some difficulty at this point in the theory of the equilibrium of fluids. We can shew clearly that certain conditions must hold for equilibrium; but it is not quite obvious that if these conditions are satisfied there will be equilibrium. Our modern writers seem to shrink from making the positive assertion of Clairaut, though perhaps sometimes it is implicitly adopted. But it is obvious that Clairaut asserts too much. Suppose for simplicity we restrict ourselves to one plane, and put X and Y as usual for the forces: it is *not sufficient* for equilibrium that $\frac{dX}{dy} = \frac{dY}{dx}$. For example take $X = \frac{y}{x^2 + y^2}$, and $Y = -\frac{x}{x^2 + y^2}$; let p denote the pressure, and ρ the density as usual. Then we get $dp = \rho \frac{ydx - xdy}{x^2 + y^2}$; and therefore $p = -\rho \tan^{-1} \frac{y}{x} + \text{constant}$. But this value of p is not admissible, for it would involve discontinuity, that is more than one value of p at the same point. See D'Alembert's *Opuscules Mathématiques*, Vol. V. page 10. In fact Clairaut's own pages 83...90 are sufficient to shew that his language is too positive.

307. The condition $\frac{dX}{dr} = \frac{dR}{dx}$ ensures that $Xdx + Rdr$ is a *complete differential.* The notion of a complete differential, and the appropriate condition, seem to have been first introduced by Clairaut himself: he refers to his memoir on the Integral Calculus in the Paris *Mémoires* for 1740.

Clairaut explains thus, in a note on his page 38, one of the symbols which he uses: "On entend par $\frac{dP}{dx}$ la différentielle de la fonction P, prise en supposant x seulement variable, et dont on a ôté les *dx*." It seems more natural to take the *differential coefficient* as the prior and simpler conception, and not the *differential*, as Clairaut here does.

308. Clairaut's fifth Chapter occupies pages 40...52; it introduces the use of *Level Surfaces*; these were first considered by Maclaurin; see Art. 248. Clairaut calls a level surface a *surface courbe de niveau;* and the space comprised between two level surfaces he calls a *couche de niveau.*

Clairaut gives the following proposition: suppose a mass of fluid divided into an infinite number of infinitesimal shells; if at any point of every shell the thickness of the shell is inversely proportional to the resultant accelerating force, the fluid will be in equilibrium. I cannot say that Clairaut's reasoning satisfies me. Indeed even with the modern methods, although it is easy to shew that when fluid is in equilibrium the thickness of the infinitesimal shells must follow the law assigned, yet to shew decisively that when this law of thickness holds the fluid must be in equilibrium seems far from easy: see Art. 306. Some remarks on Clairaut's reasoning will be found in the *Cambridge Mathematical Journal*, Vol. II. pages 18...22.

However, granting the proposition, Clairaut very ingeniously deduces the same equation as before for the free surface of a mass of fluid in relative equilibrium; and also the same condition as before connecting the forces: see Art. 306.

Another example of the strange mode of expression which we find in the book occurs on Clairaut's page 51. If we take an infinitesimal canal within an infinitesimal level shell we say in modern language that the *pressure is constant* throughout the canal; Clairaut speaks of *la liqueur..., ne pesant point.*

309. Clairaut's sixth Chapter occupies pages 52...63; it supplies examples in which the equation to the free surface of fluid in relative equilibrium is found when given forces act. In one example fluid is supposed to rotate round a vertical axis, the velocity of rotation being a function of the distance from the axis. Clairaut refers to two solutions which had already been proposed for this problem; namely, a correct solution by Daniel Bernoulli

on pages 244, 245 of his *Hydrodynamica,* and an incorrect solution by Hermann on page 372 of his *Phoronomia:* see Arts. 98 and 230.

In another example Clairaut supposes the fluid to be attracted to any number of fixed centres.

In another example Clairaut supposes the particles of fluid to attract each other with a force varying as the distance; and the fluid to rotate round an axis: in this case the free surface is that of an oblatum. Clairaut uses the known theorem that under such a law of attraction the resultant attraction varies as the distance from the centre of gravity of the whole attracting body: see Art. 12.

310. Clairaut's seventh Chapter occupies pages 63...77; it discusses a problem in fluid equilibrium proposed by Bouguer. We will give an account of the substance of the problem by the modern method.

Let x, y, z be the coordinates of any point of a fluid; let the shortest straight line be drawn from this point to a given surface; let r be the length of this straight line, and x', y', z' the coordinates of the point where it meets the given surface. Let the force acting on the fluid at (x, y, z) be along the line of r, and be denoted by f. It is required to determine the pressure at any point, and the form of the free surface.

In modern notation we have

$$\frac{1}{\rho}\frac{dp}{dx} = \frac{x'-x}{r}f, \quad \frac{1}{\rho}\frac{dp}{dy} = \frac{y'-y}{r}f, \quad \frac{1}{\rho}\frac{dp}{dz} = \frac{z'-z}{r}f.$$

Now

$$rdr = (x'-x)(dx'-dx) + (y'-y)(dy'-dy) + (z'-z)(dz'-dz);$$

and

$$(x'-x)\,dx' + (y'-y)\,dy' + (z'-z)\,dz' = 0,$$

because r is the shortest distance between (x, y, z) and the given surface; hence

$$rdr = -(x'-x)\,dx - (y'-y)\,dy - (z'-z)\,dz;$$

therefore

$$\frac{1}{\rho}dp = -fdr.$$

Hence f must be a constant, or a function of r; say $f = \phi(r)$,

and $$\frac{p}{\rho} = -\psi(r) + \text{constant},$$

where $\psi(r)$ is the integral of $\phi(r)$.

Thus the pressure at any point of the fluid mass is determined, and the form of the free surface is found by making the pressure constant.

Of course this is not Clairaut's method, as we have already remarked that he does not use a symbol for the pressure. He restricts himself to the case in which the given surface is a surface of revolution.

Clairaut considers that in order to render the hypothesis natural we must suppose there to be a *central solid mass;* for otherwise we should have some particles of fluid indefinitely close to each other, and yet acted on by forces the directions of which include a finite angle, *ce qui est choquant.*

311. Clairaut gives a second solution of the problem by a kind of general reasoning; see his page 69. He restricts himself to the case in which the given surface is a surface of revolution; and so, instead of considering normals to a surface as we did in the preceding Article, he considers normals to a given curve. Take a second curve, the points of which have a constant shortest distance from the given curve; that is, take a second curve which has the same *evolute* as the given curve: then it follows from the preceding Article that the pressure is constant throughout the second curve. Clairaut arrives, in his own way, at a result which corresponds to this; he expresses it, however, by saying that *le poids de OT doit être nul;* where OT denotes an infinitesimal canal in the form of our second curve.

312. Clairaut's eighth Chapter occupies pages 78...93; in modern language, we may say, that it is a modification of the sixth Chapter, by using polar coordinates instead of rectangular: thus confining ourselves to one plane, instead of $Xdx + Ydy$ we now get $Rdr + Trd\theta$.

The most interesting part of the Chapter is what Clairaut

calls the explanation of a species of paradox. The general equation to the free surface of the fluid is $\int Rdr + \int Trd\theta = \text{constant}$; the paradox consists in this, that Newton's principle of balancing columns gives $\int_0^r Rdr = \text{constant}$ for the equation to the free surface, which may in some cases differ from the former result.

We will omit all reference to the rotation of the fluid. Suppose, for example, that $R = r\theta^2$, and $T = r\theta$; then the two results agree: so also they agree if $R = \frac{\theta}{\sqrt{(a^2 + r\theta)}}$, and $T = \frac{1}{\sqrt{(a^2 + r\theta)}}$. But suppose that $R = \frac{r}{\sqrt{(r^2 + \theta^2)}}$, and $T = \frac{\theta}{r\sqrt{(r^2 + \theta^2)}}$; then according to Newton's principle of balancing columns we get $\sqrt{(r^2 + \theta^2)} - \theta = \text{constant}$; while the other result is $\sqrt{(r^2 + \theta^2)} = \text{constant}$.

Clairaut's explanation consists of reasoning to shew that the latter result is correct; but it does not appear to me that he is happy in his explanation. Such a force as $\frac{\theta}{r\sqrt{(r^2 + \theta^2)}}$ is inconceivable when $r = 0$; and thus to render his problem reasonable, a portion of the fluid round the origin must be supposed to become solid; and then Newton's principle of columns balancing at the centre is no longer applicable. D'Alembert objects, with justice, to Clairaut's explanation: see the *Opuscules Mathématiques*, Vol. v. pages 11 and 15.

Similar remarks to those in Art. 306 are applicable here. It is *not sufficient* for equilibrium that $Rdr + Trd\theta$ should be a perfect differential. Suppose, for instance, that this is the differential of a function $f(r, \theta)$; then if, when $r = 0$, the value of $f(r, \theta)$ still involves θ, the pressure is not the same in all directions round the origin.

Not one of Clairaut's three examples could correspond to the equilibrium of a free surface. Suppose, for instance, that $T = r\theta$; then when θ increases by 2π, we get a different value of T at the same point. But there might be equilibrium in a portion of the fluid confined, when necessary, by fixed planes.

313. Clairaut's ninth Chapter occupies pages 94...105; in this Chapter the results are extended to space of three dimensions, which in the previous Chapters had practically been applied only to space of two dimensions. Thus with the modern usual notation Clairaut finds that the free surface of the fluid in equilibrium must be such as to make the integral of $Xdx + Ydy + Zdz$ a constant; and, moreover, the following conditions must hold:

$$\frac{dX}{dy} = \frac{dY}{dx}, \quad \frac{dX}{dz} = \frac{dZ}{dx}, \quad \frac{dY}{dz} = \frac{dZ}{dy}.$$

These conditions are satisfied for such forces as occur in nature; so that Clairaut arrives substantially at this result: a mass of homogeneous fluid, under the influence of such forces as occur in nature, will be in equilibrium if Huygens's principle of the plumb-line holds at the free surface.

314. Clairaut's tenth Chapter occupies pages 105...128; it is on capillary attraction. Clairaut gives only extreme generalities. He may be said to shew that it is not impossible, and even not improbable, that the phenomena may be explained by supposing particles of fluid and particles of a solid tube to attract an adjacent particle of fluid with forces which are sensible only at a very small distance. But the Chapter is too remote from my subject to warrant me in examining it closely. Laplace devotes a paragraph to Clairaut's theory of capillary attraction in the *Mécanique Céleste*, Livre XI. § 1; Laplace's opinion is not favourable, he says: "cette théorie me paraît insignifiante...."

315. Clairaut's eleventh Chapter occupies pages 128...138; it treats of the equilibrium of fluid which is not homogeneous. In modern language, Clairaut undertakes to shew that level surfaces must be surfaces of equal density: we now know that this proposition is not necessarily true, unless $Xdx + Ydy + Zdz$ is a perfect differential. To this D'Alembert seems to refer in his *Traité...des Fluides*, second edition, page 50.

When a mass of fluid, like a planet, is not homogeneous, but yet is in equilibrium, Clairaut considers that the denser shells must be below the rarer; see his pages 134, 138, 280, 292. He does not demonstrate this condition, which is theoretically not necessary for equilibrium, though it may be essential for *stable* equilibrium.

316. Clairaut's twelfth Chapter occupies pages 139...151; it shews how we may determine the law of attraction at the surface of the Earth, from the results given by observation. By pendulum experiments we determine the force of gravity at any point on the Earth's surface; by measuring various lengths of degrees of the meridian we ascertain the form of the Earth's surface, and thus we can deduce the effect of the centrifugal force at any point: then knowing the values of gravity and of centrifugal force at any point, we can obtain the attraction at that point. But this does not determine the law of attraction within the surface of the Earth; so that on this point we must endeavour to make some natural hypothesis by the aid of the theory of fluid equilibrium.

Assuming that the Earth is a homogeneous fluid, and that the direction of attraction always passes through the centre, Clairaut gives a simple proof that the ratio of the axes must be very approximately that of 576 to 577, *whatever be the law of attraction;* see Art. 56. Hence, assuming that the ratio of the axes as determined by the French and Lapland arcs is really that of 177 to 178, it follows that the direction of attraction cannot always pass through the centre.

As an example Clairaut takes the ratio of the axes of the Earth to be that of 177 to 178; and he assumes that the diminution of gravity in passing from the pole to the equator varies as the square of the cosine of the latitude, the total diminution being $\frac{10}{2025}$ of the polar gravity: these facts depend on observations in France and Lapland. Then he shews that these data are consistent with an hypothesis of the law of force belonging to Bouguer's class: see Art. 310. This example is worked out in detail by Clairaut; but though not destitute of interest theoretically, it is of no practical value.

317. We now arrive at Clairaut's second part, which is that with which we are specially concerned. It consists of some intro-

ductory observations, followed by five Chapters. The introductory observations occupy pages 152...158.

Clairaut refers to his own former memoirs in the *Philosophical Transactions*, which we have noticed in our Chapter VI. Clairaut's researches on the figure of the Earth, considered homogeneous, arose from his desire to demonstrate Newton's postulate: see Art. 44. Clairaut's researches on the figure of the Earth, considered heterogeneous, arose from his desire to test and correct a remark made by Newton, namely, that the Earth if denser towards the centre would be more flattened than if it were homogeneous: see Art. 30.

Although the case of the homogeneous figure of the Earth could be deduced by a single substitution from the formulæ given by Clairaut for the heterogeneous figure, yet he judged it convenient to treat separately the homogeneous figure; and for this purpose to abandon his own method and follow that given in Maclaurin's *Fluxions*.

318. Clairaut's first Chapter occupies pages 158...198; it contains the theory of the homogeneous figure of the Earth or a planet. This is essentially the same theory as Maclaurin gave; but it is more easy to follow by being broken up into short sections, and printed in a more pleasing manner.

The exact values of the components of the attraction of an oblatum on a particle at its surface are given; the components being estimated parallel and perpendicular to the axis of revolution.

Clairaut holds that a rotating mass of fluid in relative equilibrium must assume the form of an oblatum; see his page 171. We have already observed that Maclaurin and Thomas Simpson in like manner asserted more than they were able to demonstrate: see Articles 249 and 283.

On his pages 188...190 Clairaut shews that the gravity varies as PG, to use our notation in Art. 153; but instead of the simple method which we adopt there, Clairaut first demonstrates the proposition of Art. 33, and then deduces the required result.

The relation which connects the ellipticity of the Earth with the value of the ratio of the centrifugal force to the attraction can be expressed exactly, or approximately in various forms, according to the notation adopted: see Arts. 262 and 283.

The following is the approximate result in Clairaut's notation: he takes the ratio of the equatorial axis to the polar axis to be that of $1 + \delta$ to 1; and he uses ϕ to express the ratio of the centrifugal force at the equator to the *gravity* there, not to the *attraction*: then

$$\phi = \frac{4}{5}\delta - \frac{2}{175}\delta^2 - \frac{8}{875}\delta^3$$

from which $$\delta = \frac{5}{4}\phi + \frac{5}{224}\phi^2 + \frac{135}{6272}\phi^3 \ldots$$

His δ is our $\frac{1}{\sqrt{(1-e^2)}} - 1$; and his ϕ is our $\frac{j}{1-j}$.

He finds $\frac{100}{28752}$ for the value of ϕ; see his page 194, from which he gets $\delta = \frac{1000}{230002}$.

319. On his pages 195...198, Clairaut applies his formula to determine the ellipticity of Jupiter; he arrives at the conclusion that the ratio of the axes is that of $100\frac{1}{2}$ to $90\frac{1}{2}$. This differs very little from Newton's final value: see Art. 29.

Modern observation gives a much smaller value to Jupiter's ellipticity than that which Newton and Clairaut derived from theory. Sir J. Herschel in his *Outlines of Astronomy*, 1849, Art. 512, states the ratio of the axes as that of 107 to 100; he adds:

And to confirm, in the strongest manner, the truth of those principles on which our former conclusions have been founded, and fully to authorize their extension to this remote system, it appears, on calculation, that this is really the degree of oblateness which corresponds, on those principles, to the dimensions of Jupiter, and to the time of his rotation.

In the edition of 1869 the ratio is changed to that of 106 to 100; but the passage just quoted remains unchanged. It is obvious that the remark cannot be accepted. For in the first place, if we consider Jupiter to be homogeneous, theory and observation are by no means in correspondence; secondly, if we suppose Jupiter not to be homogeneous, we shall be compelled to make some arbitrary hypothesis respecting the internal constitution of the planet, and cannot therefore appeal to the result as confirming in the strongest manner the truth of our principles; and thirdly, if a calculation once gave $\frac{7}{100}$ as the ratio of the difference of the axes to the minor axis, we cannot afterwards assert that the calculation gives $\frac{6}{100}$ as the ratio.

320. Clairaut's second Chapter occupies pages 198...232; it treats of the relative equilibrium of rotating homogeneous fluid which surrounds a spheroid composed of strata of varying density.

We have first a theorem respecting the attraction of a circular lamina on an external particle which is so situated that its projection on the lamina is very near the centre. Take the centre of the circle as the origin; let the axis of x pass through the projection of the attracted particle, and let h denote the distance of this projection from the centre, and k the distance of the particle from its projection; let r denote the radius of the circle, τ the thickness of the lamina, and ρ the density.

Then the attraction resolved parallel to the axis of x, estimated *towards* the origin,

$$= -\rho\tau \iint \frac{(x-h)\, dx\, dy}{\{(x-h)^2 + y^2 + k^2\}^{\frac{3}{2}}},$$

the integration being taken over the area of the circle.

Integrate first with respect to x; the limits may be denoted by $-\xi$ and ξ: thus we get

$$\rho\tau \int dy \left[\frac{1}{\{y^2 + k^2 + (\xi - h)^2\}^{\frac{1}{2}}} - \frac{1}{\{y^2 + k^2 + (\xi + h)^2\}^{\frac{1}{2}}}\right].$$

The process is *exact* up to this point. If we suppose h very small, we may expand the expression under the integral sign in powers of h; and thus we get $2\rho\tau h \int \frac{\xi dy}{(y^2 + k^2 + \xi^2)^{\frac{3}{2}}}$, that is $\frac{2\rho\tau h}{(r^2 + k^2)^{\frac{3}{2}}} \int \xi dy$. But $2 \int \xi dy$ is equal to the area of the circle; thus we obtain finally

$$\frac{\rho\tau h}{(r^2 + k^2)^{\frac{3}{2}}} \times \text{the area of the circle.}$$

The investigation would apply to a lamina which is nearly though not exactly circular, and leads to the same result.

Clairaut's own process is given in a geometrical form, but it is substantially equivalent to ours. We proceed to make use of the result.

321. Clairaut requires the approximate value of the attraction of a nearly spherical oblatum on an external particle. Let C denote the centre of the oblatum, and M the external particle. The attraction may be resolved into components along MC, and at right angles to MC. It is sufficient for Clairaut's purpose to consider the attraction along MC to be the same as if the mass of the oblatum were collected at C. To find the attraction at right angles to MC, he calculates the aggregate effect by the aid of the result in Art. 320.

Let the diagram represent the ellipse which is obtained by a meridian section of the oblatum passing through M. Through any point H in CM draw a chord at right angles to CM; the middle points of all such chords will be on a diameter. Let CK be the direction of this diameter, so that HK is the h of Art. 320, the chord itself being the intersection of a lamina at right angles to CH with the meridian plane.

Let $CH = x$, and the angle $HCK = \beta$, so that $h = x \tan \beta$. Let $CN = c$, and $CM = \gamma$. Now h is very small, because $\tan \beta$ is very small; and thus, without introducing any error to the order of accuracy which we adopt, we can use certain approximate values of the $r^2 + k^2$, and the *area*, which occur in Art. 320. We take $\pi (c^2 - x^2)$ for the *area*, and $(\gamma - x)^2 + c^2 - x^2$ for the $r^2 + k^2$. These approximations amount to neglecting the ellipticity of the oblatum; and as we have the common factor $\tan\beta$, our error is of the order of the product of $\tan\beta$ into the ellipticity.

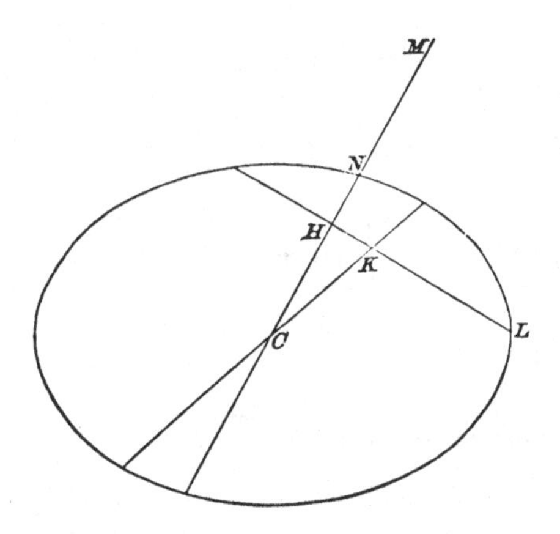

Hence by Art. 320 we have for the attraction of the whole oblatum at M in the direction at right angles to MC and towards CK,

$$\pi\rho\tan\beta\int_{-c}^{c}\frac{x(c^2-x^2)\,dx}{\{(\gamma-x)^2+c^2-x^2\}^{\frac{3}{2}}}.$$

The value of the definite integral which occurs here is $\frac{4c^5}{5\gamma^4}$; Clairaut himself obtains it by a peculiar artifice. By modern methods we may proceed thus:

Put $\gamma^2+c^2-2\gamma x=z^2$, and let $a=\gamma-c$, and $b=\gamma+c$; then we find that

$$\int_{-c}^{c}\frac{x(c^2-x^2)\,dx}{(\gamma^2+c^2-2\gamma x)^{\frac{3}{2}}}=-\frac{1}{8\gamma^4}\int_a^b(\gamma^2+c^2-z^2)(b^2-z^2)(a^2-z^2)\frac{dz}{z^2}$$

$$=\frac{1}{8\gamma^4}\int_a^b\left(1-\frac{a^2+b^2}{2z^2}\right)(b^2-z^2)(a^2-z^2)\,dz.$$

Integrate by parts, observing that

$$\int\left(1-\frac{a^2+b^2}{2z^2}\right)dz=z+\frac{a^2+b^2}{2z};$$

thus we find that the integral

$$= \frac{1}{8\gamma^4} \int_a^b \left(z + \frac{a^2 + b^2}{2z}\right) (a^2 + b^2 - 2z^2)\, 2z\, dz$$

$$= \frac{1}{8\gamma^4} \int_a^b \left\{(a^2 + b^2)^2 - 4z^4\right\} dz$$

$$= \frac{1}{8\gamma^4} \left\{(b - a)(a^2 + b^2)^2 - \frac{4}{5}(b^5 - a^5)\right\}$$

$$= \frac{c}{\gamma^4} \left\{(\gamma^2 + c^2)^2 - \frac{1}{5}(5\gamma^4 + 10\gamma^2 c^2 + c^4)\right\}$$

$$= \frac{4c^5}{5\gamma^4}.$$

Thus the required attraction is $\frac{4\pi\rho \tan \beta c^5}{5\gamma^4}$. The angle β is exactly the same as the angle between the diameter CK and the normal at its extremity, and is therefore very approximately equal to the angle between CN and the normal at N.

322. Clairaut introduces and defines the term *ellipticity of a spheroid* on his page 209; with him it denotes the ratio of the difference of the equatorial and polar diameters to the polar diameter: so that taking $2a$ for the equatorial diameter and $2b$ for the polar diameter the ellipticity is $\frac{a-b}{b}$. To the order however which is sufficient for our subject we might also define the ellipticity as $\frac{a-b}{a}$, and this is the sense in which we prefer to use the term.

323. We can now give an outline of Clairaut's investigations; we shall however change his notation for a more modern one.

Suppose the central part of the Earth solid, consisting of strata nearly spherical; and outside this let there be homogeneous fluid. Let r denote strictly the polar semiaxis of a stratum, but with sufficient approximation in many cases the radius drawn from the centre in any direction to the stratum. Let ρ denote the density and ϵ the ellipticity of this stratum; let ϵ' denote the ellipticity of the stratum which forms the boundary of the solid part. Let ρ_1 denote the density of the fluid, and ϵ_1 the ellipticity of the surface of the fluid. Let r' be the value of r at the boundary of the solid part, and r_1 the value of r at the surface of the fluid. Let ϕ_1 be the angle at the point corresponding to r_1 between the normal to the stratum and r_1. Thus the subscript 1 always indicates a value relative to the surface of the fluid.

Since the fluid is homogeneous Huygens's principle furnishes us with the necessary and sufficient condition for equilibrium. At any point of the surface of the fluid we have a central force which we will call F, and a force in the meridian plane at right angles to the radius vector towards the equator which we will call T; there is also the centrifugal force which at the equator would be jF in our usual notation, and which will be $jF \sin \lambda$ at the place considered, if λ denote the angle between the radius vector and the polar semiaxis. Hence resolving all the forces along the tangent to the meridian we have as the condition of equilibrium

$$F \sin \phi_1 - T - jF \sin \lambda \cos \lambda = 0 \ldots\ldots\ldots\ldots\ldots (1).$$

We must now develope this equation. With regard to F it will be sufficient to consider the whole mass as made up of spherical strata of varying density; and thus

$$F = \frac{4\pi}{r_1^2} \int_0^{r_1} \rho r^2 dr.$$

Next consider T. If there were a *homogeneous* oblatum of density ρ this force, by Art. 321, would be $\dfrac{4\pi\rho \tan \beta r^5}{5r_1^4}$, where r denotes the radius vector of the oblatum in the direction of r_1. For such an oblatum in which the radius vector is $r + dr$ the force would be $\dfrac{4\pi\rho}{5r_1^4}\left\{\tan \beta r^5 + \dfrac{d}{dr}(\tan \beta r^5) dr\right\}$. Hence the force arising from a shell of which ρ is the density and dr the thickness in the direction of r_1 is $\dfrac{4\pi\rho}{5r_1^4}\dfrac{d}{dr}(\tan \beta r^5) dr$.

Thus we obtain

$$T = \frac{4\pi}{5r_1^4} \int_0^{r_1} \rho \frac{d}{dr}(\tan \beta r^5) dr,$$

where ρ, now supposed variable, indicates the density of the stratum corresponding to r.

This method of treating strata of varying density occurs very frequently in our subject and should be carefully noticed.

Now by the nature of an ellipse it follows that to the order of approximation which we here retain $\tan\beta$ or $\sin\beta$ is equal to $2\epsilon\sin\lambda\cos\lambda$; and to that order ϕ_1 has the same meaning as β_1. Hence by simplifying we get from (1)

$$(2\epsilon_1 - j)\int_0^{r_1} \rho r^2 dr = \frac{2}{5r_1^2}\int_0^{r_1} \rho \frac{d}{dr}(\epsilon r^5) dr \text{..........(2).}$$

The form of (2) may be modified by separating the integral into two parts, one extending from 0 to r', and the other from r' to r_1; in the second interval the density is constant and is denoted by ρ_1. Thus

$$\int_0^{r_1} \rho r^2 dr = \int_0^{r'} \rho r^2 dr + \frac{\rho_1(r_1^3 - r'^3)}{3},$$

$$\int_0^{r_1} \rho \frac{d}{dr}(\epsilon r^5) dr = \int_0^{r'} \rho \frac{d}{dr}(\epsilon r^5) dr + \rho_1(\epsilon_1 r_1^5 - \epsilon' r'^5).$$

If we employ the second of these modifications, (2) becomes

$$2\epsilon_1\left\{\int_0^{r_1} \rho r^2 dr - \frac{\rho_1 r_1^3}{5}\right\} = j\int_0^{r_1} \rho r^2 dr + \frac{2}{5r_1^2}\int_0^{r'} \rho \frac{d}{dr}(\epsilon r^5) dr - \frac{2\rho_1 \epsilon' r'^5}{5r_1^2}.$$

If we put A for $\int_0^{r'} \rho r^2 dr$, and D for $\int_0^{r'} \rho \frac{d}{dr}(\epsilon r^5) dr$ we get, by employing also the first modification,

$$\epsilon_1 = \frac{\frac{1}{r_1^2}(6D - 6\rho_1 \epsilon' r'^5) + j(15A + 5\rho_1 r_1^3 - 5\rho_1 r'^3)}{30A + 4\rho_1 r_1^3 - 10\rho_1 r'^3} \text{.........(3).}$$

This is a very important formula in our subject; it agrees with that given by Clairaut on his page 217, allowing for a misprint with him: the investigation is substantially like his though in form rather different.

324. The general result (3) of the preceding Article is applied by Clairaut on his pages 218...222 to some special cases.

I. Suppose the whole mass homogeneous; then

$$A = \frac{\rho_1 r'^3}{3}, \quad D = \rho_1 \epsilon' r'^5;$$

and we obtain $$\epsilon_1 = \frac{5j}{4}:$$

this as far as it goes agrees with Art. 318.

II. Suppose the solid part homogeneous as well as the fluid part, but the densities of the two parts different. Let the density of the solid part be denoted by $\rho_1(1+\kappa)$; then

$$A = \frac{\rho_1 r'^3}{3}(1+\kappa), \quad D = \rho_1 \epsilon' r'^5 (1+\kappa);$$

and we obtain $$\epsilon_1 = \frac{\dfrac{6\kappa\epsilon' r'^5}{r_1^2} + 5j(\kappa r'^3 + r_1^3)}{10\kappa r'^3 + 4r_1^3}.$$

We shall find hereafter that this result reappears in the *Mécanique Céleste*, Vol. v. page 30.

Clairaut remarks that if we consider ϵ_1 to be known by observation, this formula will guide us in making suitable hypotheses as to the radius, the ellipticity, and the density of the assumed solid central part. He warns us that if we suppose κ to be negative we must remember that it is to be numerically *less than unity;* but the result shews us that this is an inadequate restriction: for if κ be negative it must not be numerically nearly equal to $\frac{4r_1^3}{10r'^3}$, and this might be much less than unity if r' were nearly equal to r_1. The truth is that if κ be *positive* the above result may be accepted without scruple; but if κ be negative we must carefully examine whether the value of ϵ_1 obtained from the formula is a small quantity.

If in the above formula for ϵ_1 we suppose $\epsilon' = 0$, and κ negative, and put $\frac{\kappa r'^3}{r_1^3} = -\lambda$, the result agrees with that obtained on page 156.

III. Suppose as a particular case of II that the solid part is to be *similar* to the whole mass, and that we require the ellipticity to be greater than it is when the whole mass is homogeneous. Then put $\epsilon_1 = \epsilon' = \frac{5j}{4}(1+p)$, where p is some positive quantity; thus we deduce

$$\kappa = -\frac{pr_1^3}{\frac{3}{2}\left(r'^3 - \frac{r'^5}{r_1^2}\right) + p\left(\frac{5}{2}r'^3 - \frac{3}{2}\frac{r^5}{r_1^2}\right)},$$

so that κ is necessarily negative.

IV. In the preceding result, suppose that the difference between r' and r_1 is infinitesimal; put $r' = r_1(1-\lambda)$, where λ is infinitesimal: then

$$\kappa = -\frac{p}{3\lambda + p},$$

so that κ differs only infinitesimally from unity. Thus we have the case of a film of fluid which surrounds a solid body of infinitesimal density; the outer and inner surfaces of the film are similar, similarly situated, and concentric oblata.

V. Instead of being a film as in IV. let us suppose the planet to be a shell of finite thickness; and let the internal part, though hard, be supposed of no density or of no attracting power: then we must solve the equation

$$\frac{3}{2}\left(r'^3 - \frac{r'^5}{r_1^2}\right) + p\left(\frac{5}{2}r'^3 - \frac{3}{2}\frac{r'^5}{r_1^2}\right) = -pr_1^3,$$

and take for $\frac{r'}{r_1}$ a positive value less than unity, if such a value should occur among the roots of the equation.

VI. Now return to the suppositions in II. If the density of the central part is to be greater than the density of the fluid, and ϵ_1 to be greater than $\frac{5j}{4}$, then ϵ' must be greater than $\frac{r_1^2\epsilon_1}{r'^2}$.

For put $\epsilon' = (\epsilon_1 + \gamma)\frac{r_1^2}{r'^2}$, and substitute in the result given in II.; thus we get

$$\epsilon_1 = \frac{5j}{4} + \frac{3\gamma\kappa r'^3}{2\kappa r'^3 + 2r_1^3};$$

and the second term will not be positive unless γ is positive.

325. Clairaut applies the last result of the preceding Article to two criticisms on Newton.

In the case of the Earth, if we wish to have ϵ_1 greater than $\frac{5j}{4}$, it is *not sufficient* merely to suppose a solid nucleus of greater density than the fluid; it is necessary to have the ellipticity of this solid nucleus greater than $\frac{r_1^2\epsilon_1}{r'^2}$: see Art. 37.

In the case of Jupiter, if we wish to have ϵ_1 less than $\frac{5j}{4}$, it is *not necessary* to suppose that the equatorial parts have been scorched by the Sun into a greater density than the other parts; it is sufficient to suppose that the solid nucleus is denser than the fluid, and that it has an ellipticity less than $\frac{r_1^2\epsilon_1}{r'^2}$: see Art. 31.

326. Clairaut shews in his pages 224 and 225, that an oblongum may be a form of relative equilibrium.

For in case II. of Art. 324, if ϵ' is negative and numerically greater than $\frac{5j(\kappa r'^3 + r_1^3)r_1^2}{6\kappa r'^5}$, then ϵ_1 is negative.

But even if ϵ' is positive, it will be possible to have ϵ_1 negative if κ be negative; Clairaut does not make this remark, to which D'Alembert seems to attach great importance; see the *Opuscules Mathématiques*, Vol. VI. page 77. The fact simply is that Clairaut's general formula contains somewhat more than he himself verbally drew from it.

327. Suppose the depth of the sea to be not greater than the height of the mountains; then Clairaut considers that we may without sensible error regard the earth as an oblatum covered with a film of water; see his page 225. In this case he takes $\epsilon_1 = \epsilon'$, and $r_1 = r'$; and so the equation (3) of Art. 323 becomes

$$10A\epsilon_1 - \frac{2D}{r_1^2} = 5Aj.$$

328. It has been objected that Clairaut ought not to have supposed $\epsilon_1 = \epsilon'$: see D'Alembert's *Opuscules Mathématiques*, Vol. VI. page 75, and Cousin's *Astronomie Physique*, page 164. If then we put $r_1 = r'$, but do not put $\epsilon_1 = \epsilon'$, the equation (3) of Art. 323 becomes

$$(10A - 2\rho_1 r_1^3)\,\epsilon_1 = \frac{2}{r_1^2}(D - \rho_1 \epsilon' r_1^5) + 5Aj.$$

As before then we may say that Clairaut's general formula contains more than he was contented to draw from it. But we must observe that if we suppose the stratum of fluid to be very thin, but do not take $\epsilon_1 = \epsilon'$, the fluid will not necessarily *cover* all the solid: either the polar parts or the equatorial parts may be left without fluid.

329. Clairaut applies his result which we have given in Art. 327 to shew that if $\epsilon = \epsilon_1 \left(\frac{r_1^2}{r^2} - u\right)$, where u is positive, and the density diminishes continually from the centre, the ellipticity will be less than when the mass is homogeneous.

For, using this expression for ϵ, we have

$$D = \int_0^{r_1} \rho \frac{d}{dr}(\epsilon r^5)\, dr = 3\epsilon_1 r_1^2 \int_0^{r_1} \rho r^2\, dr - \epsilon_1 \int_0^{r_1} \rho \frac{d}{dr}(r^5 u)\, dr$$

$$= 3\epsilon_1 r_1^2 A - \epsilon_1 (\rho r^5 u)_1 + \epsilon_1 \int_0^{r_1} r^5 u \frac{d\rho}{dr}\, dr$$

$$= 3\epsilon_1 r_1^2 A - \epsilon_1 G \text{ say,}$$

where G is some positive quantity, since by supposition $\frac{d\rho}{dr}$ is negative.

Thus we obtain,

$$10A\epsilon_1 - 6A\epsilon_1 + \frac{2}{r_1^2}\epsilon_1 G = 5Aj,$$

so that

$$\epsilon_1 = \frac{5j}{4} \frac{1}{1 + \dfrac{G}{2r_1^2 A}},$$

which is less than $\frac{5j}{4}$.

Clairaut expresses his result very awkwardly in words; he says that the spheroid will be less flattened than in the homogeneous case, unless the ellipticity of the strata diminishes from the centre to the circumference, and in a *greater ratio than the squares of the distances.* The language would imply that the squares of the distances diminish from the centre to the circumference. He should have said, *provided the product of the ellipticity into the square of the distance is never greater than at the surface.*

Clairaut on his pages 228.. 232 gives some special cases of the general result in Art. 328, by assuming special laws of density; his results are accurate, and he points out the objection to some corresponding investigations of Maclaurin: see Art. 267.

330. Clairaut's third Chapter occupies pages 233...262; it discusses the law of the variation of gravity at the surface of a spheroid of revolution composed of strata of varying density and ellipticity. Clairaut shews that the diminution of gravity in passing from the pole to the equator varies as the square of the cosine of the latitude; and he establishes the theorem which we now call *Clairaut's Theorem.* We will proceed to give some details.

331. Suppose a particle placed outside a circular lamina; when the projection of the particle on the lamina is very near the centre of the circle, the resultant attraction on the particle is very nearly the same as if the particle were at the same distance from the centre of the circle, but had its projection coincident with the centre: Clairaut shews this briefly by general reasoning. If we proceed analytically as in Art. 320, we shall find that when the particle is displaced so that its projection moves from the centre to a distance h from the centre, the attraction *perpendicular* to the lamina is not changed to the order h, while there is a *transverse* attraction produced of the order h; so that the change in the resultant attraction is of the order h^2.

The result holds also for an ellipse or any other central curve.

332. If a circular lamina, and an oval lamina which is nearly circular, have the same centre and the same plane and equal areas, they exert approximately the same attraction on a particle, the

projection of which would coincide with the common centre: Clairaut shews this briefly by general reasoning.

333. The propositions of the two preceding Articles lead Clairaut to the following general result.

Let C be the centre of an ellipsoid of revolution nearly spherical, and M an external particle; let MC cut the solid at N, and let MC produced cut the solid at n; the attraction of the solid on a particle at M is approximately the same as that of an ellipsoid of revolution of equal volume having Nn for its axis of revolution.

The original solid may be an oblatum or an oblongum; whichever it be the derived solid will be sometimes an oblatum and sometimes an oblongum, according to the position of the straight line CM.

It must be observed that the approximation holds as far as the first power of the ellipticity *inclusive;* in fact the errors in Arts. 331 and 332 are of the order of the *square* of the ellipticity.

334. Clairaut then finds the attraction of an ellipsoid of revolution which is nearly spherical on a particle which is on the prolongation of the axis of revolution. I have already adverted to the method which he uses: see Art. 165. The result is correct to the first power of the ellipticity inclusive.

335. The pages 233...243 of Clairaut's work which we have just considered were, in substance, originally published in the *Philosophical Transactions;* see Art. 164: these pages well deserve perusal as a good specimen of the ingenuity and simplicity of Clairaut's investigations.

A modern student will probably like to verify by analysis the important result in Art. 333. The simplest way perhaps is to find the *potential* of the original ellipsoid of revolution on the particle at M, and shew that it is equal to the potential of the derived ellipsoid of revolution, so far as terms of the first order inclusive.

Take C for the origin, CN for the axis of z; let the axis of y be the diameter which is conjugate to Nn in the meridian plane of the given ellipsoid which contains CM; let the axis of x be at right angles to those of y and z.

Let $CM = \gamma$; let χ denote the angle between the axes of y and z. Then the potential of the given ellipsoid is

$$\iiint \frac{\sin\chi dx\, dy\, dz}{\{x^2 + y^2 + 2y(z-\gamma)\cos\chi + (z-\gamma)^2\}^{\frac{1}{2}}};$$

the limits of the integration are determined by the equation to the given ellipsoid, which we may take as

$$\frac{x^2}{a^2} + \frac{y^2}{b^2} + \frac{z^2}{c^2} = 1.$$

Put r^2 for $x^2 + y^2 + (z-\gamma)^2$; then we may expand the term under the integral signs in the form

$$\frac{1}{r} - \frac{y(z-\gamma)\cos\chi}{r^3} + \frac{3}{2}\frac{y^2(z-\gamma)^2\cos^2\chi}{r^5} + \dots$$

The second of these terms gives zero as the result when integrated, because y is as often negative as positive. Thus if we reject the *squares* and the higher powers of the small quantity $\cos\chi$, the potential becomes

$$\sin\chi \iiint \frac{dx\, dy\, dz}{\{x^2 + y^2 + (z-\gamma)^2\}^{\frac{1}{2}}}$$

Assume $x = a\xi$, $y = b\eta$, $z = c\zeta$; then the potential can be transformed to

$$V\iiint \frac{d\xi\, d\eta\, d\zeta}{\{a^2\xi^2 + b^2\eta^2 + (c\zeta - \gamma)^2\}^{\frac{1}{2}}} \quad \dots\dots\dots\dots\dots\dots (1),$$

where V stands for $abc \sin\chi$, and the limits of integration are determined by

$$\xi^2 + \eta^2 + \zeta^2 = 1.$$

Now when we form the potential of the derived ellipsoid we obtain, if h denote the two equal semiaxes,

$$h^2c\iiint \frac{d\xi\, d\eta\, d\zeta}{\{h^2\xi^2 + h^2\eta^2 + (c\zeta - \gamma)^2\}^{\frac{1}{2}}} \qquad (2),$$

the limits being the same as before. And by the condition of equal volumes we have

$$abc \sin \chi = h^2 c \tag{3}$$

Since the original ellipsoid is nearly spherical we have

$$a = c(1 + \lambda) \quad \text{and} \quad b = c(1 + \mu)$$

where λ and μ are small, being of the order of the ellipticities.

Thus from (3) we have

$$h^2 = c^2 (1 + \lambda + \mu) \sin \chi,$$

but $\sin \chi$ being the sine of an angle, nearly a right angle, we shall find that it differs from unity by a quantity which is of the order of the *squares* of the ellipticities. Thus to our order

$$h^2 = c^2 (1 + \lambda + \mu),$$

and so we have to our order

$$h^2 = \frac{a^2 + b^2}{2}$$

Hence since $a^2 = \dfrac{a^2 + b^2}{2} + \dfrac{a^2 - b^2}{2}$, and $b^2 = \dfrac{a^2 + b^2}{2} - \dfrac{a^2 - b^2}{2}$,

we see that to our order (1) becomes

$$V \iiint \frac{d\xi \, d\eta \, d\zeta}{\left\{h^2\xi^2 + h^2\eta^2 + \dfrac{a^2 - b^2}{2}(\xi^2 - \eta^2) + (c\zeta - \gamma)^2\right\}^{\frac{1}{2}}}$$

Expand the denominator under the integral sign in powers of $\dfrac{a^2 - b^2}{2}(\xi^2 - \eta^2)$; then the term under the integral sign which involves the first power of this small quantity obviously vanishes by symmetry: so if we neglect the square of $a^2 - b^2$, the expression (1) reduces to the form (2). This is the required result.

336. We are now prepared to find the value of gravity at any point of the surface of our hypothetical Earth.

Suppose r the polar radius, $r(1 + \epsilon)$ the equatorial radius of an oblatum, where ϵ is small; let ρ be the density. We shall first determine the attraction on a particle at the distance R from the centre, the direction of R making with the polar axis an angle whose sine is s.

By Clairaut's proposition in Art. 333, we substitute instead of the oblatum, a certain ellipsoid of revolution of equal mass. Let c denote the polar semiaxis, and λ the ellipticity of this derived ellipsoid. The attraction which it exerts

$$=\frac{\text{mass of oblatum}}{R^2}\left(1-\frac{6\lambda c^2}{5R^4}\right):$$

this may be deduced from Art. 261, supposing λ positive; or it may be obtained in Clairaut's manner, to which we have referred in Art. 334, and then it will be found to hold whether λ be positive or negative.

Now to our order of approximation.

$$c=r(1+\epsilon s^2);$$

and the condition of equal masses gives

$$r^3(1+2\epsilon)=c^3(1+2\lambda)=r^3(1+3\epsilon s^2)(1+2\lambda);$$

so that
$$\lambda=\left(1-\frac{3s^2}{2}\right)\epsilon.$$

Also, supposing the attracted particle to be on a concentric and similarly situated oblatum, the dimensions of which are given by r_1 and ϵ_1, we have

$$R=r_1(1+\epsilon_1 s^2).$$

The mass of the oblatum $=\dfrac{4\pi r^3}{3}(1+2\epsilon)\rho$.

Hence the attraction of the oblatum

$$=\frac{4\pi r^3}{3r_1^{\,2}}(1+2\epsilon)(1-2\epsilon_1 s^2)\rho\left\{1-\frac{6r^2}{5r_1^{\,2}}\left(1-\frac{3}{2}s^2\right)\epsilon\right\}$$

$$=\frac{4\pi r^3\rho}{3r_1^{\,2}}\left\{1+2\epsilon-2\epsilon_1 s^2-\frac{6\epsilon r^2}{5r_1^{\,2}}+\frac{9\epsilon s^2r^2}{5r_1^{\,2}}\right\}.$$

Let us denote this for a moment by $\rho f(r)$; then for the attraction of the shell of density ρ, comprised between the surfaces which correspond to r and $r+dr$, we have $\rho\dfrac{df(r)}{dr}dr$.

Hence, if we suppose the density of each shell to be uniform, but the density to vary from shell to shell, we have for the whole attraction

$$\int_0^{r_1} \rho \frac{df(r)}{dr} dr.$$

Let $A = \int_0^{r_1} \rho r^2 dr$, $B = \int_0^{r_1} \rho \frac{d(r^3\epsilon)}{dr} dr$, $D = \int_0^{r_1} \rho \frac{d(r^5\epsilon)}{dr} dr$, then the attraction is

$$\frac{4\pi}{r_1^2}(1 - 2\epsilon_1 s^2) A + \frac{8\pi B}{3r_1^2} - \frac{4\pi}{5r_1^4}(2 - 3s^2) D.$$

We must now introduce the centrifugal force. The centrifugal force at the equator is approximately $\frac{4\pi jA}{r_1^2}$; and therefore it is $\frac{4\pi sjA}{r_1^2}$ at the point under consideration: the resolved part of this along the radius is $\frac{4\pi s^2 jA}{r_1^2}$, which must be subtracted from the *attraction* to obtain the *gravity*.

Hence the gravity at the point under consideration

$$= \frac{4\pi}{r_1^2}(1 - 2\epsilon_1 s^2) A + \frac{8\pi B}{3r_1^2} - \frac{4\pi}{5r_1^4}(2 - 3s^2) D - \frac{4\pi s^2 jA}{r_1^2}.$$

Let P denote the gravity at the pole, g the gravity at the point under consideration; then

$$P - g = \frac{4\pi}{r_1^2}\left\{(j + 2\epsilon_1) A - \frac{3D}{5r_1^2}\right\} s^2.$$

Thus $P - g$ varies as s^2; that is, the diminution of gravity in passing from the pole to the equator varies as the square of the cosine of the latitude.

Let E denote the gravity at the equator; then

$$P - E = \frac{4\pi}{r_1^2}\left\{(j + 2\epsilon_1) A - \frac{3D}{5r_1^2}\right\}.$$

Divide this by E; then on the right-hand side it will be sufficient to use $\frac{4\pi A}{r_1^2}$ for E, so that

$$\frac{P - E}{E} = j + 2\epsilon_1 - \frac{3D}{5Ar_1^2}.$$

Substitute for $\frac{D}{A}$ from Art. 327, and we have

$$\frac{P-E}{E} = \frac{5}{2}j - \epsilon_1.$$

This remarkable result is called *Clairaut's Theorem.* The fraction $\frac{P-E}{E}$ we shall call *Clairaut's fraction,* as in Art. 171, and shall denote it by v; so that we have

$$v + \epsilon_1 = \frac{5}{2}j.$$

We know by Art. 28 that $\frac{5}{2}j$ is twice the ellipticity of the earth, supposed homogeneous; and this is the form in which Clairaut himself expresses this term.

337. The assumptions on which Clairaut's demonstration of his famous theorem rests should be carefully noticed. The strata are supposed to be ellipsoidal, and of revolution round a common axis, and nearly spherical. Each stratum is homogeneous, but there is no limitation on the law by which the density varies from stratum to stratum: the density may change discontinuously if we please. It is not assumed that the strata were originally fluid; but it is assumed that the *superficial* stratum has the same form as if it were fluid and in relative equilibrium when rotating with uniform angular velocity. There is no limitation on the law by which the ellipticity varies from stratum to stratum, except that the ellipticity must be continuous, and at the surface must be such as would correspond to the relative equilibrium of a film of rotating fluid.

We shall find that D'Alembert in 1756 mistook the range of Clairaut's demonstration: see the *Recherches sur...Système du Monde,* Vol. III. page 187.

In some modern works there has been a want of strict accuracy as to the Theorem, owing perhaps to an undue regard to brevity. Thus we read in one that Clairaut established his Theorem on "the hypothesis of the Earth being a fluid mass"; and we read in another that Clairaut discovered his Theorem for "the case of a rotating fluid mass, or solid with density distributed as if fluid."

338. Clairaut on his pages 251...259 uses his theorem to support certain criticisms on Newton, David Gregory, and Maclaurin. We have already noticed these criticisms: see Arts. 30, 84, and 271.

On his pages 260...262 Clairaut in like manner uses his theorem in relation to Cassini's hypothesis that the earth was an oblongum with an ellipticity of $\frac{1}{93}$ In Art. 336 put $\epsilon_1 = -\frac{1}{93}$, and $j = \frac{1}{289}$; then we get approximately $v = \frac{1}{93} + \frac{1}{115} = \frac{1}{51}$ But, as Clairaut observes, this is a far greater value of v than pendulum observations warrant. Cassini's number however seems to have been 95 not 93: see Art. 104.

339. Clairaut's fourth Chapter occupies pages 262...296; it considers the figure of the Earth, supposed to have been originally fluid, and composed of strata of varying densities. In fact Clairaut now proposes to investigate the connection between the density and the ellipticity in order that strata of the kind considered in the preceding Chapter may be in relative equilibrium if they are fluid. A process like that of Art. 323 must now be applied to *each* stratum.

340. Suppose a shell of density ρ bounded by two concentric and similarly situated oblata; let ζ_1 be the ellipticity of the inner surface, and ζ_2 the ellipticity of the outer surface. Suppose a particle situated on the inner surface of the shell; we shall determine the attraction which the shell exerts on this particle in the direction at right angles to the radius vector from the centre of the shell to the particle. This problem is solved by Clairaut on his pages 262...265. Our solution is substantially the same.

The attraction of the shell is of course equal to the difference of the attractions of the oblatum which is bounded by the outer surface, and the oblatum which is bounded by the inner surface. We will consider these bodies separately, beginning with the larger.

The larger oblatum produces the same effect as would be produced by a similar, similarly situated, and concentric oblatum, having the particle on its surface; for the difference of these two similar, similarly situated, and concentric oblata produces no effect by Art. 13.

Hence the attraction of the larger oblatum in the assigned direction is $\frac{4\pi\rho c \tan\beta}{5}$ in the notation of Art. 321; for now $\gamma = c$. And, as in Art. 323, we have $\tan\beta = 2\zeta_2 \sin\lambda\cos\lambda$, so that the attraction becomes $\frac{8\pi\rho c\zeta_2 \sin\lambda\cos\lambda}{5}$.

In like manner the attraction of the smaller oblatum in the assigned direction is $\frac{8\pi\rho c\zeta_1 \sin\lambda\cos\lambda}{5}$.

Thus it follows that the required attraction of the shell in the assigned direction is $\frac{8\pi\rho c(\zeta_2 - \zeta_1)\sin\lambda\cos\lambda}{5}$

Now suppose that there is an infinitesimally thin shell of the density ρ; let r be the polar semiaxis of the inner surface, and ϵ the ellipticity of this surface; then $\epsilon + \frac{d\epsilon}{dr}dr$ will denote the ellipticity of the outer surface. Therefore the attraction, in the direction at right angles to the radius vector, of this shell on the inside particle is $\frac{8\pi\rho}{5} c\sin\lambda\cos\lambda \frac{d\epsilon}{dr}dr$; this is obvious from the preceding investigation.

341. We now proceed to apply a process like that of Art. 323 to any stratum.

Let there be a particle of fluid in any stratum at the distance r' from the centre; let λ be the angle between the radius vector to the particle and the polar semiaxis.

The central attraction on the particle is $\frac{4\pi}{r'^2}\int_0^{r'} \rho r^2 dr$, approximately; for the strata beyond the particle produce no central effect to the order of accuracy which we have to consider. This central attraction gives rise to a component in the meridian plane, at right angles to the radius vector, towards the pole equal to $2\epsilon' \sin\lambda\cos\lambda \times \frac{4\pi}{r'^2}\int_0^{r'} \rho r^2 dr$. We will call this a *transverse* attraction.

The transverse attraction on the particle from the strata below the particle towards the equator, is $\frac{8\pi r' \sin\lambda \cos\lambda}{5r'^5}\int_0^r \rho \frac{d}{dr}(r^5\epsilon)dr$, by Art. 323.

The transverse attraction on the particle from the strata beyond the particle towards the equator is $\frac{8\pi r' \sin\lambda\cos\lambda}{5}\int_{r'}^{r_1} \rho \frac{d\epsilon}{dr} dr$, by Art. 340.

Let ω denote the angular velocity; then the transverse component of the centrifugal force is $\omega^2 r' \sin\lambda \cos\lambda$.

Hence, as in Art. 323, equating to zero the whole transverse force, and dividing by $4\pi \sin\lambda\cos\lambda$ we obtain

$$\frac{2\epsilon'}{r'^2}\int_0^{r'} \rho r^2 dr - \frac{2}{5r'^4}\int_0^{r'} \rho \frac{d}{dr}(\epsilon r^5)\, dr - \frac{2r'}{5}\int_{r'}^{r_1} \rho \frac{d\epsilon}{dr} dr - \frac{r'\omega^2}{4\pi} = 0.$$

If as usual we denote by j the ratio of the centrifugal force at the equator at the surface of the fluid to the attraction there, we have to the order of accuracy which we require

$$r_1\omega^2 = j\frac{4\pi}{r_1^2}\int_0^{r_1} \rho r^2 dr.$$

Substituting this value of ω^2 our equation becomes

$$\frac{2\epsilon'}{r'^2}\int_0^{r'} \rho r^2 dr - \frac{2}{5r'^4}\int_0^{r'} \rho \frac{d}{dr}(\epsilon r^5)\, dr - \frac{2r'}{5}\int_{r'}^{r_1} \rho \frac{d\epsilon}{dr} dr - \frac{jr'}{r_1^3}\int_0^{r_1} \rho r^2 dr = 0.$$

This important equation occurs for the first time in Clairaut's page 273; it has ever since been permanently associated with the subject: I shall call it *Clairaut's primary equation.* Whether we leave ω^2 in the equation, or substitute for it in the manner just explained, is of no consequence.

342. If ϵ and ρ are taken so as to satisfy Clairaut's primary equation we have a *possible* constitution for the earth. Clairaut however asserts more than this on his page 265, namely that if j be very small the strata *will* be elliptical spheroids. Even Laplace has scarcely arrived at this point; he has only shewn that if the strata *are assumed* to be nearly spherical they must be elliptical spheroids.

343. Clairaut transforms his primary equation. It will not lead to any confusion if we now drop the accent from r' and from ϵ': we may then write the equation thus:

$$10\epsilon r^2 \int_0^r \rho r^2 dr - 2\int_0^r \rho \frac{d}{dr}(\epsilon r^5)\, dr - 2r^5 \int_r^{r_1} \rho \frac{d\epsilon}{dr} dr - \frac{5r^5\omega^2}{4\pi} = 0.$$

Differentiate with respect to r; thus

$$\left(20\epsilon r + 10r^2 \frac{d\epsilon}{dr}\right) \int_0^r \rho r^2 dr + 10\rho r^4 \epsilon - 2\rho \frac{d}{dr}(\epsilon r^5)$$
$$- 10r^4 \int_r^{r_1} \rho \frac{d\epsilon}{dr} dr + 2r^5 \rho \frac{d\epsilon}{dr} - \frac{25r^4\omega^2}{4\pi} = 0.$$

Simplify, and divide by $10r^4$; thus

$$\left(\frac{1}{r^2}\frac{d\epsilon}{dr} + \frac{2\epsilon}{r^3}\right) \int_0^r \rho r^2 dr = \int_r^{r_1} \rho \frac{d\epsilon}{dr} dr + \frac{5\omega^2}{8\pi}.$$

Differentiate again with respect to r; thus

$$\left(\frac{1}{r^2}\frac{d^2\epsilon}{dr^2} - \frac{6\epsilon}{r^4}\right) \int_0^r \rho r^2 dr + \rho\left(\frac{d\epsilon}{dr} + \frac{2\epsilon}{r}\right) = -\rho \frac{d\epsilon}{dr},$$

so that
$$\frac{d^2\epsilon}{dr^2} + \frac{2\rho r^2 \frac{d\epsilon}{dr}}{\int_0^r \rho r^2 dr} = \left(\frac{6}{r^2} - \frac{2\rho r}{\int_0^r \rho r^2 dr}\right)\epsilon.$$

This I shall call *Clairaut's derived equation.*

344. Clairaut puts his derived equation in another form.

Let $\frac{1}{\epsilon}\frac{d\epsilon}{dr}$ be denoted by u; so that $\frac{d\epsilon}{dr} = \epsilon u$.

Then
$$\frac{d^2\epsilon}{dr^2} = \epsilon u^2 + \epsilon \frac{du}{dr}.$$

Thus
$$\frac{du}{dr} + u^2 = -\frac{2u\rho r^2}{\int_0^r \rho r^2 dr} + \frac{6}{r^2} - \frac{2\rho r}{\int_0^r \rho r^2 dr}.$$

Put $u + \dfrac{\rho r^2}{\int_0^r \rho r^2 dr} = t$; and then we obtain

$$\frac{dt}{dr} + t^2 = \frac{r^2 \dfrac{d\rho}{dr}}{\int_0^r \rho r^2 dr} + \frac{6}{r^2}.$$

Clairaut observes that this equation falls under the case of $dy + y^n dx = X dx$, where X is a function of x; and that what we now call the separation of the variables had not yet been effected in general. Accordingly he does not propose to seek for the ellipticities which correspond to a given law of density, except in the case in which ρ varies as r^n. See his page 276.

345. Suppose then that ρ varies as r^n. We have by the preceding Article

$$\frac{dt}{dr} + t^2 = \frac{n^2 + 3n + 6}{r^2}.$$

This becomes homogeneous and easily integrable by putting a new variable instead of $\dfrac{1}{r}$; and thus we obtain

$$t = \frac{\dfrac{1}{2} + q}{r} - \frac{2qar^{-2q-1}}{ar^{-2q} + 1},$$

where $q = \sqrt{\left(n^2 + 3n + \dfrac{25}{4}\right)}$, and a is an arbitrary constant.

With this value of t we find

$$\epsilon = bar^{-n-\frac{5}{2}-q} + br^{-n-\frac{5}{2}+q},$$

where b is another arbitrary constant.

This value of ϵ then satisfies Clairaut's derived equation; we must examine if it also satisfies the primary equation. Substitute the value of ϵ, and we find after simplification that the primary equation is satisfied provided the following relation holds between the constants:

$$\left(q - \frac{1}{2} - n\right) br_1^{\frac{1}{2}+q} - ba\left(q + \frac{1}{2} + n\right) r_1^{\frac{1}{2}-q} = \frac{5j}{2} r_1^{n+3}.$$

Thus there is only one relation between two constants, and so it would appear that the solution is not determinate. Clairaut offers an explanation on this point. It has been assumed throughout that the ellipticity of the strata is small; moreover he considers that n must be negative in order that the density may diminish from the centre to the circumference, which he says the laws of hydrostatics require: see Art. 315. Hence we must have $a = 0$; for otherwise ϵ would be infinite at the centre.

Also even if n be considered positive we must have $a = 0$; in this case ϵ would be finite at the centre, but $r\epsilon$ would be infinitesimal: see Clairaut's page 281.

For a particular case suppose $n = 0$, then $q = \frac{5}{2}$; and after putting $a = 0$ we obtain $b = \frac{5j}{4}$: and then $\epsilon = b = \frac{5j}{4}$ as it should be.

346. Clairaut's derived equation may be put in the form

$$\frac{6\epsilon - r^2 \dfrac{d^2\epsilon}{dr^2}}{2r\epsilon + 2r^2 \dfrac{d\epsilon}{dr}} = \frac{\rho r^2}{\int_0^r \rho r^2 dr}$$

Then if ϵ be given as a function of r, the left-hand member of the equation becomes a known function of r; denote it by P; from this we deduce $\rho r^2 = Pe^{\int P dr}$ which gives ρ. See Clairaut's page 283.

347. The formulæ which have been investigated for the case of an infinite number of indefinitely thin strata may be applied to the case of a finite number of shells surrounding a central part, the density changing abruptly from the central part to the adjacent shell, and then from shell to shell. Clairaut considers this on his pages 286...293; taking the density constant throughout the central part, and throughout each shell.

He shews that the ellipticities increase from the centre to the surface, assuming that the densities diminish from the centre to the surface: see his page 292.

Clairaut takes for an example the case in which the whole mass is supposed to consist of two parts throughout each of which the density is constant. Let ρ_1 be the density of the outer part, ϵ_1 the ellipticity of the outer surface, r_1 the polar semiaxis; let ρ_2, ϵ_2, and r_2 be corresponding quantities for the inner part. Then in the integrations which occur in Clairaut's primary equation we have to make $\rho = \rho_2$ from $r = 0$ to $r = r_2$, and $\rho = \rho_1$ from $r = r_2$ to $r = r_1$. Put λ for $\frac{r_2}{r_1}$, and suppose $\rho_2 = \rho_1 + \sigma$. Then apply Clairaut's primary equation first to the extreme stratum of the inner part, and next to the extreme stratum of the outer part: thus we obtain after reductions

$$\epsilon_2 (10\rho_1 + 4\sigma) - 6\epsilon_1\rho_1 = 5j(\rho_1 + \sigma\lambda^3),$$

$$\epsilon_1 (4\rho_1 + 10\sigma\lambda^3) - 6\epsilon_2\sigma\lambda^5 = 5j(\rho_1 + \sigma\lambda^3).$$

From these equations we deduce

$$\epsilon_1 (\rho_1 + \sigma\lambda^3) = \epsilon_2 \left(\rho_1 + \frac{2}{5}\sigma + \frac{3}{5}\sigma\lambda^5\right),$$

$$\epsilon_2 = \frac{25j(\rho_1 + \sigma\lambda^3)^2}{(10\rho_1 + 4\sigma)(2\rho_1 + 5\sigma\lambda^3) - 18\rho_1\sigma\lambda^5}.$$

348. In his pages 294 and 295, Clairaut points out two limits for the ellipticity of a planet, assuming that the planet was originally fluid, and that the denser strata are the nearer to the centre. One limit is the ellipticity which corresponds to the case of a homogeneous mass. The other limit is that in which the attraction at any point is directed towards the centre, and varies inversely as the square of the distance from the centre, for this may be regarded as equivalent to having the density infinite at the centre: see Arts. 64 and 173.

Clairaut states on his page 296 that the theorem which we now call *Clairaut's Theorem*, holds for the case in which the earth is supposed to have been originally a fluid of the nature which he has considered. This is obvious from the demonstration already given: see Art. 336.

349. Clairaut's fifth Chapter occupies pages 296...304; it is on the comparison of the theory with observation.

Clairaut considers that the observations of the diminution of gravity in passing from the pole to the equator agree sufficiently well with his theory. But the comparison of the French and Lapland arcs gave the ellipticity apparently *greater* than $\frac{1}{230}$, whereas his theory required the ellipticity to be *less* than $\frac{1}{230}$. But, as he justly says, the comparison of these arcs was sufficient to establish the oblateness of the earth, but not to determine accurately the ratio of the axes; for the latter purpose the measurement of more distant degrees was required. He alludes to the operations in Peru, the result of which was now expected; this became known soon after the publication of Clairaut's work. The comparison of the French and Peruvian arcs would have given a smaller ellipticity, and therefore more favourable to Clairaut's Theory: see Boscovich *De Litteraria Expeditione*, page 501.

Some years later Clairaut made an attempt to explain the conflict between theory and observation as to the Figure of the Earth in an Essay which received a prize from the Academy of Toulouse; but this Essay seems never to have attracted any attention: I shall give some notice of it in Chapter XV.

350. Clairaut's work is one of the most interesting and remarkable in the literature of mixed mathematics. Laplace says in the *Mécanique Celeste*, Vol. v. page 7, after an analysis of the work:

L'importance de tous ces résultats et l'élégance avec laquelle ils sont présentés, placent cet ouvrage au rang des plus belles productions mathématiques.

In the Figure of the Earth no other person has accomplished so much as Clairaut; and the subject remains at present substantially as he left it, though the form is different. The splendid analysis which Laplace supplied adorned, but did not really alter, the theory which started from the creative hands of Clairaut.

Physical astronomy began with Newton in England; the memoirs which Maupertuis and Clairaut contributed to the *Philosophical Transactions* may be regarded as a graceful tribute to the country which gave birth to the greatest of scientific men. Newton, according to Bailly, reigned alone; but at his death, his empire, like that of Alexander, was divided: and Clairaut, D'Alembert and Euler succeeded. *Histoire de l'Astronomie Moderne*, Vol. III. page 154. Perhaps the names of Maclaurin and of Thomas Simpson ought to be recorded among the successors of Newton, but I fear it cannot be denied that on the whole his countrymen have left to foreigners the glory of continuing and extending his empire. England has produced numerous patient and able observers, but for the modern theory of physical astronomy we must chiefly study the great French writers, including among them two Italians, Lagrange and Plana, who in language have associated themselves with Laplace.

CHAPTER XII.

ARC OF THE MERIDIAN MEASURED IN PERU.

351. WE have seen in Chapter VII. that the expedition for measuring an arc of the meridian in Lapland started from Paris after that which went to Peru; nevertheless, the question as to the oblate or oblong form of the Earth was settled by the Arctic company before any result had been obtained at the Equator. In accordance with the plan of the present work, we might, therefore, leave the operations in Peru without further notice; but their extent and importance will justify us in devoting some space to a brief sketch of their course and conclusion.

352. It will be convenient to collect together the titles of the original works, accompanied with an indication of the nature of their contents. They will be arranged in the order of publication, and distinguished by Roman numerals, for the sake of easy reference.

I. La Condamine. *Relation abrégée d'un Voyage fait dans l'intérieur de l'Amérique Méridionale.* 8vo. Paris, 1745.

This gives an account of the voyage which La Condamine made down the river Amazon on his return home; it is a very interesting volume, but does not relate to our subject.

II. A Spanish translation of I., or of part of I., with some additions, seems to have been published at Amsterdam in 1745: see a note on page x. of XIII.; and also the life of La Condamine, by Biot, in the *Biographie Universelle*, republished in Biot's *Mélanges Scientifiques et Littéraires*, Vol. III.

III. La Condamine. *Lettre...sur l'Emeute populaire excitée en la Ville de Cuença......*

This seems to have been published at Paris in 1746 in octavo. It contains an account of a tumult at Cuença in 1739, which led to the death of Seniergues the surgeon of the French expedition. La Condamine encountered great trouble in carrying on the prosecution of the guilty persons.

IV. An English translation of I. was published at London in 1747. According to Biot, cited under II., there was also a Dutch translation.

V. Bouguer. In the Paris *Mémoires* for 1744, published in 1748, there is a memoir entitled *Relation abrégée du Voyage fait au Pérou....* The memoir occupies pages 249. .297 of the volume; it was read on the 14th of November, 1744. There is an account of the memoir on pages 35...40 of the historical portion of the volume.

The memoir consists of two parts. The first part relates to the voyage; and this is an abridgement of the introductory portion of IX. The second part is an outline of the operations described at full in the body of IX.

Bouguer is rather rash on his page 296; he made some observations with a common quadrant in 1738, and says: "je vis assez clairement que l'aplatissement alloit aussi loin que l'a prétendu ce grand homme [Newton]"... Thus he saw clearly what we now know did not exist. The passage does not appear to be reproduced in IX.

La Condamine was more cautious than Bouguer as to this matter. XIII. 63, XVIII. 64.

VI. Juan and Ulloa. *Relacion Historica del Viage a la America Meridional*...4 vols. 4to. Madrid, 1748. The first volume contains pages 1...404, besides Half-title, Frontispiece, Title, Preface, and Table of Contents. The second volume contains pages 405...682, besides Half-title and Title. The third volume contains pages 1...379, besides Half-title, Frontispiece, Title, Table of Contents, and Errata. The fourth volume contains pages 380...603 and i...cxcv, besides Half-title and Title. In the first and second volumes there are plates and maps, which are numbered from i. to xxi. continuously. In the third and fourth volumes there are plates and maps, which are numbered from i. to xii. continuously; and also a sheet containing the portraits of twenty-two emperors of Peru, beginning with Manco-Capac, the fabled child of the sun, and ending with Ferdinand the Sixth of Spain.

These four volumes give the account of the occupations of the two Spanish officers, and a description of the countries of Peru and Chili and of their inhabitants; they were drawn up by Ulloa. They form an interesting work, which, however, is very slightly connected with our subject.

VII. Juan and Ulloa. *Observaciones Astronomicas, y Phisicas*....4to. Madrid, 1748. Pp. xxviii + 396, besides Half-title, Frontispiece, Title, Preface, Table of Contents and Index. There are plates numbered continuously from i. to viii.; besides a map of the moon. This volume contains the detail of the geodetical and astronomical work, drawn up by Juan; it is an essential adjunct to VI., though copies of VI. are sometimes found without VII.

We will return to this volume: see Art. 362.

VIII. La Condamine. In the Paris *Mémoires* for 1745, published in 1749, there is a memoir entitled *Relation abrégée d'un voyage fait dans l'intérieur de l'Amérique Méridionale*... The memoir occupies pages 391...492 of the volume: it was read on the 28th of April, 1745. There is an account of the memoir on pages 63...73 of the historical portion of the volume.

This memoir agrees substantially with I.; but the two are not identical. A few passages occur in the memoir which are not in the book. Perhaps the book, which was published first, coincides with the discourse actually read to the Academy; and then the memoir received the slight additions before the volume for 1745 appeared. The memoir contains a plate which is not given in the book. This consists of a chart and a view of a remarkable part of the Amazon, where the river runs in a narrow channel between high rocks.

IX. Bouguer. *La Figure de la Terre...* 4to. Paris, 1749. Pp. cx + 394, besides Title, *Avertissement,* Table, and Errata.

This is the most elaborate work for our purpose to which the expedition gave rise; we will return to it: see Art. 363.

X. Bouguer. In the Paris *Mémoires* for 1746, published in 1751, there is a memoir entitled *Suite de la Relation abrégée, donnée en* 1744,... The memoir occupies pages 569...606 of the volume: it was read on the 18th of February, 1750.

This contains the geodetical measurements and the astronomical observations: it is an abridgement of the corresponding part of IX. to which Bouguer refers for full information.

XI. La Condamine. In the Paris *Mémoires* for 1746, published in 1751, there is a memoir entitled *Extrait des opérations Trigonométriques, et des observations Astronomiques...* The memoir occupies pages 618...688 of the volume: it was read on the 27th of May, 1750.

This is an abridgement of XII., which was just about to be published. La Condamine says: "J'ai usé du droit d'auteur en faisant mon extrait, et on y trouvera quelques particularités omises dans le livre même." These additions to XII., however, are small and not important.

XII. La Condamine. *Mesure des trois premiers degrés du Méridien...* 4to. Paris, 1751. Pp. 266 + x, besides Title, *Avertissement,* and Table.

This is La Condamine's account of the scientific operations. It is divided into two parts; the first part relates to the geodetical measurements, and the second to the astronomical observations. The pages 239...258 contain an important discussion of Picard's operations.

XIII. La Condamine. *Journal du Voyage fait par ordre du Roi...* 4to. Paris, 1751. Pp. xxxvi + 280 + xv, besides Title.

This is La Condamine's account of the voyage and the residence in Peru.

XIV. A French translation of VI. and VII. was published at Amsterdam and Leipsic in 1752, 2 vols. 4to. The first volume contains Frontispiece, Title, Dedication, Publisher's Advertisement, Preface, Table of Contents, and Errata, and then 554 pages of text. The second volume contains Frontispiece, Title, and Table, and then 316 pages of text, with an index for the history of Peru. This brings us to the end of the translation of VI.; and the remainder of the volume is devoted to VII.: this consists of Title, Preface, Table of Contents, 309 pages of text, and an Index. The translation has the same plates and maps as the original, except the sheet with the portraits of the emperors of Peru. The translation has in addition plans of Cape François and of Louisbourg; and also eight plates which are intended to illustrate the early history of Peru.

We learn from the Publisher's Advertisement that this translation was not allowed to be published at Paris.

The translation of VII. is very unsatisfactory; many passages are here perverted into absolute nonsense, which are quite intelligible in the Spanish original.

XV. There is an English translation of VI. I have not seen any edition except the third, which is dated 1772, and was published at London. This is in two octavo volumes. The first contains xxiv + 479 pages; the second contains 419 pages, besides the Title, Contents, and Index. There are plates and maps which are numbered from i. to vii. continuously; these reproduce on a small scale most of the illustrations of the original work.

The English translation omits the following portions of the original: the explanation of the construction and use of the sextant, Vol. I. pages 196...213; the description of the map of the western coast of South America, Vol. IV. pages 469...485; and the sketch of Peruvian history, Vol. IV. pages i...cxcv. Moreover, Ulloa in returning to Spain was taken prisoner by the English; and he complains of the barbarous treatment he received from those who captured him, Vol. IV. pages 447 and 517: these complaints are omitted in the English translation.

XVI. Bouguer. *Justification de plusieurs faits...* 4to. Paris, 1752. Pp. viii + 54, besides a double Title, and a leaf containing the *Approbation, Privilége du Roi*, and Errata.

This is an attack on La Condamine; it is of no scientific value, for it does not bear on any of the results obtained by the expedition, but only on trifling personal matters. For example, Bouguer's first twenty-one pages are spent on maintaining that the other Academicians were disposed to begin by measuring an arc of the Equator, before the orders from France were received which required them to confine themselves to an arc of the meridian. Even if Bouguer established this point, which is not certain, there cannot be any importance attached to it.

XVII. La Condamine. *Supplément au Journal Historique... Première Partie.* 4to. Paris, 1752. Pp. viii + 52, besides the Title and *Approbation.*

XVIII. La Condamine. *Supplément au Journal Historique... Seconde Partie.* 4to. Paris, 1754. Pp. 222 + xxviii, besides the Title, *Avertissement* and *Approbation.* There are also two pages containing supplements to the Errata for XII. and XIII.

In XVII. and XVIII. we have the reply of La Condamine to XVI.

XIX. Bouguer. *Lettre ... divers points d'Astronomie pratique...* 4to. Paris, 1754. Pp. 51 besides the Title and *Approbation.* This is a rejoinder to XVII. and XVIII.

XX. *Réponse de Monsieur *** à la Lettre de M. Bouguer, sur divers points...* Pp. 11.

I have seen only one copy of this publication; and that had no indication of date or place. It contains a page by an anonymous writer, which introduces a letter from La Condamine, constituting a rejoinder to XIX.

XXI. La Condamine. *Relation abrégée...* 8vo. Maestricht, 1778. Pp. xvi + 379, besides Title and *Approbation.*

This consists of a reprint of I. and II., augmented by two letters. One letter is from La Condamine, and contains a sketch of the fortunes of the members of the arctic and equatorial expeditions up to about 1773. The other letter is from one of the subordinate members of the equatorial expedition, Godin des

Odonais; it gives an account of the calamities which befell his wife on her return to Europe down the Amazon. In the *Quarterly Review*, Vol. 57, 1836, will be found a description of two modern voyages down the Amazon by English explorers, and also some notice of the sufferings of Madame Godin des Odonais.

XXII. There is a reprint of XVI. also in 4to. Paris, 1809. This is in rather smaller type than the original, and contains vi + 44 pages, besides the Title. I am at a loss to imagine what could have been the motive for reprinting this controversial piece so many years after all the persons concerned had passed away.

353. We will now give a brief account of the operations of the expedition, and the results obtained; we shall cite the pages of the original works from which our statements are derived.

The French expedition left Rochelle on the 16th of May, 1735, and arrived at Carthagena on the 16th of November; the two Spanish officers had already been waiting there for several months. The party reached Panama on the 29th of December. XIII. 3, 5, 8.

A base was measured during October, 1736, near Quito; the whole party was divided into two bands: one band measured from the north end to the south, and the other from the south end to the north. The difference between the two measurements was less than three inches in 6273 toises. XII. 5.

The geodetical angles were observed with quadrants. La Condamine's quadrant had a radius of three feet, Bouguer's about two feet and a half, Godin's not quite two feet; the Spanish officers, after their arrival in Peru, received from Paris a quadrant intermediate in size between Godin's and Bouguer's. IX. 60, XII. 13. There were two series of triangles; one was measured by Godin and Juan; the other was measured by Bouguer and Ulloa, and also by La Condamine. The two series had about half their triangles common; differing only towards the extremities of the arc. Thus three separate Trigonometrical measurements were obtained, each of which may be considered complete and independent. Every angle was observed; each person observing at least two angles of every triangle. XII. 12...15.

354. The geodetical work was carried on with great difficulty owing to the nature of the country. There was a narrow valley running nearly in the direction of the meridian, between two lofty chains of mountains. On the elevated points, which were chosen for stations, the observers suffered much from the inclemency of the weather; and they were often compelled to remain for several days or even weeks, to obtain a glimpse of the points for which they were looking, as these points were usually enveloped in mist. XIII. 52. More than once a report was current that the observers had perished. On the occasion of one very severe storm, to which they were exposed, public prayers were offered for them; or as La Condamine cautiously adds, "du moins on nous l'assura." VI. first Vol. 314, XIII. 81. The Indians caused much trouble by deserting in critical circumstances, and by incurable dishonesty. XIII. 50, 52, 72. The upper classes, who were of course Spaniards, at least by descent, seem to have received the expedition in general with politeness and kindness. XIII. 65, 75. But on the other hand we must place the tumult excited at Cuença, by which the French surgeon lost his life. Moreover, a frivolous charge of acting contrary to the orders of the king of Spain, was on one occasion brought against La Condamine; and on another occasion he was disturbed by a nocturnal visit of a police official. XIII. 26...30, 101. La Condamine seems to have had great trouble and anxiety respecting matters which ought not to have been thrown on a person fully occupied with his proper scientific work. He had at the commencement of the operations to undertake a voyage to Lima, in order to procure money for the expenses. XIII. 19...25. He had to engage in tedious proceedings at law in order to prosecute the persons who had caused the death of the surgeon. XIII. 86. He was also involved in a vexatious business connected with the erection of two pyramids to mark the extremities of the measured base, and with the inscriptions to be placed on them: these pyramids were finally destroyed by orders from Spain. La Condamine devotes a large space to the history of the pyramids, prefixing the motto "*Etiam periêre ruinæ.*" XIII. 219...280. According to the *Avertissement* to XIV. Ulloa was about to issue a history of the transactions relative to the pyramids: I do not know whether this ever appeared.

355. Near the South end of the arc a base of verification was measured. Bouguer and Ulloa measured it from South to North; La Condamine and Verguin, the draughtsman to the expedition, measured it from North to South. The two measures agreed within two inches in 5259 toises. Part of this base was measured across a shallow pool; the measuring rods floated on the surface. The calculated length of the base of verification differed from the measured length by about a toise. XII. 72, 85; XIII. 83. Godin and Juan also measured a base of verification, near the South end of the arc, but not the same as that just noticed. VII. 165; XIII. 83.

356. The astronomical part of the operations was naturally the most difficult and the most important: we must now for a time fix our attention on Bouguer and La Condamine. Any sketch will give but a faint idea of the obstacles which had to be overcome, and of the assiduity of the observers; and, indeed, there is danger lest a sketch should contain or suggest some erroneous notions.

The star ϵ of Orion was selected for observation at both ends of the arc. At the North end this star crossed the meridian to the South of the zenith, and at the South end it crossed the meridian to the North of the zenith. Thus the two zenith distances had to be found, and their sum gave the amplitude of the arc.

A sector of 12 feet radius, with an arc of 30°, had been brought from France; this was used in some observations for determining the obliquity of the ecliptic at the early part of the residence in Peru. But the arc was far longer than was necessary for the zenith distance of the selected star; and so Bouguer and La Condamine substituted a new arc. The most remarkable circumstance connected with the new arc is, that it was *not graduated*. The zenith distance of the star was known approximately; an arc was taken nearly equal to this known value, and having its chord a certain submultiple of the radius: this arc was set off on the limb of the instrument. Then the difference between this arc and the actual zenith distance of the star was determined by the aid of a micrometer. XII. 108. Some dispute arose afterwards as to the person to whom the credit of this contrivance was due. XII. 120; XVI. 36; XVIII. 111.

357. Observations were made by Bouguer and La Condamine at Tarqui, the southern station, in December, 1739, and January, 1740; and at Cotchesqui, the northern station, in February, March, and April, 1740. They appear to have been at the time contented with their results, and to have considered the object of the expedition fulfilled. XII. 165.

I do not perceive any distinct statement of the causes which led Bouguer and La Condamine to suspect the accuracy of the astronomical observations of 1739 and 1740, and in consequence to postpone their return to Europe. Perhaps La Condamine was detained by the affairs of the death of the French surgeon, and of the pyramids. They naturally wished before they left Peru to compare their result with Godin's; and Godin had not yet arrived at his conclusions. XIII. 105. However, Bouguer made more observations at Tarqui; and towards the end of 1741 he announced to La Condamine that the work which they had imagined to have been finished more than a year since must still be continued for several months: the old observations at Tarqui were to be rejected because they differed so much from the more recent observations. XIII. 128.

The untrustworthiness of the early observations seems to have been due mainly to a want of rigidity in the whole instrument, composed of radius, limb, and telescope. One unfortunate circumstance, for example, was that the radius had been constructed in two pieces for facility of transport from France; and when the instrument was to be used, the screws could not be found which were to fasten the two parts firmly together. XVIII. 42. The necessary rigidity was finally secured by the aid of strengthening bars and wire. But even after his return to France La Condamine considered that the matter was not fully explained. XVIII. 73.

It is obvious also that the optical defects of the telescope gave great trouble. The single object-glass could not bring all the different coloured rays to the same focus; and thus in the use of the micrometer there was an opening for serious error. Both

Bouguer and La Condamine treat at length on this matter, but not with perfect clearness. IX. 202...214; XII. 196...215.

358. Finally, simultaneous observations of the star were made by La Condamine at Tarqui and by Bouguer at Cotchesqui, towards the end of 1742 and the beginning of 1743. Those by Bouguer were made with a new sector of 8 feet radius, constructed under his own direction and inspection. Those by La Condamine were made with the 12 feet sector, improved successively by Bouguer and himself. XII. 185, 190. By taking simultaneous observations, the corrections for precession, nutation, and aberration, were rendered unnecessary. The aberration of light was known, but not the laws of the correction which it involved for observations of the stars. XII. 139, 220.

The amplitude of the arc was found to be about 3° 7′ 1″. La Condamine obtains 56749 toises for the length of the first degree of the meridian reduced to the level of the sea. XII. 229. Bouguer gives 56753 toises. IX. 275. Delambre recalculated the astronomical work of Bouguer and La Condamine; and fixed the amplitude at 3° 7′ 3″ He took a mean between the lengths assigned by Bouguer and La Condamine, and thus obtained, for the length of the arc reduced to the level of the sea, 176877 toises. See *Base du Système Métrique,...* Vol. III. page 133. The corresponding length of a degree is about 56737 toises.

359. We stated at the commencement of Article 356 that we confined ourselves to the proceedings of Bouguer and La Condamine. Let us now advert to the other members of the party.

Godin himself published no account of his operations; nor have I ever seen any reference to manuscripts which he may have left. Much of his work, however, was executed in association with Juan; and there is good reason to conclude that his results must have agreed substantially with those of Bouguer and La Condamine. XII. 231; XIII. 140.

The arc on which the Spanish result depends fell rather short of the arc of Bouguer and La Condamine at the southern end, but went beyond it at the northern end. The extension of the arc northwards introduced five new triangles; Juan and Ulloa were both concerned in this extension, and I presume that Godin also was with them. VII. 167, 224; XII. 231. The details connected with the triangles as observed and calculated both by Juan and Ulloa are recorded. VII. 144, 214.

For the astronomical work Godin constructed a very large sector; this is said in various places to have had a radius of 20 feet: but La Condamine correcting his former statements put it ultimately at 18 feet. VII. 272; IX. 273; XIII. 85, 99; XVI. 38; XVIII. 77.

Observations of three stars, ϵ of Orion, θ of Antinous, and α of Aquarius were made at Cuença, the southern station, in August and September, 1740, by Godin, Juan, and Ulloa. The Spanish observers were then withdrawn from their scientific occupations, and employed in the naval service, to assist in defending the country against the expected attacks of the English. Hence the observations at Pueblo Viejo, the northern station, were not made by them until April and May, 1744; Godin did not assist at these. VII. 283. The amplitude of the arc was finally settled at $3^\circ\ 26'\ 52\frac{3}{4}''$.

We do not see in the Spanish account anything corresponding to the excessive trouble which Bouguer and La Condamine experienced in their astronomical observations; we learn little more than this, that the first large sector which was made was unsatisfactory, and so another was made. VII. 271.

The Spanish result gave 56768 toises for the length of the degree of the meridian. VII. 295; XII. 234.

360. Bouguer arrived in Paris towards the end of June, 1744, about eight months before La Condamine. XIII 215. A violent controversy subsequently arose between them; and this leads us to enquire on what terms the Academicians had been during their operations. Godin separated himself from the other two in Peru. XVIII. 43. Bouguer seems to have been displeased at this, but La Condamine does not record any disapprobation. IX. 228; XII. 106: see also XVIII. 6.

La Condamine asserts that he had been on good terms with Bouguer during the ten years of the expedition, and for three years afterwards. XVII. iii: see also XVII. 28, 30; XVIII. 180, 203, 206. But on the other hand there are statements which imply that there must have been a want of perfect cordiality between these two, even in Peru. XVIII. 6, 62, 64, 143, 175, 182; XIX. 18, 49. Each of them claims to have been on good terms with Godin. XVI. 39; XVIII. 43; XIX. 38. The date of the public explosion is November 1748; the cause was the charge made by Bouguer, that his colleagues were inclined to measure a degree of the equator, instead of a degree of the meridian, until arrested by orders from France. XVIII. 67, 212. The strife extends over the series of works XVI....XX.; but even these seem to have formed but a small portion of the statements, verbal and written, which were brought before the Paris Academy. There was scarcely any exaggeration in La Condamine's complaint, that ten years of labour in the new world were followed by as many of controversy in the old. XVIII. iii, 190. The quarrel seems to me remarkable, alike for its fierceness and for the triviality of the matters in dispute. Thus, besides the measuring of an arc of the equator, to which we have already alluded, there is much contention as to the origin and value of certain suggestions in optics and practical astronomy. A sketch of the history of the quarrel, followed by a summary of the main points, is given in XVIII. 205...221. My own sympathy is on the side of La Condamine, although I consider Bouguer to have been by far the superior as a mathematician and an astronomer.

361. We may give a cursory notice to some miscellaneous points.

The equatorial expedition was suggested by Godin; see IX. iv, and Bailly's *Histoire de l'Astronomie Moderne,* Vol. III. page 11, note. Godin seems to have proposed it to the Paris Academy in 1734; but even in 1733 La Condamine had offered to measure degrees near the equator at Cayenne. IX. iv; XVII. 28; XVIII. 190. When La Condamine, on his return home, arrived at Guyana, he came to the conclusion that the country was well adapted for trigonometrical operations. XXI. 188; XIII. 194, 201. And at a later period he bitterly regretted that his original design had not been carried out, and then he would not have lost the ten most precious years of his life in preparing vexations for ten more. XVIII. 190.

Spherical trigonometry was now employed, apparently for the first time, in geodetical calculations; this improvement is claimed by Bouguer. IX. 131; X. 584; VII. 255.

To Bouguer is also due the idea of making observations with the view of determining the attraction of the mountain Chimborazo; La Condamine contributed a valuable suggestion in the practical operation. XVIII. 146.

We shall now give more details respecting the works VII. and IX.

362. The Spanish volume of observations and experiments begins with a Preliminary Discourse, which consists of a history of opinions and investigations with respect to the Figure of the Earth.

After having explained the views of Newton and Huygens, which involved the hypothesis of the rotation of the Earth, Juan says:

Assi discurrian estos grandes ingenios en la Hypotesis del movimiento diurno de la Tierra; pero aunque esta Hypothesis sea falsa, . . .

The French translation supplies the following significant note:

On doit se souvenir que l'Auteur de cet Ouvrage, ne parle pas en Mathématicien quand il suppose faux le sentiment de ceux qui affirment que la Terre tourne, mais en Homme qui écrit en *Espagne*, c'est-à-dire dans un Pays où il y a une Inquisition.

The volume is divided into nine books, which treat on the following subjects: the obliquity of the ecliptic, observations of latitude, observations of longitude, expansion and contraction of metals, barometrical experiments, the velocity of sound, the length of the degree of the meridian, pendulum experiments, navigation on the surface of the oblatum.

Juan holds that the Earth is an oblatum, and that the anomalies which seem to occur may fairly be attributed to errors of observation.

In order to obtain the ellipticity of the Earth, Juan assumes that in passing from the Pole to the Equator the seconds pendulum increases 2·16 lines. Hence by using Clairaut's theorem he obtains $\frac{1}{265}$ for the ellipticity. See his page 334. The 2·16 lines is, I presume, an arbitrary value; for although it would appear from his page 344, that this is in exact conformity with the observations of Maupertuis in Lapland, yet this must be a misprint, as we see by page 331.

An investigation is given on pages 337...345 for the rectification of the ellipse. Two infinite series are obtained, one for the length of an arc measured from the end of the minor axis, and the other for the length of an arc measured from the end of the major axis; the former is nearly correct, the latter very much less so. The mathematical process is rather clumsy; for to expand $\frac{1}{(1-x^2)^{\frac{1}{2}}}$ in powers of x, Juan in effect expands $(1-x^2)^{\frac{1}{2}}$, and then divides unity by the series; instead of simply expanding $(1-x^2)^{-\frac{1}{2}}$. To ensure tolerably rapid convergency, Juan proposes to calculate the arc from the end of the minor axis up to a certain point by his first formula, and the arc from this point to end of the major axis by his second formula. However he finally in his numerical work retains only what we should call the square of the excentricity, and it is easy to see that to this order of accuracy he might have avoided infinite series altogether, and expressed his required result in a simple finite form.

In treating on navigation Juan refers to a work by Murdoch, of which we shall give some account hereafter. Juan supplies tables of *Meridional Parts*, like Murdoch's, but much more copious, as they are calculated to every minute instead of to every degree. Juan adopts in his Tables $\frac{1}{266}$ for the ellipticity.

363. Let us now turn to Bouguer's *Figure de la Terre.*

The cx. preliminary pages give an account of the voyage and a description of the physical peculiarities of Peru, and the character of the inhabitants.

The 394 pages of text are divided into seven sections.

The first section is mainly devoted to shewing that it was advisable to determine the length of a degree of the meridian rather than the length of a degree of the equator.

The second section gives an account of the triangles, including the measurement of the base.

The third section treats of the reduction of the triangles to the plane of the horizon, and the determination of the situation of the sides with respect to the meridian.

The fourth section relates to the precautions taken with respect to the astronomical observations.

The fifth section contains the astronomical observations. The pages 227...258, however, do not belong to our subject; they relate to the observations for determining the obliquity of the ecliptic which were made during the early part of the residence in Peru.

The sixth section is thus entitled: *Qui contient diverses recherches sur la Figure de la Terre et sur les proprietés de cette Figure.*

The investigations of this section are interesting, though rather speculative than practical.

Bouguer considers the curve which represents the meridian of the Earth as unknown; but from this curve he supposes another deduced by the perpetual intersection of the normals, and he calls the deduced curve the *gravicentrique:* it is the *evolute* of the meridian curve in the language of modern mathematics.

Bouguer investigates properties of the *gravicentrique* on the supposition that the length of it measured from the equator varies as the mth power of the sine of the latitude. He specially considers the cases in which $m=2$, $m=3$, and $m=4$: see his pages 284...289. The law for the length of the *gravicentrique* is also the law for the increase of the radius of curvature of the meridian in passing from the equator to the pole.

The results of observation which had to be satisfied were the lengths of a degree of the meridian in Peru, France, and Lapland. Bouguer at first adopted the usual hypothesis of $m=2$, and obtained $\frac{1}{223}$ for the ellipticity: see his page 297. But after the French degree had been corrected, this hypothesis did not seem to him to agree with the observations; accordingly he supposed $m=4$, and obtained $\frac{1}{179}$ for the ellipticity: see his page 303. Besides the three degrees of the meridian, he also pays attention to the degree of longitude which had been measured towards the South of France.

Bouguer's hypothesis of $m=4$ is quite arbitrary. It had, however, sufficient vitality to experience the adverse criticism of Laplace, who shews that it is inconsistent with pendulum observations. *Mécanique Céleste*, Livre III. § 33.

Bouguer in his pages 319...326 explains the nature of the changes which must be made in certain tables constructed for navigation, on the hypothesis that the Earth is spherical, in order to adjust them to the actual fact.

Bouguer's seventh section is entitled *Détail des Expériences ou Observations sur la gravitation, avec des remarques sur les causes de la Figure de la Terre.*

This section contains some very interesting matter, although there is nothing as to what we usually understand by the theory of the Figure of the Earth. Bouguer says on his page 327:

> Nous n'entreprendrons point de nous élever jusqu'à une Théorie complette de la Figure de la Terre; parce que nous ne voulons rien donner s'il est possible à nos conjectures.

Bouguer describes the way in which he made his pendulum experiments; and then considers what reductions must be applied to the immediate results. He allows for the diminution of the weight of the pendulum caused by the air which it displaces; he says that this correction is now made for the first time: see his page 340. He adverts to the effect of the resistance of the air; and he states as a result which could be obtained by investigation, that the time occupied in the ascending part of an oscillation will be diminished as much as the time occupied in the descending part is increased. This we find established in the modern works on Dynamics: see Poisson's *Traité de Mécanique*, Vol. I. pages 348...361.

Bouguer treats on the diminution of attraction at different heights above the level of the sea. He finds that on a mountain at the height h above the level of the sea, the attraction is proportional to $(r - 2h)\,\Delta + \frac{3}{2}h\delta$, where r is the Earth's radius, Δ the Earth's mean density, and δ the density of the mountain. This is the first appearance of the formula, which has now passed into elementary books; see *Statics*, Art. 219.

On pages 364...394 we have an account of the observations made by Bouguer and La Condamine to determine the attraction of the mountain Chimborazo. A deviation of about $7\frac{1}{2}''$ in the situation of the plumb-line seemed to be produced; but this was much less than might have been expected. The mountain therefore must contain great cavities, or be composed of materials of comparatively small density. It is plain, however, from the account that the observations were scarcely adequate to settle the matter; nor does Bouguer himself appear to lay much stress on them.

The work of Bouguer exhibits some tendency towards unnecessary speculative refinements, and will require careful attention in order to master its complexity; but nevertheless, both on practical and theoretical grounds, it may be justly considered the most important of all which the Peruvian expedition occasioned, and as that which should be selected by a student who desires to confine himself to one of the original accounts.

364. If we consider the whole transaction we shall have abundant reason to commend the patience and devotion which the history of the expedition clearly manifests. Ten years of exile from Paris, for a Frenchman and an Academician, formed a costly sacrifice to science; and in this case the exile was aggravated by incessant labour, anxiety, and suffering. The result remains to this day one of the principal elements in the numerical facts of the subject; and while we must be grateful to the two who mainly obtained it, we may pardon them if by contests which harassed only themselves they shewed how easy it is for human infirmity to tarnish the noblest names and the brightest deeds.

CHAPTER XIII.

D'ALEMBERT.

365. THE subjects of Attraction and the Figure of the Earth engaged much of the attention of D'Alembert: in the present Chapter and a subsequent Chapter we shall consider his researches in order.

We begin with his *Traité de l'équilibre et du mouvement des fluides*. The first edition was published in 1744; the second in 1770: both are in quarto. The first edition has a Preface which occupies xxxii pages, including the Title-leaf; then a *Table des Titres;* then the text of 458 pages, followed by a page of Corrections. The second edition has an *Avertissement* which occupies a page, followed by a reprint of the preface to the first edition, and a *Table des Titres*; then the text of 476 pages. The text of the second edition is a reprint of that of the first, with some additions which furnish references to researches made by D'Alembert since the publication of the first edition of the work.

366. The only part of the edition of 1744 which directly concerns us is a section on pages 47...51, entitled *De l'équilibre des Fluides, dont la surface supérieure est Courbe.* D'Alembert says that this matter is important on account of its connexion with the question of the Figure of the Earth. Huygens had taken for the principle of equilibrium the perpendicularity of gravity at the surface. Newton used the principle of the equilibrium of central columns. Bouguer and Maupertuis shewed that both principles must hold for equilibrium. Clairaut had used the principle of canals; and had also shewn that the thickness of a level film must be inversely proportional to the resultant force at the point.

It will be seen that in this brief sketch D'Alembert names Huygens *before* Newton: see Art. 65.

367. After his brief sketch of the history of the theory of fluid equilibrium D'Alembert says on his page 48:

Les différentes Loix d'équilibre, découvertes par les Savans Geométres que nous venons de citer, paroissent être les seules auxquelles nous devions nous arrêter pour le présent, jusqu'à ce que l'Expérience, ou une connoissance plus parfaite de la nature des Fluides nous ait persuadé qu'il n'y en a point d'autres, ou peut-être nous en fasse découvrir d'autres.

It will be seen from this extract that D'Alembert knew that certain conditions were *necessary* for fluid equilibrium, but did not know what conditions were *sufficient*. He proceeds to offer certain conjectures which we now know to be inadmissible. He seems half inclined to believe that when fluid is in equilibrium the bounding surface must be plane or spherical, and the resultant force constant at all points of the surface.

D'Alembert says that one of the best methods of deciding the question, at least in part, would be to shew that the Figure of the Earth found by theory agrees with that found by actual measurement. He adds on his page 51:

......Car on ne sauroit douter que la Terre ne soit applatie vers les Pôles, après les opérations si exactes qui ont été faites au Nord, opérations confirmées par celle qu'a faite M. *Cassini de Thury* en 1740, et de laquelle il a conclu l'applatissement de la Terre, sans égard pour plusieurs mesures précedentes, d'où résultoit le contraire, et qu'apparemment il n'a pas cru assez exactes.

368. It will be convenient to notice here the additional remarks on our subject which occur in the edition of 1770; although we thus disturb the order of chronology.

On his page 36, D'Alembert objects to Clairaut's apparent belief that the laws of Hydrostatics required the denser strata of the Earth to be the nearer to the centre; D'Alembert refers to page 280 of Clairaut's work, and he might also have referred to other pages. See Art. 315.

The section which we have cited in Art. 366 is enlarged in the second edition. The names of Maclaurin and Daniel Bernoulli are mentioned as having in effect before Clairaut given the principle, that the fluid in any canal with its ends at the surface of the fluid must be in equilibrium. But D'Alembert allows that Clairaut was the first to develop the use of the principle. D'Alembert adds, with reference to Clairaut, on his page 50:

> Je crois au reste, que ce Savant s'est trompé, quand il a avancé que dans un Fluide hétérogène, les couches de différente densité devoient toutes être de niveau. Voyez à ce sujet l'art. 86 de mes *Recherches sur la cause des vents, et mon Essai sur la Résistance des Fluides*, art. 165, 166 et 167. Il est vrai que je me suis aussi trompé moi-même, en croyant que dans lé système de l'Attraction, les couches de la Terre pourroient n'être pas de niveau. C'est ce que le célèbre M. *de la Grange* a remarqué dans le second volume des Mémoires de la Société Royale des Sciences de Turin, et ce que je prouverai moi-même ailleurs plus en détail. Mais il n'en est pas moins vrai, que dans un grand nombre d'hypothèses, un Fluide peut être en équilibre, sans que les particules d'une même densité se trouvent nécessairement placées dans une couche de niveau. Quoi qu'il en soit, il est constant, suivant le Principe général dont on vient de parler [the principle of canals], que chaque couche de niveau doit être également pressée en tous ses points; et qu'ainsi l'épaisseur en chaque point doit être en raison inverse du produit de la densité par la pesanteur.

See Art. 315. D'Alembert in fact admitted his error in 1768: see his *Opuscules Mathématiques*, Vol. v. page 2. I have not found where he returns to the subject after 1770, as we might expect he would from his words above, "je prouverai moi-même..." Perhaps it really refers to what he gave in the fifth volume of the *Opuscules Mathématiques*, and was written before, though published after, that volume. There is another memoir on Fluids in the eighth volume of the *Opuscules Mathématiques*, but it does not seem to bear on this point.

369. I will notice some matters of interest which have presented themselves in reading the Articles 1...58 of D'Alembert's *Traité...des Fluides*.

In his Article 2 he criticises, and I think justly, a demonstration given by Newton, namely, the second case of Proposition 19 in the second Book of the *Principia.*

In his Article 13 there are some remarks to shew the insufficiency of two common demonstrations of the proposition that the resultant force at any point of the surface of a fluid in equilibrium must be perpendicular to the surface at that point.

The first demonstration stands thus: if the force be not perpendicular the tangential component will tend to move the particle on which it acts, and the fluid will, as it were, descend an inclined plane. D'Alembert objects that a set of equal balls might be placed, one above the other, and be in equilibrium on an inclined plane; so that if a fluid be composed of such particles it would appear that the fluid might be in equilibrium with its upper surface inclined to the horizon instead of being horizontal.

The second demonstration rests on the assumption that for equilibrium the centre of gravity should be as low as possible. D'Alembert brings forward two exceptions; in one the centre of gravity is at a maximum height, and in the other some forces act besides gravity. Thus in fact D'Alembert's objections hold against the improper extension of a certain theorem, and not against the proper enunciation of the theorem. See *Statics,* Chapter XIV.

A remark made by D'Alembert in his Article 18 deserves, I think, the attention of modern elementary writers. Suppose we have a conical vessel and a cylindrical vessel with equal bases; let them be filled with water to the same height: then the pressures on the bases will be equal. A popular mode of establishing this proposition amounts to taking the cylindrical vessel with its water, and then supposing a certain part to become solid, so as to leave a conical interior of fluid. D'Alembert says in substance that we ought not to assume that the pressure is unaltered by this solidification of part of the fluid: for suppose we solidify a complete horizontal lamina of the fluid, we can thus in effect remove from the base the pressure of all the fluid above this lamina.

I observe some modern writers adopt the reverse order; they begin with the conical vessel and afterwards dissolve the sides, instead of beginning with the cylindrical vessel and solidifying: but it may be fairly doubted if the process is more satisfactory in this way.

D'Alembert's Article 26 calls for some observations. We will give an account of his investigation in modern language.

Let a mass of fluid be acted on by a force the *direction* of which is constant, but not necessarily the *intensity*. Take the axis of x parallel to this fixed direction; let X denote the force at the distance x from the origin, p the pressure there, and ρ the density. We have then, as is well known,

$$\frac{dp}{dx} = \rho X;$$

therefore $$p = \int \rho X dx = \psi(x) \text{ say.}$$

Suppose the fluid to be enclosed in a vessel of any shape, the ends being plane figures at right angles to the axis of x. Take $\psi(x)$ such that it vanishes at one end. If $\psi(x)$ is such that it vanishes also at the other end, and is never negative, the ends may be removed without destroying the equilibrium: this is obvious. But if $\psi(x)$ can become negative, equilibrium will not hold when the ends are removed: this is also obvious. Suppose then the ends to remain.

D'Alembert says that the pressure at the end for which $\psi(x)$ vanishes will be numerically equal to the greatest negative value of $\psi(x)$. This is inaccurate. The pressure cannot indeed be less than this, but may be as much greater as we please. In fact we may take $p = C + \psi(x)$, where C is an arbitrary constant: and provided C be large enough to ensure that p is always positive, equilibrium will subsist.

The value of the pressure at the other end will then be determined by ascribing the proper value to x in the expression $C + \psi(x)$: but D'Alembert seems to say that the pressure will be $\psi(x)$.

370. The next work by D'Alembert which we have to examine is his *Réflexions sur la Cause générale des Vents.* This work

was published in 1747; it gained the prize proposed by the Berlin Academy for 1746. The work is in quarto. There is a Title-page, a Dedication, and an *Avertissement;* an Introduction of xxviii pages; then 194 pages which contain a French translation of the original essay with some additions; and lastly, 138 pages which contain the original essay in Latin. In our remarks we shall confine ourselves to the French translation.

371. The dedication is to Frederic, called the Great; and is in the usual adulatory strain of these objectionable compositions.

The introduction gives a general account of the contents of the essay, intended for the use of readers with little mathematical knowledge. Two sentences are of sufficient interest to be reproduced.

One sentence offers a curious reason for referring the winds to the action of the Sun and the Moon; it occurs on page ii. After stating that the ebb and flow of the tide are admitted to be due to this action, D'Alembert says:

......Quel que soit le principe de cette action, il est incontestable que pour se transmettre jusqu'à l'Ocean, elle doit traverser auparavant la masse d'air dont il est environné, et que par conséquent elle doit mouvoir les parties qui composent cette masse.

The other sentence relates to the difficulty which the Cartesians found in admitting that the attraction of the Sun or of the Moon could produce high water simultaneously on the meridian under the attracting body, and on the opposite meridian. D'Alembert says, with zeal amounting to anger, on his page x:

...... La preuve simple et facile que je viens de donner du contraire, sans figure et sans calcul, anéantira peut-être enfin pour toujours une objection aussi frivole, qui est pourtant une des principales de cette Secte contre la Théorie de la gravitation universelle.

372. In the work itself we first notice pages 11...17. These contain an approximate solution of what we may call a companion to Huygens's problem. D'Alembert enunciates it in the most general form, namely, where the attractive force is any function of the distance from a fixed point; but in his solution he finds it sufficient to take the force constant. See Arts. 55, 56, and 173. Let ω denote the angular velocity, f the constant central force, c the radius of the sphere which the fluid would form if there were no rotation; then assuming that $\omega^2 c$ is small compared with f, the surface will be a spheroid, and the equation to the generating curve will be

$$r = c - \frac{\omega^2 c^2}{3f} + \frac{\omega^2 c^2}{2f} \sin^2 \theta,$$

where r is the radius vector from the centre of force, and θ is the angle which r makes with the axis of revolution. This result may be easily deduced from that given in Art. 55. D'Alembert himself solves the problem by what we should now call a method of Virtual Velocities.

D'Alembert finds the volume of the solid bounded by the spheroid, the sphere of radius c, and the double cone having its vertex at the common centre, and having the semi-vertical angle θ: see his page 15. The result in our notation is $\frac{2\pi\omega^2 c^4 \cos\theta \sin^2\theta}{3f}$; this may be easily verified. In this expression some of the volume is estimated negative if θ be so great that we get beyond the value for which the sphere and the spheroid intersect.

373. We have no concern with the discussions on the motion of a fluid, to which D'Alembert now proceeds, so that we pass on to pages 33...45 of his work.

D'Alembert determines the form of relative equilibrium of a thin layer of fluid spread over a solid spherical mass; taking the action of the fluid itself into account, and supposing uniform rotation.

D'Alembert requires the attraction of a homogeneous oblatum, which is nearly spherical, on a particle situated at any point of its surface. This he obtains by three steps.

(1) He quotes a theorem given by Maclaurin in his *Essay on the Tides*, by which the attraction on a particle at any point is known, when it is known for a particle at the pole and for a particle at the equator. See Art. 244.

(2) He has an approximate investigation for finding the attraction on a particle at the pole. This was originally given by Clairaut, but D'Alembert does not refer to him. See Art. 233.

(3) He has an approximate investigation for finding the attraction on a particle at the equator. He mentions Daniel Bernoulli in connexion with this; but the principle is the same as in the investigation for the particle at the pole, first given by Clairaut.

374. We will now furnish in modern language, and in our own notation, an equivalent to D'Alembert's process. Suppose s the radius, and σ the density of the central sphere, and ρ the density of the fluid. We may consider that there is an oblatum of density ρ, and also a sphere of density $\sigma - \rho$.

Let the ellipticity of the oblatum be ϵ, which is supposed small; let x and z be the coordinates of a point parallel respectively to the major and minor axes of the generating ellipse; then the attractions of the oblatum in these directions will be, by Art 261, respectively

$$\frac{4\pi\rho}{3}\left(1-\frac{2\epsilon}{5}\right)x \text{ and } \frac{4\pi\rho}{3}\left(1+\frac{4\epsilon}{5}\right)z.$$

Put the first in the form $\frac{4\pi\rho}{3}\left(1+\frac{4\epsilon}{5}\right)x-\frac{8\pi\rho\epsilon}{5}x$. Then on the whole we have a force *towards* the centre, the value of which is the product of the distance into $\frac{4\pi\rho}{3}\left(1+\frac{4\epsilon}{5}\right)$; together with the force $\frac{8\pi\rho\epsilon}{5}x$ parallel to the major axis outwards from the minor axis.

Thus we see that we can avail ourselves of the solution of the companion to Huygens's problem, provided we add $\frac{8\pi\rho\epsilon}{5}$ to the ω^2, and use the proper value of the central force. This central force at the distance r will be $\frac{4\pi\rho}{3}\left(1+\frac{4\epsilon}{5}\right)r+\frac{4\pi}{3}(\sigma-\rho)\frac{s^3}{r^2}$.

Hence, as by Art. 55 we have $\epsilon = \frac{j}{2}$, we now obtain

$$\epsilon = \frac{\frac{r}{2}\left(\frac{8\pi\rho\epsilon}{5} + \omega^2\right)}{\frac{4\pi\rho}{3}\left(1 + \frac{4\epsilon}{5}\right) r + \frac{4\pi}{3}(\sigma - \rho)\frac{s^3}{r^2}};$$

therefore
$$\epsilon = \frac{\frac{r\omega^2}{2}}{\frac{4\pi\rho}{3}\left(1 + \frac{4\epsilon}{5}\right) r + \frac{4\pi}{3}(\sigma - \rho)\frac{s^3}{r^2} - \frac{4\pi\rho r}{5}}.$$

For an approximation we reject $\frac{4\epsilon}{5}$ in comparison with unity in the denominator; and indeed our investigation is not accurate enough to justify us in retaining this term: thus

$$\epsilon = \frac{\frac{\omega^2}{2}}{\frac{8\pi}{15}\rho + \frac{4\pi}{3}(\sigma - \rho)\frac{s^3}{r^3}}$$

D'Alembert's own process is ruder and he has $\frac{s}{r}$ instead of our $\frac{s^3}{r^3}$ in our notation.

As yet we have not introduced the condition that the layer of fluid is thin; suppose it so thin that s may be taken equal to r in the denominator: thus

$$\epsilon = \frac{\frac{\omega^2}{2}}{\frac{4\pi\sigma}{3}\left(1 - \frac{3\rho}{5\sigma}\right)} = \frac{\eta}{1 - \frac{3\rho}{5\sigma}},$$

where η is what would be the ellipticity if the attraction of the fluid itself were entirely neglected.

375. On his page 40 D'Alembert proceeds to some remarks on the Figure of the Earth; for these he had prepared us on his page 10, saying, "...où je démontre plusieurs vérités fort paradoxes sur cette matiere." The remarks amount in substance to

the two obvious statements that the value just found for ϵ is very large if 3ρ is nearly equal to 5σ, and will be negative if 3ρ is greater than 5σ. If ϵ is not numerically small, our approximations do not hold. If ϵ is negative and numerically small our supposed oblatum is really an oblongum.

D'Alembert seems to consider it rather singular that an oblongum should be a possible form for the surface. See his page 41.

376. D'Alembert next considers the case in which the nucleus is not a sphere but an oblatum; the process is less satisfactory than that in Art. 374, because we have now to deal with the attraction of an oblatum on an external particle. Suppose, however, that the layer of fluid is very thin; let the ellipticity of the solid oblatum be small, and denote it by ϵ'. Then we see that we shall obtain an approximation to the required result by adding $\frac{8\pi}{5}(\sigma-\rho)\epsilon'$ to ω^2; so that

$$\epsilon = \frac{\frac{\omega^2}{2} + \frac{4\pi}{5}(\sigma-\rho)\epsilon'}{\frac{4\pi\sigma}{3}\left(1-\frac{3\rho}{5\sigma}\right)} = \frac{\eta + \frac{3}{5}\left(1-\frac{\rho}{\sigma}\right)\epsilon'}{1-\frac{3\rho}{5\sigma}}.$$

377. The result just obtained is one to which D'Alembert seems to have attached great importance. It must be observed, however, that it is only a particular case of a general formula given by Clairaut. Take the final result of Art. 323: in the integrals represented by A and D let the density be constant, and denote it by σ. Thus

$$A = \frac{\sigma r'^3}{3}, \quad D = \sigma\epsilon' r'^5;$$

therefore,

$$\epsilon_1(10\sigma r'^3 + 4\rho_1 r_1^3 - 10\rho_1 r'^3) = \frac{6}{r_1^2}(\sigma-\rho_1)\epsilon' r'^5 + 5j(\sigma r'^3 + \rho_1 r_1^3 - \rho_1 r'^3);$$

this is in fact given in Case II. of Art. 324. We have here then the more accurate form: if we now suppose that the difference between r' and r_1 may be neglected, we obtain

$$\epsilon_1(10\sigma - 6\rho_1) = 6\epsilon'(\sigma-\rho_1) + 5j\sigma,$$

which agrees with D'Alembert's result; it is more simple but less accurate than the immediately preceding form. D'Alembert himself subsequently obtained the more accurate form: see his *Recherches...Système du Monde,* Vol. III. page 225. Clairaut was content with somewhat less than he might have deduced from his own formula; see Art. 328.

378. The value of ϵ obtained in Art. 376 may be negative; it will be negative if the numerator is positive and $\frac{3\rho}{5\sigma}$ is greater than unity. D'Alembert says on his page 42,

... Donc si la Terre étoit un Sphéroide allongé, il ne seroit pas absolument nécessaire d'avoir recours pour expliquer ce Phenoméne, à un noyau intérieur allongé. Car il pourroit se faire que ce noyau fût applati, et que la Terre fût allongée vers les Pôles.

This remark is probably aimed at Clairaut; see Boscovich *De Litteraria Expeditione...* page 464: we have, however, shewn in Art. 326, that Clairaut might have drawn the same inference if he pleased. But Clairaut had a conviction of the propriety of assuming the Earth to be densest at the centre; and thus he would naturally neglect any hypothesis which was inconsistent with this conviction.

With respect to the formula of Art. 376, D'Alembert remarks that if $5\sigma - 3\rho = 0$, and also $\eta + \frac{3}{5}\left(1 - \frac{\rho}{\sigma}\right)\epsilon' = 0$, then ϵ may have any value we please, provided only it be small: he repeats this remark in his *Recherches ... Système du Monde,* Vol. III., page 190.

379. D'Alembert makes a statement at the top of his page 44 which I do not verify. He proposes to estimate the force on the fluid in the direction of a tangent at any point of the meridian of the nucleus. Let f denote the force to the centre, θ the angle between the axis and the radius vector to the point, then the required force is the product of $\sin\theta\cos\theta$ into

$$r\left\{\omega^2 + \frac{8\pi\rho}{5}\epsilon + \frac{8\pi(\sigma-\rho)}{5}\epsilon'\right\} - 2f\epsilon',$$

that is $$2f(\epsilon-\epsilon')\sin\theta\cos\theta,$$

that is $$2f\sin\theta\cos\theta\left\{\frac{\frac{r\omega^2}{2f}+\frac{3}{5}\left(1-\frac{\rho}{\sigma}\right)\epsilon'}{1-\frac{3\rho}{5\sigma}}-\epsilon'\right\},$$

that is $$\frac{\frac{r\omega^2}{2}-\frac{2f\epsilon'}{5}}{1-\frac{3\rho}{5\sigma}}\,2\sin\theta\cos\theta.$$

D'Alembert omits the term $\frac{2f\epsilon'}{5}$. In fact the force along the tangent must vanish if $\epsilon'=\epsilon$; but D'Alembert's expression would never allow it to vanish.

380. We proceed to pages 151...158 of the *Réflexions sur... Vents*, which contain some new and interesting matter relating to attractions. D'Alembert obtains, in effect, formulæ for determining the attraction at any point of the surface of an ellipsoid which is nearly spherical. He first *states* what the results are for points at the ends of the three axes; he does not give his investigation, which was probably of the kind which he attributed to Daniel Bernoulli: see Art. 373. Let the three semi-axes be r, $r-\beta$, $r-\gamma$, where β and γ are small: it is easy to shew by this method that the attraction at the end of the first axis is $\frac{4\pi r}{3}-\frac{8\pi}{15}(\beta+\gamma)$. If, for greater symmetry, we denote the semi-axes by $r-\alpha$, $r-\beta$, $r-\gamma$, where α, β, γ are small, the attraction at the end of the first axis is $\frac{4\pi}{3}(r-\alpha)-\frac{8\pi}{15}(\beta-\alpha+\gamma-\alpha)$, that is $\frac{4\pi}{3}\left(r-\frac{\alpha+2\beta+2\gamma}{5}\right)$. In order to express the attraction at *any point* of the surface, D'Alembert uses, in effect, the property that the attraction perpendicular to a principal plane of the ellipsoid varies as the distance from that plane. This, he says, follows from the principles given in Maclaurin's *Essay on the Tides*. Maclaurin himself did not explicitly go beyond the case of ellipsoids of revolution; but D'Alembert's extension was very obvious.

Let x, y, z be the coordinates of any point on the surface of the ellipsoid referred to the axes as axes of coordinates; let X, Y, Z be the attractions parallel to these axes: then

$$X = \frac{x}{r-\alpha} \frac{4\pi}{3} \left(r - \frac{\alpha + 2\beta + 2\gamma}{5}\right) = \frac{4\pi x}{3} \left(1 + \frac{4\alpha - 2\beta - 2\gamma}{5r}\right),$$

and similar expressions hold for Y and Z.

381. Then D'Alembert shews that an ellipsoid of homogeneous fluid, differing very little from a sphere, cannot be in equilibrium under its own attraction; in fact, the resultant force will not be at right angles to the free surface. D'Alembert's demonstration is laborious, but sound, if we use the correction of a mistake furnished by himself in his *Opuscules Mathématiques*, Vol. I. page 252. The modern method would be to form the condition which involves the direction cosines of the resultant force and of the normal to the surface. This condition is

$$X \div \frac{x}{(r-\alpha)^2} = Y \div \frac{y}{(r-\beta)^2} = Z \div \frac{z}{(r-\gamma)^2},$$

that is, approximately,

$$\frac{X}{x}\left(1 - \frac{2\alpha}{r}\right) = \frac{Y}{y}\left(1 - \frac{2\beta}{r}\right) = \frac{Z}{z}\left(1 - \frac{2\gamma}{r}\right)$$

This condition is not fulfilled.

D'Alembert some years later supposed that he had demonstrated the relative equilibrium of a rotating ellipsoid of fluid to be impossible; see his *Recherches...Système du Monde*, Vol. III. page 256: but he forgot that the so-called centrifugal force must also be considered. We know now by Jacobi's Theorem that such relative equilibrium is possible.

Further, D'Alembert's demonstration shews that a fluid ellipsoid *which is nearly spherical* cannot be in equilibrium under its own attraction; but it does not shew that this result holds for *every* ellipsoid. This is however the case; for in the demonstration of Jacobi's Theorem we shall find that the angular velocity has a value which cannot vanish.

382. On his page 156, D'Alembert proceeds to the case in which a solid homogeneous ellipsoid is surrounded by a thin stratum of fluid of different density in equilibrium. The mistake already referred to influences this investigation; and moreover D'Alembert misinterprets his results, and infers that if the solid part is a solid of revolution *it must be a sphere,* and that the density of the solid part must be *exactly* $\frac{3}{5}$ of the density of the fluid. This contradicts his own investigation in pages 40...44 of the work: see Art. 375. However, in his *Opuscules Mathématiques,* Vol. I. pages 253...255, he corrects his errors, and is more successful.

Let σ be the density of the solid, ρ the density of the fluid; let ϵ_1 and ϵ_2 be the ellipticities of the two principal sections of the solid, ζ_1 and ζ_2 the corresponding ellipticities of the two sections of the external fluid surface. D'Alembert obtains an approximate result which we may thus express

$$\frac{\epsilon_1}{\zeta_1} = \frac{\epsilon_2}{\zeta_2} = \frac{\sigma - \rho}{\frac{5}{3}\sigma - \rho}.$$

So far he is correct, but he adds that the solid figure and the external figure are *semblables,* which is not admissible: to make the figures *like* we should require $\frac{\epsilon_1}{\zeta_1}$ and $\frac{\epsilon_2}{\zeta_2}$ both to be equal to unity.

383. It will be instructive to notice the principle involved in D'Alembert's treatment of this problem: I will give it in substance though not in his form.

I use as before $r - \alpha$, $r - \beta$, $r - \gamma$ for the semi-axes of the external figure; and $r - \alpha'$, $r - \beta'$, $r - \gamma'$ for those of the solid part. We may then consider that we have a body with the former semi-axes, of the density ρ, and also a body with the latter semi-axes of the density $\sigma - \rho$.

For the former body we may take as before

$$X = \frac{4\pi\rho x}{3}\left(1 + \frac{4\alpha - 2\beta - 2\gamma}{5r}\right),$$

and similar expressions for Y and Z.

For the latter body we take

$$\frac{4\pi(\sigma-\rho)x}{3}\left(1+\frac{4\alpha'-2\beta'-2\gamma'}{5r}\right),$$

and two similar expressions. This amounts to supposing the second body enlarged in size until it just passes through the attracted point; that is in fact we introduce a thin ellipsoidal shell of density $\sigma-\rho$. But no sensible error is thus produced; for the action of this shell is in *amount* only of the first order; and is in *direction*, as we now know, accurately along the normal to its outer surface. Hence the shell would supply a force along the tangent plane to the fluid surface which would be only of the second order; and so for our purpose may be neglected. D'Alembert leaves his readers to think this point out for themselves, but in a later work he supplied a hint: see his *Opuscules Mathématiques*, Vol. VI. page 226.

Thus we take for the whole attraction parallel to the axis of x

$$\frac{4\pi x}{3}\left\{\rho\left(1+\frac{4\alpha-2\beta-2\gamma}{5r}\right)+(\sigma-\rho)\left(1+\frac{4\alpha'-2\beta'-2\gamma'}{5r}\right)\right\}$$

Call this X_1; and let Y_1 and Z_1 have similar meanings.

We know that for equilibrium we must have

$$\frac{X_1}{x}\left(1-\frac{2\alpha}{r}\right)=\frac{Y_1}{y}\left(1-\frac{2\beta}{r}\right)=\frac{Z_1}{z}\left(1-\frac{2\gamma}{r}\right)$$

This leads by easy reduction to

$$\frac{\alpha-\beta}{\alpha'-\beta'}=\frac{\alpha-\gamma}{\alpha'-\gamma'}=\frac{\sigma-\rho}{\frac{5}{3}\sigma-\rho}$$

D'Alembert then shews that if the whole mass revolve round one of the axes with uniform angular velocity relative equilibrium may subsist.

Take the axis of x as that of revolution; let ω be the angular velocity: then we must put $-\omega^2 y$ to what we called Y_1, and $-\omega^2 z$ to what we called Z_1. This will be found to lead to

$$(\alpha'-\beta')(\sigma-\rho)=\left(\frac{5}{3}\sigma-\rho\right)(\alpha-\beta)-\frac{5\omega^2}{8\pi},$$

and

$$(\alpha'-\gamma')(\sigma-\rho)=\left(\frac{5}{3}\sigma-\rho\right)(\alpha-\gamma)-\frac{5\omega^2}{8\pi}.$$

384. The next work by D'Alembert which we have to examine is his *Recherches sur la Précession des Equinoxes*...

This work was published in 1749; it is in quarto. The Title, Dedication and Introduction occupy xxxviii pages; then follows a table of Contents, and then the text of 184 pages.

There is a German translation of this work in octavo, by Dr G. K. Seuffert, published at Nürnberg, 1857.

385. We are concerned only with Chapter IX. of the work, which is entitled *Conséquences qui résultent de la Théorie précédente par rapport à la figure de la Terre;* this occupies pages 95...105.

By comparing his theory of Precession with observation, D'Alembert obtained the following numerical relation

$$\frac{\int_0^1 \rho \frac{d(r^5\epsilon)}{dr} dr}{\int_0^1 \rho \frac{dr^5}{dr} dr} = \frac{1}{324}.$$

The notation will be understood from what has been said before: see Art. 323.

This very important result remains almost unchanged in the modern theory; the fraction $\frac{1}{324}$ being replaced by 00326, which differs little from it: see Résal, *Traité Elémentaire de Mécanique Céleste*, page 226.

386. D'Alembert combines his own result with one given by Clairaut on his page 226: it is that which occurs in our Art. 327; denoting r_1 by unity, we may write it thus:

$$10\epsilon_1 \int_0^1 \rho \frac{dr^3}{dr} dr - 6 \int_0^1 \rho \frac{d(r^5\epsilon)}{dr} dr = 5j \int_0^1 \rho \frac{dr^3}{dr} dr \;\ldots\ldots\ldots (1).$$

Now D'Alembert, relying on the measures in Lapland and Peru, takes $\epsilon_1 = \frac{1}{174}$; and so the result in Art. 385 may be written thus:

$$\int_0^1 \rho \frac{d(r^5\epsilon)}{dr} dr = \epsilon_1 \frac{174}{324} \int_0^1 \rho \frac{dr^5}{dr} dr \;\ldots\ldots\ldots\ldots\ldots\ldots (2).$$

Assume $$\int_0^1 \rho \frac{dr^5}{dr} dr = k \int_0^1 \rho \frac{dr^3}{dr} dr \ldots\ldots\ldots\ldots\ldots\ldots (3).$$

Then from (1), (2), and (3) we obtain $\epsilon_1 = \dfrac{5j}{10 - \dfrac{174 \times 6k}{324}}$.

Now we shall shew presently that k is less than $\frac{5}{3}$; so that ϵ_1 is less than $\dfrac{5j}{10 - \dfrac{1740}{324}}$, that is less than $\dfrac{\frac{5}{289}}{10 - \dfrac{1740}{324}}$. This he says makes ϵ_1 less than $\frac{1}{256}$, which is inconsistent with the value $\epsilon_1 = \frac{1}{174}$, given by observation.

Instead of 256 we might put 267.

Thus D'Alembert infers that the Earth cannot be composed of solid elliptic strata, which is the hypothesis on which the result quoted from Clairaut was obtained. We know now that ϵ_1 cannot be so great as $\frac{1}{174}$; and thus the contradiction which D'Alembert points out no longer exists.

387. We shall now shew, as we have stated, that k is less than $\frac{5}{3}$. We have to shew that

$$3 \int_0^1 \rho \frac{dr^5}{dr} dr \text{ is less than } 5 \int_0^1 \rho \frac{dr^3}{dr} dr,$$

where the symbols denote positive quantities. D'Alembert spreads the demonstration over six pages. He makes three cases; that in which ρ always decreases as r increases from 0 to 1, that in which ρ always increases, and that in which ρ sometimes decreases and sometimes increases. But the required result can be obtained instantaneously. We have to shew that

$$\int_0^1 \rho r^4 dr \text{ is less than } \int_0^1 \rho r^2 dr,$$

or that $$\int_0^1 \rho r^2 (r^2 - 1)\, dr \text{ is negative};$$

and this is obvious, for every element of the last integral is negative.

388. We may also shew that if ρ always decreases as r increases from 0 to 1, then

$$\int_0^1 \rho \frac{dr^5}{dr} dr \text{ is less than } \int_0^1 \rho \frac{dr^3}{dr} dr.$$

Integrate by parts: let ρ_1 be the value of ρ at the surface. Then we have to shew that

$$\rho_1 - \int_0^1 \frac{d\rho}{dr} r^5 dr \text{ is less than } \rho_1 - \int_0^1 \frac{d\rho}{dr} r^3 dr,$$

or that $$\int_0^1 \frac{d\rho}{dr} r^3 (1 - r^2)\, dr \text{ is negative};$$

and this is obvious, for $\frac{d\rho}{dr}$ is negative by supposition, so that every element of the last integral is negative.

389. D'Alembert's page 101 is not intelligible to me. I imagine he means to say that perhaps some person will be able to shew that if ρ increases constantly from the centre $\int_0^1 \rho \frac{dr^5}{dr} dr$ is less than $\left(\frac{5}{3} - \beta\right) \int_0^1 \rho \frac{dr^3}{dr} dr$, where β is some positive quantity. This we have shewn in Art. 388, where $\frac{5}{3} - \beta$ is equal to unity, so that $\beta = \frac{2}{3}$.

390. D'Alembert then considers on his pages 103...105, whether the facts and the theory will agree on the supposition that the Earth consists of a solid elliptic mass covered with a thin layer of fluid. We must observe that the layer here is to be of finite thickness though thin; the case of an *infinitesimal* layer was in fact that which was dismissed as untenable in Art. 386.

D'Alembert assumes without any adequate investigation that the action of the fluid on the solid will not affect the Precession. See on this point Résal, *Traité Elémentaire de Mécanique Céleste*, pages 353...356.

As in Art. 376, we have

$$\epsilon\left(1-\frac{3\rho}{5\sigma}\right)=\eta+\frac{3}{5}\left(1-\frac{\rho}{\sigma}\right)\epsilon';$$

here ϵ is the ellipticity of the exterior surface of the fluid, and ϵ' the ellipticity of the solid nucleus. Thus

$$\frac{1}{174}\left(1-\frac{3\rho}{5\sigma}\right)=\frac{1}{578}+\frac{3}{5}\left(1-\frac{\rho}{\sigma}\right)\epsilon';$$

therefore
$$\frac{\rho}{\sigma}=\frac{\frac{5}{3}\left(\frac{1}{174}-\frac{1}{578}-\frac{3}{5}\epsilon'\right)}{\frac{1}{174}-\epsilon'}.$$

If we take ϵ' less than $\frac{1}{256}$ we find $\frac{\rho}{\sigma}$ to be positive; the number $\frac{1}{256}$ is that which presented itself in Art. 386; but it appears to me quite arbitrary to introduce it here. D'Alembert, however, has no misgiving: see his page 105.

391. D'Alembert gives the following inequality on his page 99:

If x is a proper fraction, 2 is greater than $x^3(5-3x^2)$. He establishes it easily by taking the differential coefficient of $x^3(5-3x^2)$.

We can establish it by common Algebra. For

$$\begin{aligned}2-x^3(5-3x^2)&=2(1-x^5)-5x^3(1-x^2)\\&=(1-x)\{2(1+x+x^2+x^3+x^4)-5x^3(1+x)\}\\&=(1-x)\{2(1+x)(1-x^3)+x^2(2-x-x^2)\};\end{aligned}$$

this is necessarily positive.

The last expression may be put also as

$$(1-x)^2\{2(1+x)(1+x+x^2)+x^2(2+x)\},$$

that is as
$$(1-x)^2\{2+4x+6x^2+3x^3\}.$$

392. The next work by D'Alembert which we have to examine is his *Essai d'une Nouvelle Théorie de la Résistance des Fluides.*

This work was published in 1752; it is in quarto. The Title, Dedication, Introduction, and Title of Contents occupy xlvi pages; the text occupies 212 pages.

The work was composed in competition for a prize proposed by the Academy of Berlin. The Academy instead of awarding the prize requested the candidates to give supplements shewing the agreement of their theories with experiments. D'Alembert seems to have been not quite satisfied with this proceeding; he resolved to abstain from a new competition, and to publish his essay at once. He adds, on his page xl:

Je souhaite par l'intérêt que je prends à l'avancement des Sciences, que les Juges nommés par cette illustre Compagnie, et qui n'ont pas sans doute proposé cette question sans s'assurer si la solution en étoit possible, trouvent pleinement de quoi se satisfaire dans les Ouvrages qui leur seront envoyés pour le concours.

393. The second Chapter of the book is entitled *Principes généraux de l'équilibre des Fluides*; it occupies pages 13...18.

D'Alembert first adverts to the *principle of Canals;* he deduces Clairaut's condition with respect to *curved* canals from Maclaurin's with respect to *straight* canals. To a modern reader the principle seems sufficiently evident without any remark.

394. D'Alembert establishes an important result which can be best explained by the aid of the modern equations for fluid equilibrium. Confining ourselves for simplicity to the case of forces in one plane we have

$$\frac{dp}{dx} = \rho X, \qquad \frac{dp}{dy} = \rho Y;$$

from these it follows that

$$\frac{d}{dx}(\rho Y) = \frac{d}{dy}(\rho X):$$

D'Alembert demonstrates this condition; for the particular case in which ρ is constant it was already known, as we have seen in Art. 306. D'Alembert considers his own demonstration simpler than any which had yet been given.

D'Alembert himself does not use the symbol p or speak of the pressure of the fluid. It will however be interesting and instructive to give the essence of his investigation in modern language.

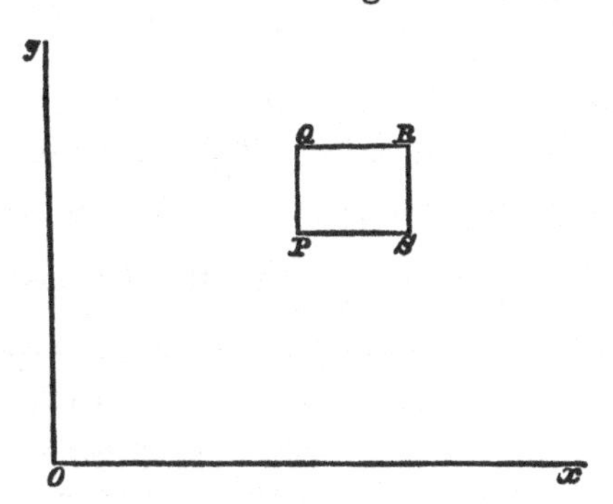

Let the coordinates of any point P be x and y; let the coordinates of an adjacent point R be $x+h$ and $y+k$. Complete the rectangle $PQRS$, having its sides parallel to the axes.

Let ρ be the density at P, let ρ_1 be the mean density along PQ, and ρ_2 the mean density along PS.

Let p be the pressure at P; then the pressure at Q will ultimately be $p+\rho_1 Yk$, and the pressure at S will ultimately be $p+\rho_2 Xh$. Now we may form two expressions for the pressure at R, one obtained by passing from Q to R, and the other obtained by passing from S to R. The former expression is ultimately

$$p+\rho_1 Yk+\rho_2 Xh+\frac{d}{dy}(\rho_2 Xh)k,$$

and the latter is

$$p+\rho_2 Xh+\rho_1 Yk+\frac{d}{dx}(\rho_1 Yk)h;$$

equate these and we obtain ultimately

$$\frac{d}{dx}(\rho_1 Y)=\frac{d}{dy}(\rho_2 X),$$

that is

$$\frac{d}{dx}(\rho Y)=\frac{d}{dy}(\rho X).$$

This mode of giving as it were a physical interpretation to the condition just obtained might be called D'Alembert's hydrostatical principle; though it is not very clearly put by himself. We may say verbally that the principle amounts to this: the change of pressure in passing from one given point of a fluid in equilibrium to another is independent of the path by which we proceed.

395. An Appendix entitled *Réflexions sur les loix de l'Equilibre des Fluides* occupies pages 190...212 of the work.

D'Alembert gives on his pages 190...194 another demonstration of the equation $\frac{d}{dx}(\rho Y) = \frac{d}{dy}(\rho X)$; this demonstration is sound but complex: he gives it, he says, because it will supply the opportunity for some important remarks on the laws of the equilibrium of fluids. The remarks do not seem to me of great importance; but the reader can judge for himself from the account which will now be given of them.

396. D'Alembert says on his page 195, in effect, that if with previous writers on this subject we suppose the density to be constant throughout every level surface we arrive at the equation $\frac{dY}{dx} = \frac{dX}{dy}$ instead of that in Art. 394: this appears to him to require explanation. Along a surface of equal density we have $\frac{d\rho}{dx}dx + \frac{d\rho}{dy}dy = 0$; if this surface is also a level surface we have $Xdx + Ydy = 0$; hence $Y\frac{d\rho}{dx} = X\frac{d\rho}{dy}$, and the equation of Art. 394 reduces to $\frac{dY}{dx} = \frac{dX}{dy}$. So far he is right, but he adds a remark which is quite erroneous; changing his notation to that which we have used, his words are:

Mais il faut remarquer que l'équation $\frac{dX}{dy} = \frac{dY}{dx}$ n'a lieu dans ce cas que pour les couches......auxquelles la direction de la pesanteur est perpendiculaire, au lieu que l'équation $\frac{d}{dx}(\rho Y) = \frac{d}{dy}(\rho X)$ a lieu généralement pour telle couche qu'on voudra...

This is a strange error: from the nature of the equation $\frac{dY}{dx} = \frac{dX}{dy}$ it is quite independent of direction.

397. D'Alembert says on his page 197, that the equation of Art. 394 supposes ρ, X, and Y to be functions of x and y: but he does not see why we should be restricted to this hypothesis. He proceeds to something which he considers more general, but which is really not so; in fact he supposes that X and Y are functions of x, y, and ζ, where ζ is itself a definite function of x and y: but it is obvious that this is practically identical with the usual hypothesis. I found after I had written this that Lagrange had made an equivalent remark in the *Miscellanea Taurinensia*, Vol. II. page 282. D'Alembert himself also subsequently admitted that this introduction of ζ was superfluous: see his *Opuscules Mathématiques*, Vol. VIII. page 16.

398. D'Alembert makes an erroneous statement on his page 199, namely, that if the pressure be equal at all points of the bounding surface the force must be equal at all points: we know that this is not necessarily the case. Indeed D'Alembert himself says on his page 201:

... A l'égard du principe de l'égalité des forces, il est évident que s'il étoit admis, toutes les Théories qu'on a données de la Figure de la Terre, en la considérant comme un Fluide, et en ayant égard à l'attraction des parties, et à la rotation de l'Axe, devroient être regardées comme fausses.

399. D'Alembert returns to the matter which we noticed in Art. 367; and seems still half persuaded of the truth of the absurd opinion stated there. However he converts himself from his error by the aid of an important principle which he had formerly given. The following is the substance of his argument: it is obvious that a fluid may be in motion without having its surface plane or spherical; and it follows from what we now call *D'Alembert's Principle* that if any motion is known we know also the forces which would maintain the system in equilibrium in the configuration which it has at any instant; thus forces do exist which would maintain a fluid in equilibrium and give to the surface a form which is neither plane nor spherical.

400. D'Alembert seems to attach great importance to the fact that if a fluid be in equilibrium the surfaces of equal density

are not necessarily level surfaces. We know now, with the usual notation, that if $Xdx + Ydy + Zdz$ is a perfect differential, the surfaces of equal density will be level surfaces; moreover for such forces as occur in nature this condition is satisfied: hence for such cases as occur in nature it is true that the surfaces of equal density are level surfaces. But D'Alembert's statement is correct, that surfaces of equal density are not necessarily level surfaces. See Arts. 315 and 368.

401. We will give briefly the example which D'Alembert discusses, translating his process into modern language.

Suppose s the distance of a point from the origin, and θ the angle which s makes with a fixed straight line. Let S denote the force along s, and T that at right angles to s; and let σ denote the density.

Then the usual equations for the equilibrium of a fluid are

$$\frac{dp}{ds} = \sigma S, \qquad \frac{dp}{sd\theta} = \sigma T,$$

where p denotes the pressure. Therefore

$$\frac{d}{d\theta}(\sigma S) = \frac{d}{ds}(\sigma s T) \quad \text{.........................} \quad (1).$$

This condition in fact agrees with what D'Alembert himself deduces from the principle of canals.

Now let us assume that the fluid is arranged in strata of equal density; let the curve of equal density be determined by the equation

$$s = r + \alpha\rho Z \quad \text{..............................} \quad (2),$$

where r is a parameter which particularises the curve we consider, ρ is a function of r, and Z a function of θ; and α is a very small quantity, the square of which will be neglected.

Also suppose that

$$-S = \rho' + \alpha\rho'' Z', \quad \text{and} \quad T = \alpha\rho''' Z'' \quad \text{..............} \quad (3),$$

where ρ', ρ'', and ρ''' are functions of r; and Z' and Z'' are functions of θ. The notation is kept very close to D'Alembert's, though not exactly the same.

Now (1) may be written

$$S\frac{d\sigma}{d\theta}+\sigma\frac{dS}{d\theta}=\sigma T+sT\frac{d\sigma}{ds}+\sigma s\frac{dT}{ds} \ldots\ldots\ldots\ldots\ldots(4).$$

The condition that σ is constant along the curves determined by (2) gives

$$\frac{d\sigma}{d\theta}+\frac{d\sigma}{ds}\frac{ds}{d\theta}=0,$$

that is,

$$\frac{d\sigma}{d\theta}+\alpha\rho\frac{dZ}{d\theta}\frac{d\sigma}{ds}=0.$$

Then (4) becomes

$$-\left(S\alpha\rho\frac{dZ}{d\theta}+sT\right)\frac{d\sigma}{ds}=\sigma T-\sigma\frac{dS}{d\theta}+\sigma s\frac{dT}{ds}.$$

Substitute from (3), and neglect the square of α; thus

$$\alpha\left(\rho\rho'\frac{dZ}{d\theta}-s\rho'''Z''\right)\frac{d\sigma}{ds}=\alpha\sigma\rho'''Z''-\sigma\frac{dS}{d\theta}+\alpha\sigma sZ''\frac{d\rho'''}{dr} \ldots\ldots\ldots(5).$$

Here $\frac{dS}{d\theta}$ means the differential coefficient of S with respect to θ, supposing s constant; and so it is found by combining

$$-\frac{dS}{d\theta}=\left(\frac{d\rho'}{dr}+\alpha Z'\frac{d\rho''}{dr}\right)\frac{dr}{d\theta}+\alpha\rho''\frac{dZ'}{d\theta},$$

and

$$0=\frac{ds}{d\theta}=\left(1+\alpha Z\frac{d\rho}{dr}\right)\frac{dr}{d\theta}+\alpha\rho\frac{dZ}{d\theta}.$$

Hence, neglecting the square of α,

$$-\frac{dS}{d\theta}=-\alpha\rho\frac{d\rho'}{dr}\frac{dZ}{d\theta}+\alpha\rho''\frac{dZ'}{d\theta};$$

also if we neglect the square of α we may put $\frac{d\sigma}{dr}$ for $\frac{d\sigma}{ds}$ in (5). Then, dividing by α, we obtain

$$\left(\rho\rho'\frac{dZ}{d\theta}-r\rho'''Z''\right)\frac{d\sigma}{dr}=\sigma\rho'''Z''-\sigma\rho\frac{d\rho'}{dr}\frac{dZ}{d\theta}$$
$$+\sigma\rho''\frac{dZ'}{d\theta}+\sigma rZ''\frac{d\rho'''}{dr} \ldots\ldots\ldots\ldots(6).$$

402. We will make some remarks on the equation (6). D'Alembert himself by transposition puts it in this form:

$$\frac{dZ}{d\theta}\frac{d}{dr}(\sigma\rho\rho') - Z''\frac{d}{dr}(\sigma r\rho''') = \sigma\rho''\frac{dZ'}{d\theta} + \sigma\rho'\frac{d\rho}{dr}\frac{dZ}{d\theta}.$$

D'Alembert obtains this result by the method which we have exemplified in Art. 394. In modern language we may say that he passes from one point of the fluid to another by two different routes; and thus he obtains two expressions for the change of pressure, which can be equated. But as he does not use the word *pressure*, or the symbol p, his method is somewhat obscure. In the diagram of Art. 394, we see that

the increase of pressure from P to Q + increase from Q to R
= increase from P to S + increase from S to R.

With D'Alembert the equivalent statement takes the less natural form,

the increase of pressure from Q to R − increase from P to S
= increase from S to R − increase from P to Q.

Instead of the words *increase of pressure from P to Q*, D'Alembert uses such words as *force of the column PQ along PQ*; and these seem scarcely intelligible. D'Alembert attempts to enunciate this case of his hydrostatical principle in words in his *Recherches...Systême du Monde*, Vol. III. page 226, where he says:

... il faut supposer la différence de pesanteur de deux couches de niveau infiniment proches, égale à la différence de pesanteur de deux couches verticales infiniment proches,...

An enunciation, partly in words and partly by symbols, is also given by Lagrange; see the *Miscellanea Taurinensia*, Vol. II. page 285.

We may remark that D'Alembert's notation might be rendered at once simpler and more general. Instead of ρZ, where ρ is a function of r and Z a function of θ, put V, where V is a function of both r and θ; also put V' instead of $\rho'' Z'$, and V'' instead of $\rho''' Z''$. Then the equation at the beginning of this Article may be written

$$\frac{dV}{d\theta}\frac{d}{dr}(\sigma\rho') - \frac{d}{dr}(\sigma r V'') = \sigma\frac{dV'}{d\theta}$$

In his *Opuscules Mathématiques,* Vol. v. page 6, D'Alembert returns to the example of Art. 401. There he takes ρ' to be a function of s instead of r; or, which comes to the same thing to his order of approximation, he puts instead of the first of equations (3)

$$-S = \rho' + \frac{d\rho'}{dr}\alpha\rho Z + \alpha\rho'' Z';$$

hence we have an additional term $\alpha\frac{d\rho'}{dr}\rho\frac{dZ}{d\theta}\sigma$ on the right-hand side of (5): and finally, instead of (6), we obtain

$$\rho\rho'\frac{dZ}{d\theta}\frac{d\sigma}{dr} - Z''\frac{d}{dr}(\sigma r\rho''') = \sigma\rho''\frac{dZ'}{d\theta}.$$

403. I am not sure that I understand D'Alembert's continuation after the point which we reached at the end of Art. 401; but I think that it is substantially equivalent to the following.

Assume that the surfaces of equal density are level surfaces; then the force along the tangent to the curve considered must vanish. Thus we obtain to our order of approximation

$$\frac{\rho\rho'}{r}\frac{dZ}{d\theta} = \rho''' Z''.$$

Now ρ' and ρ are functions of r only, and Z and Z' are functions of θ only; so we must have

$$\frac{\rho\rho'}{r} = C\rho''', \quad Z'' = C\frac{dZ}{d\theta} \quad\text{.....................(7)},$$

where C is some constant.

Substituting in (6) we obtain

$$\left\{C\left(\rho''' + r\frac{d\rho'''}{dr}\right) - \rho\frac{d\rho'}{dr}\right\}\frac{dZ}{d\theta} + \rho''\frac{dZ'}{d\theta} = 0;$$

which, as before, leads to

$$\frac{dZ'}{d\theta} = B\frac{dZ}{d\theta}, \quad C\left(\rho''' + r\frac{d\rho'''}{dr}\right) - \rho\frac{d\rho'}{dr} = -B\rho'' \text{ ......(8)},$$

where B is some constant.

Thus we have the four equations (7) and (8) holding in place of the single equation (4).

From the first of (8) we have

$$Z' = BZ + B',$$

where B' is some constant.

From the first of (7) and the second of (8) we get

$$\frac{d(\rho'\rho)}{dr} - \rho\frac{d\rho'}{dr} = -B\rho'',$$

so that

$$-B\rho'' = \rho'\frac{d\rho}{dr}.$$

Thus, finally,

$$s = r + \alpha\rho Z, \quad -S = \rho' - \frac{\alpha}{B}\rho'\frac{d\rho}{dr}(BZ + B'), \quad T = \frac{\alpha\rho\rho'}{rC}C\frac{dZ}{d\theta},$$

that is,

$$s = r + \alpha\rho Z, \quad -S = \rho' - \alpha\rho'\frac{d\rho}{dr}(Z + B_1), \quad T = \frac{\alpha\rho\rho'}{r}\frac{dZ}{d\theta},$$

where B_1 is some constant.

These results are of course less general than the single equation (4).

404. D'Alembert finishes the Appendix with some matter which is very closely connected with our subject. He says on his pages 208 and 209:

Je remarquerai à cette occasion, qu'il me semble qu'on n'a point encore résolu d'une maniére assez générale le Problême de la figure de la Terre, dans l'hypothese que l'attraction soit en raison inverse du quarré des distances, et que la Terre soit composée d'un amas de Fluides de différentes densités.

Accordingly, D'Alembert proposes his more general solution of the problem of the Figure of the Earth. It would not be advisable to devote much space to shew that D'Alembert's additions to Clairaut's investigations are worthless; but as we have already given the principal formulæ which are necessary, we shall be able with brevity to justify this opinion. D'Alembert himself refers, as we shall do, to Clairaut, for some formulæ which are necessary.

We adopt Clairaut's hypothesis that the Earth consists of ellipsoidal fluid strata of varying density and ellipticity. Let ρ denote the density; take the known equation of Art. 401,

$$\frac{d}{d\theta}[\rho S] = \frac{d}{ds}[\rho s T] \quad\text{.................(1).}$$

Here the quantities are supposed to be expressed in terms of θ and s; and we use the square brackets to indicate this. But suppose that s is changed into $r(1+\epsilon\sin^2\theta)$; then we have to transform (1) suitably.

$$\frac{d}{d\theta}[\rho S] = \frac{d}{d\theta}(\rho S) + \frac{d}{dr}(\rho S)\frac{dr}{d\theta},$$

and $\frac{dr}{d\theta}$ is to be found on the supposition that s is constant, so that to our order of approximation it is equal to $-2r\epsilon\sin\theta\cos\theta$.

Thus rejecting the square of ϵ we have from (1)

$$\frac{d}{d\theta}(\rho S) - \frac{d}{dr}(\rho S)\, 2r\epsilon\sin\theta\cos\theta = \frac{d}{dr}(\rho r T).$$

Let ϕ denote the angle between the radius vector and the tangent to the ellipse at the point considered; so that to our order $\cos\phi = 2\epsilon\sin\theta\cos\theta$. Hence

$$\frac{d}{d\theta}(\rho S) = \frac{d}{dr}(\rho r T) + r\cos\phi\frac{d}{dr}(\rho S)$$

$$= \frac{d}{dr}\rho(rT + r\cos\phi S) - 2\sin\theta\cos\theta\rho S\frac{d(r\epsilon)}{dr}.$$

Thus $$\frac{d}{d\theta}(\rho S) + 2\rho\sin\theta\cos\theta S\frac{d(r\epsilon)}{dr} = \frac{d}{dr}(\rho r Q),$$

where Q stands for $T + \cos\phi S$, that is, for the whole force along the tangent; and the first term may be written $\rho\frac{dS}{d\theta}$. Hence finally

$$\rho\frac{dS}{d\theta} + 2\rho\sin\theta\cos\theta S\frac{d(r\epsilon)}{dr} = \frac{d}{dr}(\rho r Q) \quad\text{.........(2).}$$

This equation substantially coincides with that which D'Alembert uses; but he does not sufficiently explain his process.

405. We have now to give the values of Q and S. I shall use the following notation:

$$\Upsilon(r) \text{ for } \int_0^r \rho r^2 dr,$$

$$\Omega_0(r) \text{ for } \int_0^r \rho \frac{d\epsilon}{dr}\, dr,$$

$$\Omega_3(r) \text{ for } \int_0^r \rho \frac{d\,(r^3\epsilon)}{dr}\, dr,$$

$$\Omega_5(r) \text{ for } \int_0^r \rho \frac{d\,(r^5\epsilon)}{dr}\, dr.$$

Let b be the extreme value of r, that is the value of r at the surface; and let ω be the angular velocity. Then it will be found that

$$Q = \sin\theta\cos\theta \left\{ -\frac{8\pi\epsilon\Upsilon(r)}{r^2} + \frac{8\pi\Omega_5(r)}{5r^4} + \frac{8\pi r}{5}[\Omega_0(b) - \Omega_0(r)] + \omega^2 r \right\}$$

$$-S = \frac{4\pi(1 - 2\epsilon \sin^2\theta)}{r^2}\Upsilon(r) + \frac{8\pi}{3r^2}\Omega_3(r)$$

$$+ 4\pi(3\sin^2\theta - 2)\left\{ \frac{\Omega_5(r)}{5r^4} - \frac{\Omega_0(b) - \Omega_0(r)}{15} 2r \right\} - \omega^2 r \sin^2\theta.$$

The value of Q is found as in Art. 341, or Clairaut's page 273. The value of $-S$ is found as in Art. 336, or Clairaut's page 247: it is only necessary to add to what is there given the central attraction which arises from the matter which may be said to be external to the attracted point, and thus we obtain the term which involves Ω_0 in the manner the term involving Ω_5 was obtained.

Hence $-\dfrac{dS}{d\theta}$

$$= \sin\theta\cos\theta \left\{ -\frac{16\pi\epsilon\Upsilon(r)}{r^2} + \frac{24\pi\Omega_5(r)}{5r^4} - \frac{16\pi r}{5}[\Omega_0(b) - \Omega_0(r)] - 2\omega^2 r \right\}.$$

Thus (2) becomes to our order of approximation

$$\rho\left\{ \frac{\Upsilon(r)}{r^2}\frac{d(r\epsilon)}{dr} - \frac{2\epsilon}{r^2}\Upsilon(r) + \frac{3\Omega_5(r)}{5r^4} - \frac{2r}{5}[\Omega_0(b) - \Omega_0(r)] - \frac{\omega^2 r}{4\pi} \right\}$$

$$= \frac{d}{dr}\rho\left\{ \frac{\epsilon\Upsilon(r)}{r} - \frac{\Omega_5(r)}{5r^3} - \frac{r^2}{5}[\Omega_0(b) - \Omega_0(r)] - \frac{\omega^2 r^2}{8\pi} \right\} \quad\text{.........(3).}$$

Let $K = \frac{\epsilon}{r^2}\Upsilon(r) - \frac{\Omega_5(r)}{5r^4} - \frac{r}{5}[\Omega_0(b) - \Omega_0(r)] - \frac{\omega^2 r}{8\pi}$; then multiply by r^4 and differentiate; then divide by r^4 and differentiate again. Thus we obtain

$$\frac{d^2\epsilon}{dr^2} + \frac{2\rho r^2}{\Upsilon(r)}\frac{d\epsilon}{dr} - \left\{\frac{6}{r^2} - \frac{2\rho r}{\Upsilon(r)}\right\}\epsilon = \frac{r^2}{\Upsilon(r)}\frac{d}{dr}\left\{\frac{1}{r^4}\frac{d}{dr}(Kr^4)\right\} \quad\ldots\ldots\ldots(4).$$

Moreover (3) may be written

$$\frac{d}{dr}(Kr\rho) = \rho\left\{\frac{d(r\epsilon)}{dr}\frac{\Upsilon(r)}{r^2} - \frac{4\epsilon\Upsilon(r)}{r^2} + \frac{\Omega_5(r)}{r^4} + 2K\right\};$$

multiply by $\frac{r^4}{\rho}$ and differentiate: then we obtain

$$\frac{d^2\epsilon}{dr^2} + \frac{2\rho r^2}{\Upsilon(r)}\frac{d\epsilon}{dr} - \left\{\frac{6}{r^2} - \frac{2\rho r}{\Upsilon(r)}\right\}\epsilon$$

$$= \frac{1}{r^3\Upsilon(r)}\left[\frac{d}{dr}\left\{\frac{r^4}{\rho}\frac{d(Kr\rho)}{dr} - 2Kr^4\right\}\right] \quad\ldots\ldots\ldots(5).$$

Comparing (4) and (5) we obtain $\frac{d}{dr}\left(\frac{Kr^5}{\rho}\frac{d\rho}{dr}\right) = 0$; therefore $\frac{Kr^5}{\rho}\frac{d\rho}{dr} = M$ a constant; and so the right-hand side of (4) becomes

$$\frac{r^2}{\Upsilon(r)}\frac{d}{dr}\left\{\frac{1}{r^4}\frac{d}{dr}\left(\frac{M\rho}{r}\frac{dr}{d\rho}\right)\right\}.$$

Thus D'Alembert considers he has found a more general result than had hitherto been given; for we know that Clairaut's *derived equation* agrees with (4) when the right-hand side is changed to zero: see Art. 343.

But D'Alembert himself admits, that at the external surface there can be no tangential force, and so K must vanish there; see the last line of his page 211. This would suggest $M=0$; but D'Alembert wishes to avoid this, and so he says it will be sufficient to have $\frac{d\rho}{dr}$ infinite at the external surface.

The error involved is very serious even for D'Alembert: such a strange result should have led him to review his process. If we develope the right-hand side of (3) we have one term involving

$\frac{d\rho}{dr}$, and another involving ρ; the latter term is exactly the same as we have on the left-hand side of (3). Thus (3) becomes simply, in D'Alembert's notation, $Kr\frac{d\rho}{dr}=0$; thus either $K=0$, or $\frac{d\rho}{dr}=0$; in the latter case the density is constant: in both cases the level surfaces are surfaces of equal density.

In fact, as we stated in Art. 400, we know that for such forces as occur in nature the level surfaces must be surfaces of equal density; this was pointed out by Lagrange in some observations on D'Alembert's misconception: see the *Miscellanea Taurinensia*, Vol. II. page 285.

406. D'Alembert himself briefly admitted and corrected his error in his *Opuscules*, Vol. V. page 4: my remarks were written before I had arrived at this admission; and I have ventured to retain them. It is curious to notice the complacent satisfaction with which D'Alembert, up to the period of the admission of his error, regarded his efforts to improve the important result which I call Clairaut's derived equation: see the *Recherches....Systême du Monde*, Vol. II. page 290, and Vol. III. pages xxxvi and xxxvii; and also the article *Figure de la Terre* in the original *Encyclopédie*.

407. We might have deduced equation (3) of Art. 405 from equation (6) of Art. 401. Return to the notation of Art. 401, using σ for the density. We have

$$\frac{dZ}{d\theta}=\frac{dZ'}{d\theta}=Z''=2\sin\theta\cos\theta;$$

and thus equation (6) becomes

$$\frac{d}{dr}\sigma(\rho\rho'-r\rho''')=\sigma\rho''+\sigma\rho'\frac{d\rho}{dr};$$

and
$$a\rho=r\epsilon,\quad \rho'=4\pi\left\{\frac{\Upsilon(r)}{r^2}+\text{small terms}\right\},$$

$$a\rho''=4\pi\left\{-\frac{2\epsilon\Upsilon(r)}{r^2}+\frac{3\Omega_5(r)}{5r^4}-\frac{2r}{5}\left[\Omega_0(b)-\Omega_0(r)\right]-\frac{\omega^2 r}{4\pi}\right\},$$

$$a\rho'''=4\pi\left\{\frac{\Omega_5(r)}{5r^4}+\frac{r}{5}\left[\Omega_0(b)-\Omega_0(r)\right]+\frac{\omega^2 r}{8\pi}\right\}.$$

Substitute these values in the above equation, and it will be found to agree with (3) of Art. 405.

408. In considering the writings of D'Alembert on our subject up to the present point, we find but little of importance. Not only do they fail to add anything to what Clairaut had given, but they do not even reach the same level. It seems to me that D'Alembert had not taken the trouble to study a work which far surpassed all his own efforts in the same direction.

409. The next work by D'Alembert is entitled *Recherches sur différens points importans du Systême du Monde.* This work forms three parts or volumes in quarto. The first and the second parts were published in 1754; and the third part in 1756.

The first part contains the Title, Preliminary Essay, Table of Contents, and Corrections in lxviii pages; then the text of 260 pages: there is one plate.

The second part contains the Title and Table of Contents in vi pages; then the text of 290 pages: there are three plates.

The third part contains the Title, Preface, Table of Contents, and the *Privilege du Roi* in xlviii pages; then the text and Corrections of 263 pages: there are two plates.

410. There is nothing in the first part with which we are concerned.

In the second part we have on pages 201...209, *Remarques sur la figure de la Terre, qui résulte de la Précession des Equinoxes;* and on pages 265...290, we have a Chapter entitled *De la Figure de la Terre.*

411. D'Alembert, on his pages 201...209, returns to the subject of the information which the theory of the Precession of the Equinoxes gives with respect to the theory of the Figure of the Earth. He first substantially repeats the matter of which we have given an account in Arts. 385 and 386. He then says, on his page 204:

Je dois cependant avouer qu'un grand Geométre a cru pouvoir concilier tout, en supposant que la Terre soit un solide Elliptique, dont la différence des Axes soit $= \frac{1}{200}$, et qui renferme au-dedans de lui un noyau sphérique dont la densité soit à celle du Sphéroide comme 10 est à 1, et dont le rayon soit au rayon de l'Equateur comme 3 à 5.

D'Alembert here alludes to a memoir by Euler on the Precession of the Equinoxes, published in the Berlin *Mémoires* for 1749; see page 315 of the memoir: Euler does not support his suggestion by any theory connected with our subject.

D'Alembert shews that the above supposition is inadmissible. Take a formula obtained in Art. 374, namely

$$\epsilon = \frac{\frac{r\omega^2}{2}}{\frac{4\pi\rho r}{3} + \frac{4\pi(\sigma-\rho)s^3}{3r^2} - \frac{4\pi\rho r}{5}};$$

let j denote, as usual, the ratio of the centrifugal force at the equator to the attraction there, so that

$$r\omega^2 = j\left\{\frac{4\pi\rho r}{3} + \frac{4\pi(\sigma-\rho)s^3}{3r^2}\right\};$$

therefore
$$\epsilon = \frac{\frac{j}{2}\left\{\rho + (\sigma-\rho)\frac{s^3}{r^3}\right\}}{\rho + (\sigma-\rho)\frac{s^3}{r^3} - \frac{3\rho}{5}}.$$

Now let us suppose $\sigma = 10\rho$, and $s = \frac{3}{5}r$, so that $\frac{(\sigma-\rho)s^3}{\rho r^3} = 2$ very nearly. Thus

$$\epsilon = \frac{j}{2}\frac{1+2}{1+2-\frac{3}{5}} = \frac{5j}{8};$$

this value of ϵ is smaller than observation will allow. It will be observed that D'Alembert assumes that the ellipticity of the external surface is the same as if the outer part were *fluid*. it is not obvious whether Euler contemplated this in his hypothesis that the Earth consisted of two *solid* parts.

412. We now pass to pages 265...290 of the volume. On pages 265...274, D'Alembert considers how the figure of the Earth

may be found by geographical operations. He suggests in fact that we should assume for the radius vector a series with unknown coefficients involving cosines of multiples of the colatitude. Then by measuring the lengths of degrees of the meridian in various latitudes we find the corresponding values of the radius of curvature: and thus we obtain equations for determining the unknown coefficients in the assumed expression for the radius vector.

413. D'Alembert also suggests that observations of the Moon's parallax may be employed for information as to the figure of the Earth: but he admits that practically this method would be of little value.

414. In pages 275...290, D'Alembert indicates a method for calculating the attraction of a spheroid on a particle at the surface. Suppose Q a point of the surface, C the point which may be called the centre of the spheroid. D'Alembert proposes to consider the spheroid as composed of two parts; one part being the sphere on CQ as radius, and the other part the difference between the sphere and the spheroid. He shews how the approximate value of the attraction of the second part may be conveniently calculated.

It is obvious that the principle of this method is the same as that which has since been developed by Laplace. D'Alembert gives only an outline of his method here; he works it out in detail in the third volume of the *Recherches...Systême du Monde*. We shall recur to it in our Article 424.

415. We now arrive at the third volume of the *Recherches ...Systême du Monde*. Here pages xix...xlii and 107...260 are devoted to the Figure of the Earth.

416. In pages xix...xlii D'Alembert gives some introductory remarks on the subject, the purport of which is to shew the uncertainty as to the actual facts. It was possible to doubt whether the Earth was a figure of revolution; granting it to be such, it was possible to doubt whether the northern and the southern hemispheres were exactly alike; and granting that they were

exactly alike, it was possible to doubt whether the figure was that of an ellipsoid of revolution.

D'Alembert refers to six measured lengths which had to be considered in testing any theory; five of these were arcs of meridians, namely, those in Lapland, Peru, France, the Cape of Good Hope, and Italy: one was an arc of longitude, in latitude 43° 32′. As to a degree of the meridian in France, three lengths had been proposed; Picard gave 57060 toises; the Academicians of the North corrected it to 57183 toises; and subsequently it was put at 57074 toises: see Art. 236.

D'Alembert found it impossible to assign such a value of the ellipticity as would harmonise the six measured lengths.

417. The following points of interest may be noticed in the introductory remarks by D'Alembert.

On page xxxii he says that a hemispherical mountain a league high ought to make a pendulum deviate more than 1′ from the vertical; but the high mountains in Peru scarcely produced a variation of 7″. It is easy to verify his calculation, supposing the density of the mountain equal to the mean density of the Earth. For the facts as to the mountains in Peru see Bouguer's *Figure de la Terre*, pages 364...394.

D'Alembert in a note on his page xl suggests, that in such a mountainous country as Italy, the direction of the plumb-line may have been disturbed, and thus an error produced in the measured length of a degree.

D'Alembert refers to the figure of Jupiter as suggesting by analogy what the figure of the Earth may be; but I do not understand all that is said on this matter. The following passage occurs on pages xxxv and xxxvi.

Car les observations nous prouvent que la surface de Jupiter est sujette à des altérations sans comparaison plus considérables et plus fréquentes que celle de la Terre; or si ces altérations n'influoient en rien sur la figure de l'équateur de Jupiter, pourquoi la figure de l'équateur de la Terre seroit-elle altérée par des mouvemens beaucoup moindres?

I do not know what changes in Jupiter he refers to here.

Again he suggests that we should determine by observation whether the figure of Jupiter is precisely that which theory would assign; but I cannot see any practical value in the method which he proposes. He states it thus on his page xli:

Pour cela il suffiroit de mesurer le parallele à l'équateur de Jupiter, qui en seroit éloigné de 60 degrés; si ce parallele se trouvoit sensiblement égal ou inégal à la moitié de l'équateur, le méridien de Jupiter seroit elliptique ou ne le seroit pas.

It seems to me that supposing the observation could be made with great accuracy it would afford but little information; if the parallel were not exactly half of the equator, we should know that the meridian could not be circular: but we could not in any case pronounce what the figure must be from merely knowing the value of this parallel.

418. We now proceed to the text on pages 107...260.

A brief introduction commences the discussion. D'Alembert proposes to examine the figure of the Earth, first astronomically, so far as observations make it known, and then physically by theory.

419. In the first three Chapters D'Alembert considers whether we can by direct observations determine if certain hypotheses which are usually made are strictly true. Thus, for example, we usually assume that the plane which contains the axis of the Earth and any given place will also contain the vertical line at that place: this amounts practically to assuming that the Earth is a figure of revolution. D'Alembert shews that, strictly speaking, this hypothesis may be untrue; for observations made at any given place would not enable us to decide that the vertical did or did not lie exactly in the plane containing the place and the axis of the Earth. Again, we define the vertical direction at any place as that of falling bodies; and we know that this direction is perpendicular to the surface of fluid at rest at the place: but this direction will not be necessarily perpendicular to the surface of the solid Earth at the place. Now D'Alembert shews that if the angle between these two directions is very small we shall not be able to detect it by observations.

I do not give any detailed account of these Chapters, since the propositions are of such a kind that they readily commend themselves as reasonable. The processes of D'Alembert require attention to understand them; but they will be found to present no very serious difficulty.

420. D'Alembert's fourth Chapter is entitled *De la Figure de la Terre dans les hypothèses ordinaires.* This is of the same character as the portion of the second volume of the *Recherches* which we described in Art. 412.

421. D'Alembert's fifth Chapter is entitled *Des parallaxes en tant qu'elles dépendent de la figure de la Terre.* The Earth being not a sphere the parallax of the Moon will vary with the place of observation; D'Alembert investigates formulæ for the parallax: but these investigations belong rather to Plane Astronomy than to Physical Astronomy.

422. We now pass to D'Alembert's second Section, which is entitled *De la figure de la Terre considérée physiquement.*

423. The first Chapter, on pages 166...177, contains the investigation of certain integrals which will be used in the sequel.

Thus, to take the first, required

$$\int \frac{dt}{\sqrt{(1-t^2)}} \cdot \frac{(1-t^2)^n}{(k^2+t^2-k^2t^2)^{n+1}},$$

n being a positive integer.

D'Alembert assumes $k^2+t^2-k^2t^2 = \dfrac{1}{s+1}$; and then the integral becomes

$$-\frac{1}{2(1-k^2)^n} \int \frac{s^n ds}{\sqrt{(s-k^2s-k^2s^2)}}.$$

D'Alembert requires the integral between the limits 0 and 1 of t; to these limits correspond $\dfrac{1-k^2}{k^2}$ and 0 for s. He easily obtains the required result by ordinary methods: we will verify by assuming $\sin^2\theta = \dfrac{k^2 s}{1-k^2}$, which reduces the integral to

$$\frac{1}{k^{2n+1}}\int_0^{\frac{\pi}{2}} \sin^{2n}\theta d\theta,$$

and the value is

$$\frac{1}{k^{2n+1}} \cdot \frac{(2n-1)(2n-3)\ldots 1}{2n(2n-2)\ldots 2} \cdot \frac{\pi}{2}.$$

D'Alembert arrives at the same result on his page 170; he apparently gives twice this value, but he has really taken the integral twice over.

On his page 171, he professes, I think, to investigate the integral

$$\int \frac{dt}{\sqrt{(1-t^2)}} \cdot \frac{(1-t^2)^{\frac{n}{2}}}{(k^2+t^2-k^2t^2)^{\frac{n}{2}+1}},$$

where n is an odd positive integer; but his printing is not very distinct. This integral transforms as before into

$$-\frac{1}{2(1-k^2)^{\frac{n}{2}}}\int \frac{s^{\frac{n}{2}}\, ds}{\sqrt{(s-k^2s-k^2s^2)}}.$$

It is unnecessary for his purpose to take any notice of the numerical factor which is here outside the integral sign; and so he omits it.

He gives three times, namely on his pages 174, 176, and 177, the following result:

$$\int_0^{2r} \frac{x(2rx-x^2)^2}{(2rx)^{\frac{3}{2}}}\, dx = \frac{16 \times 8r^3}{5 \times 7 \times 9}.$$

424. D'Alembert's second Chapter, on pages 178...199, is entitled *De l'attraction d'un sphéroïde sur les corpuscules placés à sa surface; et de la figure qui en résulte pour ce sphéroïde.*

We begin with a general formula for the attraction of a spheroid on a particle at the surface, resolved tangentially; we shall follow D'Alembert as to principle, but we shall simplify the mere analytical work.

Let there be a point Q on the surface of a spheroid, let s be the distance of Q from a fixed point which we may call the centre of the spheroid; let θ be the angular distance of Q from the pole. It is required to find the attraction of the spheroid at Q, resolved tangentially.

We assume that the spheroid is a figure of revolution. We may suppose that the spheroid consists of a sphere of radius s, and an additional shell: see Art. 414. We assume that the shell is at every point so thin that it may be treated as if it were condensed on the surface of the sphere of radius s. It is obvious that we need only consider the shell when we seek the tangential attraction.

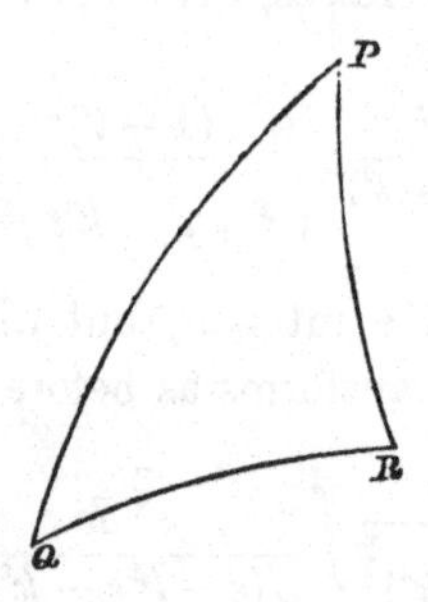

Let R be any other point on the surface of the spheroid; and let its polar co-ordinates be s' and θ'. Let P be the pole; put μ for the angle which QR subtends at the centre, and ψ for the angle PQR.

The element of spherical surface at R may be denoted by $s^2 \sin\mu \, d\mu \, d\psi$; and thus the element of mass of the shell may be denoted by $(s'-s)\, s^2 \sin\mu \, d\mu \, d\psi$, taking the density as unity. The distance from Q is $2s \sin\frac{\mu}{2}$. We first take the resolved part of the attraction along the tangent to QR at Q; and then we resolve this along the tangent to QP at Q.

Thus we obtain

$$(s'-s)\, s^2 \sin\mu \, d\mu \, d\psi \, \frac{\cos\frac{\mu}{2}}{\left(2s \sin\frac{\mu}{2}\right)^2} \cos\psi,$$

that is
$$\frac{(s'-s)\cos^2\frac{\mu}{2}\cos\psi}{2\sin\frac{\mu}{2}}\,d\mu\,d\psi.$$

If we integrate this expression between the limits 0 and π for μ, and 0 and 2π for ψ, we obtain the tangential attraction at Q *towards* the pole.

425. Now suppose, with D'Alembert, that
$$s' = r + r\alpha\,(A + B\cos\theta' + C\cos^2\theta' + D\cos^3\theta'),$$
where α is a very small constant, and r, A, B, C, D are any constants: we might suppose these constants connected by the relation $A+B+C+D=0$, and then r would be the polar semi-axis of the spheroid. However we will not use this supposition. Substitute this value of s' in the expression of the preceding Article: then we see that the tangential attraction reduces to
$$\frac{\alpha r}{2}\int_0^\pi\int_0^{2\pi}\frac{\cos^2\frac{\mu}{2}\cos\psi}{\sin\frac{\mu}{2}}(B\cos\theta' + C\cos^2\theta' + D\cos^3\theta')\,d\mu\,d\psi\,;$$
and
$$\cos\theta' = \cos\theta\cos\mu + \sin\theta\sin\mu\cos\psi.$$

We shall determine separately the values of the three parts of which the integral is composed.

The term involving B reduces to
$$\frac{\alpha rB}{2}\int_0^\pi\int_0^{2\pi}\frac{\cos^2\frac{\mu}{2}\cos\psi}{\sin\frac{\mu}{2}}\sin\theta\sin\mu\cos\psi\,d\mu\,d\psi\,;$$
this
$$= \alpha r B\sin\theta\,\pi\int_0^\pi\cos^3\frac{\mu}{2}\,d\mu = \frac{4\pi}{3}\alpha rB\sin\theta.$$

The term involving C reduces to
$$\alpha rC\int_0^\pi\int_0^{2\pi}\frac{\cos^2\frac{\mu}{2}\cos\psi}{\sin\frac{\mu}{2}}\cos\theta\cos\mu\sin\theta\sin\mu\cos\psi\,d\mu\,d\psi\,;$$
this
$$= 2\alpha rC\sin\theta\cos\theta\,\pi\int_0^\pi\cos^3\frac{\mu}{2}\cos\mu\,d\mu = \frac{8\pi}{5}\,\alpha rC\sin\theta\cos\theta.$$

The term involving D reduces to

$$\frac{arD}{2}\int_0^\pi\int_0^{2\pi}\frac{\cos^3\frac{\mu}{2}\cos\psi}{\sin\frac{\mu}{2}}(3\cos^2\theta\cos^2\mu\sin\theta\sin\mu\cos\psi + \sin^3\theta\sin^3\mu\cos^3\psi)\,d\mu\,d\psi;$$

this

$$=arD\sin\theta\cos^2\theta\,3\pi\int_0^\pi\cos^3\frac{\mu}{2}\cos^2\mu\,d\mu + arD\sin^3\theta\frac{3\pi}{4}\int_0^\pi\cos^3\frac{\mu}{2}\sin^2\mu\,d\mu$$

$$=\frac{76\pi}{35}arD\sin\theta\cos^2\theta+\frac{16\pi}{35}arD\sin^3\theta.$$

Thus the whole tangential attraction towards the pole is

$$4\pi ar\left(\frac{B}{3}\sin\theta+\frac{2C}{5}\sin\theta\cos\theta+\frac{19D}{35}\sin\theta\cos^2\theta+\frac{4D}{35}\sin^3\theta\right)$$

426. Let there be a solid sphere of radius r and density σ, surrounded by a thin fluid stratum of density σ'; and let the radius of the external surface of this stratum be the s' of Art. 425. We propose to enquire if this fluid will remain in a state of relative equilibrium when rotating with uniform angular velocity.

The attraction towards the centre may be taken as $\frac{4\pi r\sigma}{3}$; the resolved part of this tangentially towards the pole is found to the order we require by multiplying by $\frac{ds'}{s'\,d\theta'}$, using θ instead of θ' in the result. Let j denote the ratio of the centrifugal force at the equator to the attraction there; then $\frac{4\pi r\sigma}{3}j\sin\theta\cos\theta$ *from* the pole is the tangential action of the centrifugal force. Thus equating to zero the whole tangential force we get

$$4\pi ar\left(\frac{B}{3}\sin\theta+\frac{2C}{5}\sin\theta\cos\theta+\frac{19D}{35}\sin\theta\cos^2\theta+\frac{4D}{35}\sin^3\theta\right)\sigma'$$

$$-\frac{4\pi r\sigma}{3}j\sin\theta\cos\theta-\frac{4\pi ar}{3}(B\sin\theta+2C\sin\theta\cos\theta+3D\sin\theta\cos^2\theta)\,\sigma=0.$$

Divide by $\sin\theta$; then equate to zero the coefficients of the various powers of $\cos\theta$. Thus we obtain

$$\frac{B\sigma'}{3}+\frac{4D\sigma'}{35}-\frac{B\sigma}{3}=0 \text{ .... } (1).$$

$$\frac{2\alpha C\sigma'}{5}-\frac{j\sigma}{3}-\frac{2\alpha C\sigma}{3} \text{ .... } (2).$$

$$\frac{15D\sigma'}{35}-D\sigma=0 \text{ .... } (3).$$

From (3) we get $\sigma=\frac{3}{7}\sigma'$; then from (2) we get $\alpha C=\frac{5j}{4}$; and then from (1) we get $B=-\frac{3D}{5}$

D'Alembert has a wrong equation instead of (1), and so his value of B is wrong; he corrects the error in his *Opuscules Mathématiques*, Vol. VI. page 230.

It is remarkable, as D'Alembert says on his page 181, that the value of C is independent of B and D, and is numerically the same as it would be if we made $\sigma'=\sigma$, and therefore B and D zero, but with the opposite sign.

427. D'Alembert shews that the equation $s'=r+\alpha r(A+B\cos\theta')$ represents a circle; supposing α so small that its square may be neglected. He states that on the same supposition the equation $s'=r+\alpha r(A+B\cos\theta'+C\cos^2\theta')$ represents an ellipse. See his pages 181...183. It is easy to verify these propositions.

428. D'Alembert proceeds to another case of relative equilibrium on his page 183. He first states the value of the attraction towards its centre, produced by an oblatum of small excentricity on an external particle. Suppose the polar semiaxis to be r, and the equatorial semiaxis $r(1+\alpha)$ where α is very small; let δ be the distance of the attracted particle from the centre of the oblatum, θ the angle between the polar semiaxis and the direction of δ. Then he says that the value of the attraction towards

the centre is

$$\frac{4\pi r^3}{3\delta^2}+\frac{8\pi r^3\alpha}{3\delta^2}+\frac{4\pi r^5\alpha}{5\delta^4}-\frac{12\pi r^5\alpha\cos^2\theta}{5\delta^4};$$

he says that this can be obtained by methods given further on, or by other means.

We may easily verify this statement. If M be the mass of an oblatum, R the polar semiaxis, e the excentricity; then the attraction on a particle at the distance δ from the centre on the polar axis produced, is by Art. 261, approximately

$$\frac{M}{\delta^2}\left(1-\frac{3e^2R^2}{5\delta^2}\right).$$

Then use the theorem given by Clairaut, Art. 333; we have consequently $R=r(1+\alpha\sin^2\theta)$, and also $R^3(1-e^2)^{-1}=r^3(1+\alpha)^2$; so that $e^2=2\alpha-3\alpha\sin^2\theta$. With these values of R and e we shall verify D'Alembert's statement.

429. Now suppose the Earth to consist of a solid oblatum of density σ, surrounded by a thin layer of fluid of density σ'; as an equivalent supposition we may take two coexistent oblata, the lesser of density $\sigma-\sigma'$, and the larger of density σ'.

Let the polar and equatorial radii of the lesser oblatum be $r(1-\beta)$ and $r(1-\beta)(1+\alpha')$ respectively; and let those of the larger be r and $r(1+\alpha)$: we suppose α, α', and β so small that squares and products may be neglected.

Let P denote the gravity of a particle at the pole, and ϖ the gravity of a particle at the equator; the particle being supposed to be on the outer surface. We shall find, by Art. 428, that

$$P=\frac{4\pi r\sigma}{3}-4\pi r\beta(\sigma-\sigma')+\frac{16\pi r(\alpha-\alpha')\sigma'}{15}+\frac{16\pi r\alpha'\sigma}{15},$$

$$\varpi=\frac{4\pi r\sigma}{3}-4\pi r\beta(\sigma-\sigma')-\frac{8\pi r\alpha\sigma}{3}+\frac{52\pi r(\alpha-\alpha')\sigma'}{15}$$

$$+\frac{52\pi r\alpha'\sigma}{15}-\frac{4\pi r\sigma j}{3};$$

therefore,

$$\frac{P-\varpi}{\varpi}=2\alpha+j-\frac{9(\alpha-\alpha')\sigma'}{5\sigma}-\frac{9\alpha'}{5} \qquad (1).$$

But, by Art. 376, we have $\alpha = \dfrac{6\alpha'(\sigma - \sigma') + 5j\sigma}{10\sigma - 6\sigma'}$ .... (2).

Substitute in (1) the value of α' found from (2): thus we get

$$\frac{P - \varpi}{\varpi} = \frac{5j}{2} - \alpha \text{ .... (3).}$$

Substitute in (3) for α from (2); thus

$$\frac{P - \varpi}{\varpi} = \frac{5j}{4} + \frac{3(\sigma - \sigma')(5j - 4\alpha')}{2(10\sigma - 6\sigma')} \text{ .... (4).}$$

These results agree with D'Alembert's on his page 186, but the notation is different.

It is obvious from (4) that if $\sigma - \sigma'$ and $5j - 4\alpha'$ are both positive, then $\dfrac{P - \varpi}{\varpi}$ is greater than $\dfrac{5j}{4}$; also if $\sigma - \sigma'$ and $5j - 4\alpha'$ are both negative, and $10\sigma - 6\sigma'$ is positive, then $\dfrac{P - \varpi}{\varpi}$ is greater than $\dfrac{5j}{4}$. Also if $\sigma - \sigma'$ and $5j - 4\alpha'$ are of contrary signs, and $10\sigma - 6\sigma'$ is positive, then $\dfrac{P - \varpi}{\varpi}$ is less than $\dfrac{5j}{4}$.

430. It will be observed that the preceding investigation depends on that which we have noticed in Art. 376, and which is not altogether satisfactory, although D'Alembert seems to have been very fond of it. We may also remark that if the layer of fluid is to surround the body completely, there must be a certain condition satisfied, namely, $1 - \beta + \alpha'$ must be less than $1 + \alpha$: D'Alembert does not advert to this, but it is not of much importance.

431. D'Alembert on his pages 187 and 188 makes some remarks on Clairaut. D'Alembert here admits that Clairaut had already obtained the result (3) of Art. 429; but D'Alembert says that Clairaut's demonstration was limited to the case in which α is greater than $\dfrac{5j}{4}$. D'Alembert also states that Clairaut supposed the strata nearer to the centre to be the denser, and also supposed

that α and α' could only differ by a quantity infinitesimal compared with α or α'.

But these remarks are quite inapplicable. Clairaut believed the strata nearer to the centre to be the denser; but he did not introduce this belief in such a manner as to restrict his investigations. Clairaut does not limit himself to the case in which α is greater than $\frac{5j}{4}$: D'Alembert seems to have assumed that the quantity denoted by D in Art. 327 is necessarily positive, which it is not. Finally, Clairaut does not assume that the difference between α and α' is infinitesimal compared with α and α', when the fluid is of finite thickness, but only when this thickness is infinitesimal: see Art. 328.

D'Alembert certainly added nothing to the investigations given by Clairaut of the theorem which bears his name: in fact, D'Alembert criticised these investigations before he had taken the trouble to understand them.

432. It is curious to see D'Alembert devote a whole paragraph on his pages 188 and 189 to a very elementary piece of Algebra. If we have given that $12\alpha'(\Delta - 1)$ is greater than $15N(\Delta - 1)$, we must not infer that $12\alpha'$ is greater than $15N$, unless we know that $\Delta - 1$ is positive.

D'Alembert repeats on his page 190 a remark which he had made at an earlier date: see Art. 378.

433. D'Alembert investigates on his pages 191...197 the values of some definite integrals which are useful in the sequel, namely, various cases of $\int_0^{2r} \frac{x^p dx}{(n^2 + 2nx + 2rx)^{\frac{3}{2}}}$ and $\int_0^{2r} \frac{x^p dx}{(n^2 - 2nx + 2rx)^{\frac{3}{2}}}$, obtained by ascribing to p various positive integral values. For example

$$\int_0^{2r} \frac{dx}{(n^2 + 2nx + 2rx)^{\frac{3}{2}}} = \frac{2r}{n(r+n)(2r+n)},$$

and

$$\int_0^{2r} \frac{dx}{(n^2 - 2nx + 2rx)^{\frac{3}{2}}} = \frac{2}{n(2r-n)}.$$

We suppose that $2r$ is greater than n. We observe that the second of these two examples cannot be deduced from the first by changing the sign of n.

434. D'Alembert makes some remarks on his pages 198 and 199 on the attraction of a spherical shell. He takes r for the radius of the shell, and $\frac{4\pi r^2}{\delta^2}$ for the attraction on a particle outside the shell at a distance δ from the centre: thus he does not introduce any factor to represent the thickness of the shell. When the particle is inside the shell the attraction is zero. He adds:

De-là il me semble qu'on peut conclure que l'attraction d'une surface sphérique sur un point placé sur cette surface même, n'est pas 4π, comme il paroît qu'on l'a crû jusqu' à présent, mais seulement 2π.

If it be necessary to put the idea into words, it would be better to say that the attraction of a spherical film on a particle which forms part of the film is 2π.

D'Alembert recurs to the subject of the attraction of a spherical film in the article *Gravitation* of the original *Encyclopédie* and in the first volume of his *Opuscules Mathématiques.*

435. D'Alembert illustrates his remarks on the attraction of a spherical film by the following statement on his page 199

Les Géometres ne sont pas tout-à-fait étrangers à ces sortes de paradoxes, d'une quantité qui s'évanouit tout d'un coup sans disparoître par degrés. Ainsi la courbe $y = \sqrt{ax} + \sqrt[4]{a^3(b+x)}$ qui est du 8e degré tant que b n'est pas $= 0$, perd subitement plusieurs branches lorsque $b = 0$, parce que l'équation du 8e degré se réduit alors au 4e. Voyez les *Mémoires de l'Académie de Berlin* 1749, page 146. Dans le premier cas, cette courbe a un diametre ; dans le cas de $b = 0$, elle n'en a plus.

This illustration does not seem to me very good: it may justly be maintained that the above equation when properly understood is of the 8th degree, even when $b = 0$.

436. D'Alembert's third Chapter, on pages 200...213 is entitled *Problêmes nécessaires pour généraliser les recherches précédentes.* This Chapter consists of various definite integrals which are required by D'Alembert in his process for calculating the attraction of a spheroid. These definite integrals depend

mainly on the values of $\int_r^{-r} \frac{u^p du}{(\delta^2 + r^2 - 2\delta u)^{\frac{3}{2}}}$, when for p we put in succession 0, 1, 2, 3, 4, 5, 6. Here δ and r are constants: the integrals present different values according as δ is greater or less than r.

D'Alembert puzzles his readers by taking 1 and -1 as the limits of u on his pages 201, 202, and 208; but except on these pages the limits are those which I have stated, namely r and $-r$. His results are correct, allowing for a few obvious misprints.

437. D'Alembert's fourth Chapter, on pages 214...246 is entitled *Usages des Problêmes précédens, pour déterminer l'attraction du sphéroïde sur un corpuscule quelconque.*

He determines the attraction which a certain spheroid of revolution exerts on a particle, external or internal, at right angles to the radius vector, and along the radius vector; these he calls respectively the *horizontal* and *vertical* attractions.

He states the results, having previously given the values of certain definite integrals which are required.

We will explain how these results may be verified; the method we shall adopt is that which we have already used in Art. 424.

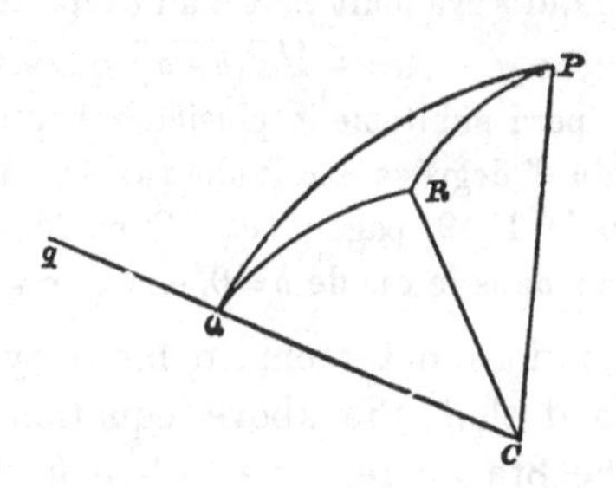

Let C denote the centre of the spheroid, CP the semi-axis of revolution, Q any point on the surface having for its polar co-ordinates s and θ. Produce CQ to any point q. It is required to find the *horizontal* attraction on a particle at q. Let $Cq = \delta$.

Let R be any point on the surface having for its polar co-ordinates s' and θ'. We suppose that the spheroid consists of a sphere of radius s and an additional shell.

Let the angle PQR be denoted by ψ, and the angle QCR by μ.

The element of the shell at $R = s^2 (s' - s) \sin\mu \, d\mu \, d\psi$.

The distance $Rq = (s'^2 + \delta^2 - 2s'\delta \cos\mu)^{\frac{1}{2}}$.

The resolved attraction of the element in the plane RCq at right angles to Cq is therefore

$$\frac{s' \sin\mu \, s^2 (s' - s) \sin\mu \, d\mu \, d\psi}{(s'^2 + \delta^2 - 2s'\delta\cos\mu)^{\frac{3}{2}}};$$

and resolving along the plane PCq we get

$$\frac{s' \sin\mu \cos\psi \, s^2 (s' - s) \sin\mu \, d\mu \, d\psi}{(s'^2 + \delta^2 - 2s'\delta\cos\mu)^{\frac{3}{2}}}.$$

We have to integrate this between the limits 0 and π for μ, and 0 and 2π for ψ; then we obtain the horizontal attraction at q *towards* P.

We suppose with D'Alembert that s' has the value given in Art. 425.

We shall obtain by effecting the integrations, neglecting the square of α,

$$\pi\alpha \sin\theta \left\{\frac{4Br^4}{3\delta^3} + \frac{8Cr^5}{5\delta^4}\cos\theta + \frac{4Dr^4}{5\delta^3} - \frac{12Dr^6}{35\delta^5} + \frac{12Dr^6}{7\delta^5}\cos^2\theta\right\}.$$

In like manner if q be between C and Q instead of on CQ produced, and Cq be called δ as before, we obtain for the horizontal attraction

$$\pi\alpha \sin\theta \left\{\frac{4Br}{3} + \frac{8C\delta}{5}\cos\theta + \frac{4Dr}{5} - \frac{12D\delta^2}{35r} + \frac{12D\delta^2}{7r}\cos^2\theta\right\}.$$

These expressions must be multiplied by a factor to represent the density, if the density is not unity.

When $\delta = r$ these expressions both coincide, as they should do, with that given in Art. 425.

438. The attraction at q in the direction at right angles to the meridian plane of q will be zero, since the spheroid is supposed a figure of revolution. D'Alembert himself makes the remark on his page 216. He adds however that this can also be seen by calculation; and he gives some calculations, which I do not find to be intelligible.

439. In Art. 437 we have investigated expressions for the horizontal attraction of the spheroid supposed homogeneous. D'Alembert deduces on his page 218 the attraction of such a spheroid on an included particle when the spheroid is composed of indefinitely thin shells of varying density: the process is the same as we have already found was used by Clairaut. See Arts. 323 and 336.

440. In order to obtain the whole action along the *tangent* to the meridian curve at any point, we must as in Art. 426 add to the horizontal attraction the resolved part of the vertical attraction along the tangent, and also the resolved part of the centrifugal force.

441. Next we proceed to find the *vertical* attraction on the particle at q.

Suppose the particle outside the spheroid. The vertical action of the sphere of radius s

$$=\frac{4\pi s^3}{3\delta^2}=\frac{4\pi r^3}{3\delta^2}(1+3\alpha A+3\alpha B\cos\theta+3\alpha C\cos^2\theta+3\alpha D\cos^3\theta).$$

We must now determine the vertical action of the shell. As in Art. 437 we find that this is

$$\int_0^{\pi}\int_0^{2\pi}\frac{(\delta-s'\cos\mu)\,s^2\,(s'-s)\sin\mu\,d\mu\,d\psi}{(s'^2+\delta^2-2s'\delta\cos\mu)^{\frac{3}{2}}}.$$

By effecting the integrations we obtain for the whole vertical attraction

$$\frac{4\pi r^3}{3\delta^2}(1+3\alpha A+\alpha C)-\frac{4\pi r^5\alpha C}{5\delta^4}+\pi\alpha\cos\theta\left\{\left(\frac{B}{3}+\frac{D}{5}\right)\frac{8r^4}{\delta^3}-\frac{48Dr^6}{35\delta^5}\right\}$$
$$+\pi\alpha\cos^2\theta\,\frac{12\,Cr^5}{5\delta^4}+\pi\alpha\cos^3\theta\,\frac{16\,Dr^6}{7\delta^5}.$$

In like manner if the attracted particle be inside the spheroid, the whole vertical attraction is

$$\frac{4\pi\delta}{3}+\frac{8\pi\alpha C\delta}{15}+\pi\alpha\cos\theta\left\{-\frac{4Br}{3}-\frac{4Dr}{5}+\frac{36D\delta^2}{35r}\right\}$$
$$-\pi\alpha\cos^2\theta\frac{8C\delta}{5}-\pi\alpha\cos^3\theta\frac{12D\delta^2}{7r}.$$

For a point on the surface we must put in either of these results $\delta=r(1+\alpha A+\alpha B\cos\theta+\alpha C\cos^2\theta+\alpha D\cos^3\theta)$: it will be found that each of them becomes then

$$\frac{4\pi r}{3}\left(1+\alpha A+\frac{2\alpha C}{5}\right)+\pi\alpha\cos\theta\frac{8Dr}{35}-\pi\alpha\cos^2\theta\frac{4Cr}{15}-\pi\alpha\cos^3\theta\frac{8Dr}{21}.$$

These expressions must be multiplied by a factor to represent the density, if the density is not unity. Then as in Art. 439 we can obtain the vertical attraction for a spheroid composed of indefinitely thin strata of varying density.

442. D'Alembert now discusses the relative equilibrium of homogeneous fluid surrounding a solid nucleus composed of strata of varying density: see his page 222. The problem is thus an extension of that in Art. 426, and it is solved in the same manner. There is no difficulty in the hydrostatical part of the problem; for since the fluid is homogeneous it is sufficient for equilibrium that the tangential action at every point of the surface should be zero.

If we equate to zero this tangential action, we obtain a result of the form

$$\sin\theta\,(M_0+M_1\cos\theta+M_2\cos^2\theta)=0,$$

where M_0, M_1, and M_2 are independent of θ. This leads, as in Art. 426, to three equations

$$M_0=0,\quad M_1=0,\quad M_2=0.$$

D'Alembert gives these three equations on his pages 222...225.

443. We must be careful as to the notation, since many symbols are required. D'Alembert leaves his notation to explain itself, and it is not very inviting. I shall use the subscript 1 to denote values relating to the external boundary of the fluid; and I shall use ρ as a general symbol for the density. Then the three

equations are

$$0=\frac{4\pi a}{3\delta^3}\int_0^{r_1}\rho\frac{d}{dr}(Br^4)\,dr+\frac{4\pi a}{35\delta^5}\int_0^{r_1}\rho\frac{d}{dr}D\,(7r^4\delta^2-3r^6)\,dr$$

$$-\frac{4\pi aB_1}{\delta^2}\int_0^{r_1}\rho r^2dr,$$

$$0=\frac{8\pi a}{5\delta^4}\int_0^{r_1}\rho\frac{d}{dr}(Cr^5)\,dr-\frac{8\pi aC_1}{\delta^2}\int_0^{r_1}\rho r^2dr-\omega^2r_1$$

$$0=\frac{12\pi a}{7\delta^5}\int_0^{r_1}\rho\frac{d}{dr}(Dr^6)\,dr-\frac{12\pi aD_1}{\delta^2}\int_0^{r_1}\rho r^2\,dr.$$

These equations may be developed. I shall use the subscript 0 to denote values relating to the internal boundary of the fluid. Then between the limits r_0 and r_1 the density is constant, by hypothesis; I shall denote it by σ. Also for δ we must put r_1 to the order which we wish to retain.

I will now express the second of the three equations in the modified form which arises from the use of the notation just explained; the other two equations may be similarly expressed.

In this equation ω denotes the angular velocity; and I shall put as usual j for $\dfrac{\omega^2r_1}{\dfrac{4\pi}{r_1^2}\displaystyle\int_0^{r_1}\rho r^2dr}$. Thus we have

$$\frac{2a\sigma}{5r_1^4}(r_1^5C_1-r_0^5C_0)+\frac{2a}{5r_1^4}\int_0^{r_0}\rho\frac{d}{dr}(Cr^5)\,dr$$

$$=\frac{2C_1a+j}{r_1^2}\left\{\frac{\sigma}{3}(r_1^3-r_0^3)+\int_0^{r_1}\rho r^2dr\right\}.$$

This equation corresponds with D'Alembert's on page 225; he puts it so as to express a in terms of the other quantities. He takes $r_1=1$ and $C_1=1$, which he may do; but then by mistake or misprint he *also* takes $C_0=1$, which he ought not to do. This equation also exactly corresponds with equation (3) in Art. 324; the aC of the present Article is the $-\epsilon$ of that Article.

444. D'Alembert passes on his page 225 to the problem in which the entire spheroid is fluid, and is composed of indefinitely thin strata of varying densities. He treats this problem according

to his own peculiar views of hydrostatical principles. He arrives at three general equations, each of which presents itself in a *primary* and in a *derived* form, like Clairaut's equation of Arts. 341 and 343.

D'Alembert's peculiar views lead him astray, and the consequence is what we have already seen in Art. 405, namely, that the results which he obtains are much more complicated than they should have been.

For instance the third equation, with the notation of Art. 405, is presented thus in its derived form by D'Alembert:

$$\frac{d^2D}{dr^2} + \frac{2\rho r^2}{\Upsilon(r)}\frac{dD}{dr} - \left\{\frac{12}{r^2} - \frac{2\rho r}{\Upsilon(r)}\right\} D = \frac{r^3 N}{\Upsilon(r)}\frac{d}{dr}\left\{\frac{1}{r^6}\frac{d}{dr}\left(\frac{\rho}{r}\frac{dr}{d\rho}\right)\right\},$$

where N is a constant. But the correct form is that in which the right-hand member is zero.

His second equation is precisely the same as (5) of Art. 405, with C instead of ϵ; the error and the correction are the same as we have already indicated in that Article.

In like manner the derived form of D'Alembert's first equation is similarly embarrassed with a superfluous term. The K which occurs on his page 231 should be zero. D'Alembert admitted his errors in the fifth volume of his *Opuscules Mathématiques*, page 5.

445. These differential equations for C and D, when written correctly with zero on the right-hand side, are cases of the general equation, which Laplace's functions must satisfy, in Laplace's Theory of the Figure of the Earth. This general equation is

$$\frac{d^2Y_i}{dr^2}\frac{\Upsilon(r)}{r^i} + \frac{2\rho}{r^{i-2}}\frac{dY_i}{dr} + Y_i\left\{\frac{2\rho}{r^{i-1}} - \frac{i(i+1)\Upsilon(r)}{r^{i+2}}\right\} = 0.$$

If we put 2 for i we arrive at the same differential equation for Y_2 as for D'Alembert's symbol C; and if we put 3 for i we arrive at the same differential equation for Y_3 as for D'Alembert's symbol D.

Laplace shews that Y_3 must be zero. If we put $D=0$ in the differential equation for D'Alembert's symbol B, we find that this is the same as the above when 1 is put for i.

446. D'Alembert makes some remarks on the integration of the differential equations which have been obtained; see his pages 231...234. By transformation he arrives at an equation which he says is integrable in several cases; he gives three cases: they are however unintelligible to me.

447. On his pages 234...244 D'Alembert extends the calculation of the horizontal and vertical attractions which we have noticed in Arts. 437...441: he introduces two new terms into the expression for the radius vector of the attracting body, namely, $ar(E\cos^4\theta' + F\cos^5\theta')$.

I have found on going over the calculation that there are numerous misprints or errors in his results.

448. On his page 245 D'Alembert takes the case of a spheroid composed of two fluids of different densities; he says that the figures of the upper and lower strata must be determined by the law of the perpendicularity of the action to each of the strata. He adds:

Car dans le cas où les couches voisines different entr'elles sensiblement par la densité, et ont une épaisseur finie, la pesanteur doit être perpendiculaire à chacune. Voyez l'Appendice de mon *Essai sur la résistance des fluides.*

The statement he makes here about the conditions of equilibrium is true; but the reference to the *Essai sur la résistance des Fluides* is very remarkable: for the doctrine maintained in the *Essai* is precisely the reverse of that which is here affirmed in the *Recherches.* We read in the *Essai* on page 206:

Supposons maintenant que le Fluide soit composé de plusieurs couches différemment denses, et dont la différence de densités soit finie; je dis que le Fluide pourra encore être en équilibre, quoique les surfaces qui séparent ces différentes couches ne soient point de niveau...

Suppose p the pressure at any point of the surface bounding fluids of different densities. Let S be the force, if any, resolved along a tangent to the surface. Then proceeding along an element of this tangent we should have in one fluid $dp = \rho S ds$, and in the other fluid $dp = \rho' S ds$, where ρ and ρ' are unequal. But these values of dp must be equal; therefore S must $= 0$.

This assumes that there is no discontinuity in the forces acting at the common surface. In the remarks on page 206, D'Alembert's *Essai*, which follow and support the words we have quoted, he allows a discontinuity to occur in the forces.

449. D'Alembert's fifth Chapter, on pages 247...260 is entitled *De l'attraction d'un sphéroïde qui n'est pas un solide de révolution.* This Chapter is not important; it merely indicates how we ought to proceed, and shews that in some cases the integrations could be effected.

450. On his page 256 he makes a mistake to which I have drawn attention in Art. 381. He says:

J'ai fait voir, par exemple, dans mes *Recherches sur la cause des vents* art. 84. n°. 10. qu'un sphéroïde elliptique, homogene et fluide, tournant autour de son axe, ne pouvoit subsister, si les méridiens n'étoient pas tous égaux et semblables;...

451. On his page 258 he alludes to the case in which we require the attraction, not of a whole spheroid, but of a segment of a spheroid. Then on his page 259 he takes for special consideration the case of a semi-spheroid; but his first paragraph is unintelligible to me: in his second paragraph he asserts that the attraction along the radius of a semi-spheroid is half the attraction of the whole spheroid, which, however, is not necessarily true of *any* semi-spheroid, though it would be true if the whole spheroid were cut *symmetrically* into two halves.

452. Let us now appreciate the contributions to our subject which D'Alembert made in his *Recherches...Systême du Monde.*

The method of estimating the attraction of a spheroid by resolving the body into a sphere and a thin additional shell, which is here systematically employed, is very valuable.

Assuming that the radius vector of a spheroid is

$$r + \alpha r(A + B\cos\theta' + C\cos^2\theta' + D\cos^3\theta' + E\cos^4\theta' + F\cos^5\theta')$$

where α is very small, he gives expressions for the resolved attractions on any particle, external or internal, the spheroid being either homogeneous or composed of indefinitely thin strata of varying density. The calculations are laborious; and though D'Alembert's results are not free from error, yet they furnish useful information.

Retaining the terms in the radius vector as far as $D\cos^8\theta'$ inclusive, D'Alembert gave the equations which must be satisfied by B, C, D, supposed variable, to ensure the relative equilibrium of a fluid mass. His equations are encumbered with terms which are really non-existent; but still in their derived forms the remarkable similarity between them to which we have drawn attention in Art. 445 is made apparent. I consider it to be quite possible that this similarity may have struck the attention of Legendre and Laplace, and thus contributed to the construction of the general equation.

As I have already hinted, D'Alembert himself over estimated the value of the conclusions that he drew from his peculiar views of Hydrostatics. In the preface to the third volume of these *Recherches....Systême du Monde*, page xxxvi, he states that hitherto the Theory of the Figure of the Earth had been restricted to verifying the agreement of the elliptic figure with the laws of Hydrostatics; and then adds, "j'ai trouvé de plus, et je le démontre dans cet Ouvrage, qu'il y a une infinité d'autres figures qui s'accordent avec ces loix, surtout si on ne suppose pas la Terre entierement homogene." This, however, as we now know is unsatisfactory. For instance, D'Alembert indeed arrives at an equation which his symbol D must satisfy, as we saw in Art. 444; but he does not solve the equation, and so shew that D is a real quantity: on the contrary, Laplace, in fact, shews that D must be zero.

CHAPTER XIV.

BOSCOVICH AND STAY.

453. THE present Chapter will contain an account of the contributions made by Boscovich to our subject, together with a notice of the poem by Stay to which Boscovich added copious explanations.

454. In 1750 two Jesuits, Maire and Boscovich, began to measure an arc of the meridian in the Papal States. The account of the survey appeared at Rome in 1755, under the title *De Litteraria Expeditione per Pontificiam Ditionem.* The volume is in quarto; it consists of Title, Dedication, Preface, and Index in xxii pages, and the text in 516 pages: there are three pages of Errata, and four Plates. A French translation was published at Paris in 1770.

The dedication is to Benedict XIV., by whose command the survey was executed: behind the cloud of incense raised by the authors, we may discern the figure of a sagacious and enlightened Pontiff.

455. The book is divided into five parts. The first gives the history of the proceedings, the second the calculations for the determination of the length of a degree of the meridian, the third the correction of the map of the district, the fourth an account of the instruments employed, the fifth a treatise on the Figure of the Earth. The second and third parts are by Maire; the others are by Boscovich.

I shall not enter into any examination of the practical operations recorded in the volume; they have been criticised by

De Zach in his *Correspondance Astronomique,* Vol. VI.: see however, Airy's Article on the *Figure of the Earth,* in the *Encyclopædia Metropolitana,* page 207.

456. The fifth part of the book is that which we have to examine. This occupies pages 385...516, and is entitled *De Figura Telluris determinanda ex æquilibrio, et ex mensura graduum.* After a few introductory sentences, the treatise is divided into two Chapters: the first, extending to page 481, relates to the Figure of the Earth, as deduced from the theory of fluid equilibrium; the second relates to the Figure of the Earth as determined by the measure of degrees.

457. It must be observed that before the publication of the book, Boscovich had issued various dissertations, bearing more or less on our subject: these seem to have been academical exercises which he delivered in his character of professor at the Roman College. I have not seen any of these dissertations. Boscovich refers to them generally on pages xviii. and 386 of the book: from the latter page it appears that few copies of the exercises were printed, and of these the larger part perished. Probably the treatise reproduces all that was valuable with respect to our subject in the previous publications. The dates and titles of some of these dissertations are given in the pages of the work which I have recorded after them:

1738. De Telluris figura, 23.

1739. De figura Telluris, 395, 399, 445, 447, 487. Perhaps we may infer from the last three lines on page 445, that this was reprinted in a subsequent year.

1741. De Inæqualitate Gravitatis, 23.

1742. De Observationibus Astronomicis, 23, 475.

1748. De Maris Æstu, 390.

De Lege virium in natura existentium, 416. The date is not stated, but it is said *exposui nuper.*

I give the titles as I find them: it is possible however, that there may be only *one* dissertation instead of the two which appear dated 1738 and 1739.

458. The work of Boscovich, to which we now proceed, may be described in general as out of date, even when first published: it is chiefly written in an antiquated geometrical fashion, which one would have thought little likely to be adopted for this subject after the appearance of Clairaut's treatise. The Latin seems to me much more elaborate than is usual in the scientific literature of the period: this might perhaps have been expected from an Italian and a Jesuit.

459. Up to his page 417, Boscovich considers the case of homogeneous fluid attracted to a fixed point by a force which is any function of the distance, and rotating with uniform angular velocity round an axis through the fixed point: the analytical solution of this problem is very short and simple, as we have seen in Art. 56. Boscovich gives correct but tedious geometrical constructions, and devotes special attention to two cases, namely, that in which the force is constant, and that in which it varies as the distance: in this way he contrives to fill thirty pages.

Boscovich gives on his page 399 a good elementary investigation like that on Clairaut's page 143: see Art. 297.

460. A strange mistake occurs on pages 411 and 412. Boscovich has assumed a value for the radius of the equator, and has found as usual that the ratio of the centrifugal force at the equator to the attraction there, is that of 1 to 289. He adds:

......Si gradus æquatoris fuerit major, vel minor, in eadem ratione duplicata major, vel minor erit sinus versus arcus similis, adeoque et vis centrifuga, et proinde in eadem ratione duplicata minuendus erit posterior proportionis numerus.

But the word *duplicata* ought to be omitted: moreover, corresponding to the words *vel minor*, the words *vel augendus* should be inserted after *minuendus*.

461. On his page 417, Boscovich says that he will investigate the Figure of the Earth on the Newtonian hypothesis of gravity, and will illustrate in the first place Maclaurin's solution: Boscovich refers to Maclaurin's *Prize Essay on the Tides*, and not to the more complete investigations contained in the *Fluxions*.

Thus from page 417 to page 448, Boscovich may be said to reproduce in substance Maclaurin's discussion of the relative equilibrium of a mass of homogeneous rotating fluid. We shall only have to notice a few matters which present some novelty.

462. Boscovich begins with a demonstration of a theorem in Conic Sections which forms the fourth Corollary to the first Lemma in Maclaurin's Essay. Boscovich considers his own demonstration, which is geometrical, more simple and more elegant than an analytical demonstration, which he ascribes to Calandrinus, printed in what we call the Jesuits' edition of the *Principia.* Boscovich does not remark that Clairaut had already given a very good demonstration by the method of projections: see Clairaut's page 159.

463. On his page 424, Boscovich enunciates the following theorem:

Si in massa quadam fluida particulæ omnes ejusmodi viribus animatæ sint, ut assumpto intra eam puncto quocumque, bini quicumque canales rectilinei ducti inde ad superficiem extimam in æquilibrio sint, ea massa erit in æquilibrio.

In his demonstration he shews that if at every point rectilinear columns are in equilibrium, so also are curvilinear canals of every form, and that a particle at the surface has no tendency to move. The part relating to curvilinear canals is the most interesting: this, however, had already been formally treated by D'Alembert in his *Essai sur la Résistance des Fluides*, page 15.

On his page 432, Boscovich supplies in fact a demonstration of what Maclaurin contented himself with affirming in the words "in like manner it is shewn": see Art. 245.

464. Boscovich has to compare the attractions of an oblatum on a particle at the pole and at the equator respectively. After remarking on page 435, that Newton had shewn how to calculate the attraction at the pole, Boscovich adds:

......sed pro puncto posito in æquatore rem nequaquam perfecit, verum crassa quadam æstimatione invenit utcumque pro ellipsoide data, et parum abludente a sphæra. Mac-Lavrinus multo sane elegantius accuratissime, et felicissime rem perfecit tam pro puncto posito in polo, quam pro puncto posito in æquatore ;...

However, Boscovich says that he will himself adopt a method which is nearly the same as Bernoulli's; it is the method, really due to Clairaut, which we have noticed in Arts. 165 and 233. Boscovich professes to use Geometry alone: but the Geometry consists chiefly in denoting the length of every straight line by two capital letters instead of a single small letter: this strange notion of Geometry has survived to our own times in the University of Cambridge.

465. Boscovich arrives at the usual result for the attraction of the excess of an oblatum over the inscribed sphere, on a particle at the pole; and with some enthusiasm he says on his page 438, Et ea quidem est elegantissima, et simplicissima expressio ejus vis.

From this result he deduces very briefly and easily the attraction of the excess of a sphere over the inscribed oblatum on a particle at the equator: see his page 439.

Hence finally he arrives at the equation which we have frequently given in our notation; namely $\epsilon = \frac{5j}{4}$: see his page 441.

466. A digression on pages 442...447 is devoted by Boscovich to Hermann. I have already noticed Hermann's *Phoronomia*, and I presume this is the work Boscovich has in view; but it does not seem so obvious to me as to Boscovich, that Hermann held Newton and David Gregory to be wrong: see Art. 95. Boscovich says on his page 442:

Et quidem Hermannus censuit, hanc ipsam suam Ellipsim esse illam, quæ in Newtoniana gravitatis theoria debeat obvenire, ac Gregorium, et Newtonum ipsum culpandos existimavit, quod ii id ipsum non viderint, et plusquam duplo majorem justo compressionem Telluri tribuerint, quam ipsa illorum principia postularent. At Hermannus ipse in eo erravit sane quamplurimum,...

The *ipsum* after *Newtonum* marks Boscovich's opinion of Hermann's audacity.

The digression is interesting because Boscovich allows that he was himself for a time to some extent misled by Hermann. Boscovich in 1739 was thus induced to suspect that the oblatum, which Newton had assumed without demonstration, was not a possible form for relative equilibrium; but in the following year Maclaurin's demonstration settled the matter, and then Boscovich was led to investigate the cause of Hermann's error: accordingly he points out what he considers to be the *erroris primus fons,* and the *alter ejusdem erroris fons.*

It may be doubted whether Boscovich himself was quite clear on the subject; he appears to fall into the mistake which has been pointed out in Art. 33, for he does not introduce the important condition involved in the words *resolved along the radius:* see his page 443. But in his commentary on Stay's poem, at a rather later date, he is quite correct: see his Article 244 on pages 371 and 372 of Vol. II.

467. On his page 448, Boscovich gives an elegant investigation of the diminution of gravity in passing from the pole to the equator. But by gravity he really means gravity resolved *along the radius,* which is not strictly the same as the gravity which is measured by observations: see Art. 34.

468. Boscovich now proceeds to consider the Figure of the Earth when it is not supposed to be homogeneous. He assumes that there is a spherical homogeneous nucleus surrounded by fluid which is also homogeneous, but not of the same density as the nucleus: to this discussion he devotes his pages 448...457. The investigation is tedious, but was probably considered by the author to be a choice specimen of his geometrical methods. Although the whole discussion was quite superfluous after the publication of Clairaut's treatise, yet there is one matter of principle in which Boscovich is rather superior to Maclaurin. As we have already stated, when Maclaurin supposed the earth to be fluid, but not homogeneous, he did not demonstrate that the whole mass would be in equilibrium; see Arts. 264, 267, 269. Boscovich shews by his language that he saw this difficulty; he says on his page 458:

......Idcirco ego, ut methodum canalium tuto adhiberem, massam solidam prius ad homogeneitatem adduxi, amandata in centrum redundante materia, tum dissolvi.

We will give a notion of Boscovich's method. Suppose we have to consider the case in which there is a homogeneous fluid surrounding a solid spherical nucleus; and let the density in the nucleus be a function of the distance from the centre. Reduce the density of the nucleus to that of the fluid, and put a force at the centre, producing an attraction equal to that of the excess of the solid nucleus above an equal volume of fluid. Then suppose the nucleus to become fluid. If the additional force at the centre attracted as the distance from the centre, we should thus obtain a problem which has been fully discussed by Maclaurin; for he has considered forces varying as the distance from the axis and from the plane of the equator, besides the attraction of the fluid: see Art. 245. Boscovich then by a supplementary investigation has to allow for the difference between his supposed force at the centre which attracts as the distance, and the real force which would attract inversely as the square of the distance.

469. Boscovich obtains a result which, as he says, had previously been given by D'Alembert and Clairaut: see Art. 377.

Boscovich points out that the result differs from one which Daniel Bernoulli had given in his Essay on the Tides, and which had been criticised by D'Alembert. I do not stay to discuss the point, as it does not strictly belong to our subject but to that of the Tides. See D'Alembert's *Réflexions sur...des Vents*, page 56; Laplace's *Mécanique Céleste*, Vol. v. page 150; and page 8 of Lubbock's work mentioned in Art. 233.

It may however be observed that Boscovich seems to have supposed that D. Bernoulli's result ought to have coincided with his own, although the circumstances of the problems differ in a very important respect. In D. Bernoulli's problem the fluid is exposed to the attraction of a distant body, and this attraction does not reduce to a single force tending to the centre and varying as the distance, which is the case that Boscovich considers.

470. In his pages 459...465 Boscovich discusses the result which, as we have stated in Art. 469, he had obtained in agreement with D'Alembert and Clairaut. Boscovich shews that in certain cases the external surface is an oblongum, not an oblatum; it appears however from his page 463, that he held the oblongum to be in modern language an *unstable* form. See Art. 378.

471. In his pages 466...468, Boscovich demonstrates *Clairaut's Theorem*, on the same hypotheses as to the constitution of the Earth which had been used to obtain the result of Art. 469. He draws some inferences from the theorem in his pages 469...471.

472. Boscovich now proceeds to the subject of the variation of gravity as tested by experiments with the pendulum. He suggests local inequalities as the cause of the observed irregularities. He calculates the effect which would be produced on the plumb-line by the attraction of a sphere of the mean density of the Earth, of a geographical mile in radius, for various positions of the sphere; see his pages 472...474.

One of his results is that such a sphere as we have mentioned, if placed just beneath the surface of the Earth in addition to the matter already there, would increase the length of the pendulum by one-eighth of a line. Then he says that if for the depth of eight geographical miles the density at the pole is twice that at the equator the length of the pendulum at the pole will be a line longer. This, he says, follows from what has been demonstrated: but there seems to be some mistake. If r be the radius, and ρ the density, of a sphere, the attraction at the surface is $\frac{4\pi\rho r}{3}$. Now if the density at the pole is changed from ρ to 2ρ throughout the depth h, the additional result is approximately the same as would be produced by the attraction of an infinite plate of thickness h: and so it is $2\pi\rho h$. Suppose $h = 8r$; then the result becomes $16\pi\rho h$: this is *twelve* times the former result $\frac{4\pi\rho r}{3}$. Accordingly instead of an increase of *one* line in the length of the pendulum we obtain an increase of $\frac{12}{8}$ of a line, that is, of a line and a half.

473. Boscovich refers to the curious opinion expressed by Newton to which we drew attention in Art. 31; Boscovich says on his page 475:

Newtonus censuit prope æquatorem debere densitatem esse potius majorem in partibus nimirum a Sole quodammodo veluti tostis. Ego contra, cum tam multa corpora dilatentur caloris vi, et vi frigoris adstringantur, opinor debere potius rariora ibi esse corpora ob id ipsum. Sed externi caloris, et frigoris vis ad tantam altitudinem infra superficiem non pertingit, ut effectum sensibilem edat in partem utramlibet.

474. Boscovich notices the fact that according to observations made by Bouguer and La Condamine, the attraction of a large mountain in Peru was much less than it ought to have been, supposing its density equal to the mean density of the Earth: see Art. 363. Boscovich offers a conjecture in explanation of this fact; he says on his page 475:

......Verum montes quidem plerique, ut ego arbitror, effecti sunt intumescentibus interni caloris vi stratis superficiei proximis; quod quidem si ita contigit, nihil ibi materiæ accedit, et vacuus inter viscera hiatus compensat omnem illam apparentem materiæ in montem assurgentis congeriem.

475. Boscovich observes that a greater effect might be produced on the pendulum by a large tract of raised land than by a single mountain. He refers to a problem on this point which he had given in his dissertation *De Observationibus Astronomicis*, 1742. The problem is the following: cut a slice from a sphere by two parallel planes, one passing through the centre; bisect the slice by a plane perpendicular to the circular ends: then find the attraction, resolved parallel to the planes of the circular ends, of one of the halves on a particle situated at that point of the half which was originally the centre of the sphere. Boscovich states, without investigation, an approximate result for the case in which the thickness of the slice is very small compared with the radius of the sphere: but this result is incorrect. In his commentary on Stay's poem, Vol. II., page 382, he gives a correct investigation. If we wish to confine ourselves to the order of approximation which is sufficient for his numerical application, we may replace the slice of a sphere by the slice of a cylinder. Let a

be the radius of the slice of the cylinder, h the thickness, ρ the density. Then the required result is easily found to be

$$2\rho h \int_0^a \frac{dr}{\sqrt{(r^2+h^2)}},$$

that is
$$2\rho h \log \frac{a+\sqrt{(a^2+h^2)}}{h}.$$

If we suppose h very small compared with a, we get approximately $2\rho h \log \frac{2a}{h}$, that is $\rho h \left(2 \log \frac{a}{h} + 2 \log 2\right)$; this agrees closely with $\rho h \left(2 \log \frac{a}{h} + 1{\cdot}389\right)$, which is Boscovich's result in his Commentary on Stay's poem.

476. Boscovich makes a curious suggestion on his page 477. He proposes to have a pendulum in a tower by the sea shore, at some place in England or the opposite continent, where the water may be raised by the high tide 50 feet above the level of the low tide. He considers that if the density of the sea is equal to the mean density of the Earth, a deviation of about 2″ will be produced in the direction of the pendulum. By having a long pendulum and using a microscope, he thinks the deviation might be observed, and thus some notion obtained of the mean density of the Earth. See some remarks on this suggestion in De Zach's work, *L'attraction des montagnes*, page 17.

477. In his pages 477...481, Boscovich cites some observations of pendulums, and draws inferences from them: he had recently made some observations at Rome, in conjunction with La Condamine, with the pendulum which had been used in America and at the Cape of Good Hope.

478. We now reach the second Chapter of Boscovich's treatise; this relates to the Figure of the Earth, as determined by the measure of degrees.

He begins with some general explanation as to what is meant by a degree, and an osculating circle; see his pages 481...486: these present nothing of interest except a curious mistake. Let s denote an arc of a curve measured from some fixed point, ρ the radius of curvature at the variable extremity of the arc, and ϕ the inclination of ρ to a fixed straight line: then we know that $\frac{ds}{d\phi} = \rho$. By the length of a degree, we mean the value of $\int \rho d\phi$ taken between limits which differ by the circular measure of a degree. Thus the length must be equal to $\rho_1 \frac{\pi}{180}$, where ρ_1 is some value of ρ which lies between the least and the greatest of the values which occur within the range of integration, ρ being supposed always finite. This statement follows from the first principles of the Integral Calculus; Boscovich, however, denies the universal truth of it, for he says on his page 484:

> Fieri itidem potest, ut arcus unius gradus plurimum differat a gradu circuli osculantis curvam ubique intra eum arcum, quod quidem tum accidere potest, cum curvatura pergendo ab altero ejus extremo ad alterum primo quidem perpetuo crescit, tum perpetuo decrescit, vel vice versa.

479. On his page 487, Boscovich seems to adopt a definition which has not been used by others. If $2a$ and $2b$ are the major and minor axes respectively of an ellipse, we call $\frac{2b^2}{a}$ the latus rectum: Boscovich seems to call $\frac{2b^2}{a}$ the latus rectum with respect to the major axis, and $\frac{2a^2}{b}$ the latus rectum with respect to the minor axis.

In his pages 487...493, Boscovich gives various geometrical constructions relating to the ellipse and its radius of curvature; he says on page 488:

> Exhibebo autem solutiones diversas ab iis, quas simplicissimas sane, et admodum elegantes, ac geometricas itidem exhibui in mea dissertatione illa de Figura Telluris.

Thus he seems to have been very well pleased with some of his own work; for I presume we are to consider the demonstrations in the book at least as good as those which had appeared in the dissertation.

A property of the ellipse may be noticed which he demonstrates on his page 489. The normal at any point P of an ellipse meets the minor axis at G; from P a perpendicular PM is drawn to the minor axis meeting at Q the circle which is described on the minor axis as diameter; from G a straight line is drawn parallel to CQ, meeting MP produced at N; then GN is equal to half the latus rectum *with respect to the minor axis*, and MP is a mean proportional between MQ and MN.

480. Boscovich obtains on his page 494 an approximate formula, which determines the ellipticity of the Earth from the lengths of a degree of the meridian at the pole and the equator; Boscovich refers to Maupertuis, who had previously obtained the formula: see Maupertuis's *Figure de la Terre*...page 130.

Boscovich however considers that the exact theorem is more elegant, namely, that the lengths of a degree of the meridian at the equator and the pole are respectively as the inverse cubes of the corresponding diameters.

Boscovich shews on his pages 495 and 496 that the diminution of the length of a degree of the meridian from the pole to the equator varies as the square of the cosine of the latitude: hence the ellipticity may be found by measuring arcs of the meridian.

481. Boscovich now proceeds to consider the actual measures of a degree of the meridian in various places. He says that there are only five measures which are accurate; namely, those in France, in Lapland, in Peru, at the Cape of Good Hope, and his own in the Papal States: see his page 497.

He holds that the value of Picard's degree may now be considered perfectly settled *post mutationes quatuor*. It is not certain what is meant by *four changes;* in Art. 236, *four* different values are given, and these are also recorded by Boscovich himself in his commentary on Stay's poem, Vol. II., page 392. But if there were four changes, there must have been five different values: perhaps then we are to include a result obtained by J. Cassini, which was between 30 and 50 toises less than Picard's own: see *De la Grandeur et de la Figure de la Terre*, page 286.

Boscovich alludes to Norwood's measure, and gives a few lines to Snell's measure; he considers them both unsatisfactory: see Arts. 68 and 105.

482. Boscovich on his pages 499...503 deduces the ellipticity of the Earth by ten different binary combinations of the five arcs; but he finds that the results are very discordant. One combination actually brings out a negative ellipticity; namely, the combination of the Roman arc with the African. The other combinations give various values of the ellipticity, the greatest being $\frac{1}{128}$, and the least about one-tenth of this. The mean ellipticity is $\frac{1}{255}$; but if the two combinations be rejected which differ very much from the rest, the mean ellipticity is $\frac{1}{195}$.

Boscovich has some troublesome misprints on his page 501; the ellipticities deduced from his sixth and tenth combinations are quite wrong: and the numbers which he gives in his following Article to denote the mean excesses of the polar degree above the equatorial are a third of the true values.

483. Boscovich says on his page 501 that some persons had tried to conciliate the results by forcing the observations:

Nonnulli, ut nuperrime Eulerus in schediasmate, cujus summam quandam mihi humanissimè communicavit hìc Romæ præsens, dum hæc scribo, Condaminius, observationibus vim inferunt, ut omnia concilient. Et is quidem gradum Lapponiensem, Africanum, Quitensem, mutatione adhibita hexapedarum 19 in singulis, conciliat cum ellipsi Newtoniana, sed Gallicus Piccardi gradus corrigendus illi est hexapedis 169, quem idcirco sibi maximè suspectum esse profitetur, et novas in Gallia mensuras desiderat. At id quidem errorem exposcit intolerabilem sane in gradu cum ingenti cura definito a peritissimis viris.

It seems absurd to suppose that an error extravagantly the greatest should occur in the arc which must have been the best determined of all at the epoch. We shall recur to Euler's speculations in Chapter XV.

484. On his pages 502...506, Boscovich discusses Bouguer's hypothesis that the increment of the length of a degree of the meridian in proceeding from the equator to the pole varies as the fourth power of the sine of the latitude: see Art. 363. Boscovich considers that the African arc overturned this hypothesis. But then it should be observed that the African arc presented much difficulty when compared with the others.

Boscovich observes in a despairing tone: "Quocumque te vertas, nihil certum, sibi constans, et regulare occurrit."

He gives on his pages 507...510 reflections on the state of knowledge of the subject: he sums up his opinions vigorously on his page 508 as to what had been established. Instead of the inquiry respecting the Figure of the Earth from the measures of degrees being finished, he considers that it had scarcely been commenced. Still some valuable results had been obtained: the hypothesis of an attraction directed to a fixed point was excluded, and the compression at the poles was extremely probable, though the amount of this compression was uncertain.

485. On his page 510, Boscovich says that the observations were not inconsistent with the hypothesis of a nucleus in which the density, in modern language, is a function of the distance from the centre. He makes two statements, as to what Clairaut had established, which seem not strictly accurate.

One statement is this: assuming that Clairaut's fraction is greater than $\frac{5j}{4}$, *then the density of the nucleus must be greater than the mean density of the Earth,* but the ellipticity less than $\frac{5j}{4}$. If Clairaut's fraction is greater than $\frac{5j}{4}$, the ellipticity must be less than $\frac{5j}{4}$, by Art. 336. But the statement that the density of the nucleus *must be greater* than the mean density of the Earth does not seem justified by anything in Clairaut: the nearest approach to it is in the second criticism of Art. 325, but this obviously falls short of the statement.

Again Boscovich proceeds thus:

......Invenit autem ejusmodi fractionem majorem revera esse, et affirmavit ellipticitatem *minorem* erui e graduum mensura; unde intulit, ea duo conciliari non posse, nisi assumatur certa nuclei ipsius ellipticitas.

The word *minorem*, which I have put in Italics, must be a misprint for *majorem*; see Art. 349. Then for all that follows *intulit* there seems not sufficient authority; the criticism in the first paragraph of Art. 325 is somewhat short of this.

486. Boscovich considers that more observations of pendulums and more measurements of degrees are required; he admits that this would involve great labour and expense, but he adds, "at nihil est, quod Astronomorum patientia, et munificentia Regum superare non possit:" see his pages 511 and 512. Since his time the endurance of Astronomers and the liberality of Sovereigns have been largely exercised in the subject.

He repeats on his page 513 that the fact of the compression at the poles might be admitted; but the amount of the compression, and the true Figure of the Earth, were still quite uncertain.

He finishes by giving on his pages 514...516 approximate solutions of the problem to determine the Figure of the Earth, assumed to be an oblatum, from two measured degrees, one or both of which might be of longitude.

487. In forming an estimate of the treatise we must remember that the author had prescribed to himself the condition of supplying *geometrical* investigations; so the Differential Calculus was not to be introduced. We must consider the treatise rather as the work of a professor for the purposes of instruction, than of an investigator for the advancement of science; and then we may award the praise that the task proposed is fairly accomplished. It would have been more desirable to study Clairaut's work than to be confined to Boscovich's geometrical methods: but the experience of our own university shews us that it is possible to find the methods used for teaching occasionally some years in arrear of those used for investigation.

Although the mathematical processes seem a little out of date, yet Boscovich's treatise reveals, I think, great knowledge and judgment in Natural Philosophy.

488. Boscovich has an unpleasant habit of giving hints as to matters which will be found in other parts of his book, without supplying exact references; I have observed many passages of this kind, and have not always been able to determine with certainty to what he is pointing. Thus we have on page 392, "de qua fortasse aliquid alibi infra;" on page 413, "videbimus;" on page 448, "porro videbimus;" on page 455, "ut infra patebit;" on page 466, "ut infra videbimus;" on page 506, "ut vidimus;" on page 507, "ut innui etiam;" on page 508, "supra innui." None of these allusions, however, are to matters of great importance; but there is a passage of more interest on page 386:

......Expediam autem, quod ad eam gravitatis legem pertinet, sive Tellus homogenea sit, in quo argumento felicissime sane Mac Laurinus se gessit, sive diversam in diversis distantiis densitatem habeat, de quo casu multo aliter ego quidem sentio, quam summi etiam nostræ ætatis viri senserint, quorum calculos laborare omnino censeo, cum Geometria duce ad conclusiones delabar prorsus contrarias eorum conclusionibus.

I cannot see anything in the treatise which corresponds to "de quo...conclusionibus." Boscovich seems to dissent from only one person, namely Daniel Bernoulli, and D'Alembert had previously objected to the same thing: see Art. 469.

489. Boscovich himself gave an abstract of his treatise in the Bologna *Commentarii*, Vol. IV. 1757, pages 353...396. This supplies nothing of importance to our subject except three separate sentences, which I quote, because I do not understand them.

With reference to the arc in Lapland, Boscovich says on page 389:

......et in Lapponia, adhibita huic postremo illa correctione, quæ adhibita est etiam a Bouguerio, et præter quam alias adhibendas non esse, ut ut ab alio nuper adhibitas, demonstrari facile potest.

Bouguer's correction is that for refraction; I do not know what the other corrections are, nor by whom they were proposed.

After drawing an inference from Clairaut's theorem, Boscovich says on page 392:

......quod ipsum cum ego in eo opusculo diserte affirmaverim, et Clerautii theorema ipsum ex mea theoria deduxerim, ipso Clerautio nominato, miratus sane sum in opusculo nuper in Hetruria edito, me contra Clerautium hac ipsa in re adduci testem pro homogeneitate, et hoc ipsum meum indigitari opusculum.

I do not know to what book Boscovich here alludes.

Boscovich, as we shall see in our account of his commentary on Stay's poem, devised a curious method of treating discordant observations, so as to obtain the best result from them. It appears that he was now in possession of the method, and he makes a numerical application of it, though he does not give any explanation. He says on his page 392:

Invenio illud, quod in memorato volumine nequaquam quæsiveram...

I do not feel certain as to the meaning of these words, but I suppose the *memoratum volumen* to be his own treatise, of which he is giving an abstract: and then he seems to say that he had now solved the problem of the advantageous combination of observations which had not been considered by him at the time of the publication of his treatise.

The two serious misprints relating to the ellipticity, which occur on page 501 of the treatise, are reproduced on page 391 of the memoir: see Art. 482.

490. We now proceed to Stay's poem, to which Boscovich supplied a commentary. The title of the poem is, *Philosophiæ Recentioris a Benedicto Stay...versibus traditæ Libri X. cum adnotationibus, et supplementis P. Rogerii Josephi Boscovich...*

This work consists of three octavo volumes, published at Rome, the first volume in 1755, the second in 1760, and the third in 1792. We have here a treatise on Natural Philosophy in Latin hexameters, extending to more than twenty-four thousand lines. Each volume contains copious notes; and to the first and second volumes elaborate supplementary dissertations are added: these are all by Boscovich. The long interval between the publication of the second and third volumes was caused by the journeyings and incessant occupations of Boscovich, which hindered him from completing his share of the work; and he died before he had drawn up the intended supplementary dissertations for the third volume.

The number of students interested both in Natural Philosophy and in Latin Verse could scarcely ever have been large; and is probably less now than formerly. Cambridge, I hope, has never been destitute of men of such tastes, but it is curious that the University Library does not possess a complete copy of the famous work by Stay and Boscovich.

491. Dugald Stewart, in his well-known *Dissertation*, after speaking in the highest terms of Boscovich, says:

Italy is certainly the only part of Europe where mathematicians and metaphysicians of the highest rank have produced such poetry as has proceeded from the pens of Boscovich and Stay. It is in this rare balance of imagination, and of the reasoning powers, that the perfection of the human intellect will be allowed to consist; and of this balance a far greater number of instances may be quoted from Italy, (reckoning from Galileo downwards,) than in any other corner of the learned world. *Works edited by Hamilton*, Vol. I. page 424.

If I might venture to give an opinion, founded on such portions of Stay's work as I have read, I should say that it is rather versification than poetry, displaying technical skill rather than imagination. The subject, however, was not very favourable to his genius; and sometimes his lines contrast unfavourably with the simple but elegant notes of his commentator. Boscovich, however, had a high opinion of the text which he explained, for he speaks of it as *operis sane immortalis;* see the *De Litteraria Expeditione*, page 390: the French translation reduces this to *ouvrage digne de l'immortalité.*

492. The work is furnished with a preface by Boscovich, and with a letter to Benedict Stay from his brother Christopher Stay. The letter refers to Bacon and to Newton; see page xxix. While Newton's devout character is praised, the wish is gently expressed that he had known religion in its purity as well as its power.

The part of the poem which concerns us consists of the latter half of the fourth Book and the former half of the fifth Book. We may say in general terms that we have an account of the results obtained by theory as to Attractions and the Figure of the Earth, and also of the operations carried on for measuring the dimensions of the Earth.

493. It may be satisfactory to the reader to have some specimens of Stay's verses.

A passage in Book IV. beginning with line 1638 is interesting. Stay illustrates the fact that although attraction is exerted by every particle of matter, yet the disturbing effect of mountains or great buildings on a falling body vanishes in comparison with the downward action of the whole earth; he finishes thus:

> Inter saxa quidem, glebasque, herbasque virentes
> Mutua vis hæc est, et ligna, et dura metalla;
> Tellus tota tamen longe, longeque trahendo
> Prævalet, absorbetque leves has undique vires
> Ingens, atque illos conatus præpedit omnes,
> Ut Sol, cum radios Cælo jaculatur ab alto,
> Non extincta licet stellarum lumina velat.

I will take next a passage beginning at line 1941 of Book IV.; Stay has explained Newton's method of determining the Figure of the Earth, and then he proceeds to shew where it was defective, and to state that Maclaurin supplied the defect.

> At reperire suo num motu Terra diurno
> Illam debuerit, quam coni segmina prima
> Proscissi dant, induere, et circumdare formam,
> Æque etiam si densa, fluensque fuisset, ut unda,
> Inclite Vir, porro non hoc accepimus a te
> Inter munera magna, quibus nos undique ditas;
> Fors voluisti, alii ut quid tantis addere possent;
> Sic alios Rex sæpe suis ditescere gaudet
> Thesauris, atque in vulgus diffundere dona,
> Postquam ipse immensam fuerit largitus opum vim.
> Hoc donum, Laurine, tuum est; stupuere docentem
> Multa Caledoniis Mortales te quoque in oris.
> Inter multa tamen longe hoc præstantius unum est:
> Illam nempe doces formam a Tellure fuisse,
> Gyros agglomerat dum circa se, subeundam,
> Si liquida, et molem foret æque densa per omnem,
> Atque, polos inter, medias attollier oras
> Mensura circum, dixi qua nuper, eadem
> Propterea debere, atque hinc quoque crescere eodem
> Ordine, quo dixi, paulatim pondera rerum,
> Inque polos illas gravitati accedere vires.

As another specimen I will take a passage beginning at line 712 of Book v.; it is part of the description of the operations of Maupertuis and his friends in Lapland:

Postquam flumineo mensura est cognita dorso
Illa prior; montes tum qua ratione adeundi?
Undique præruptis silvæ stant montibus altæ
Verbera ventorum tantùm frangentia ramos
Perpessæ, nunquam flammas, diramque bipennem,
Obstructæ nivibus, mortali fors pede nunquam
Tentatæ; jam sunt nudanda cacumina, Cæloque
Illæ ostentandæ rupes, jam montis ad imam
Radicem aerii, Kittim dixere Coloni,
Hærendum est; illic fabricanda patentia sursum
Pastorum de more mapalia, suspicerentur
Unde faces Cæli, et sublimes verticis ulnæ,
Et sunt multa locis aptanda, movendaque multis
Instrumenta gravi molimine, Dædalus ille
Præsertim multa quæ fecerat arte Britannus,
Uranie cujus tantùm est munita labore;
Ipse gradus, graduumque dedit cognosse per arcum
Particulas senas decies in quolibet uno,
Atque harum totidem quoque fragmina particularum,
Quæ non, convexis nisi vitris, cernere, tantùm est.
Nimirum, genus hoc, arcte conclusa supellex,
Ne quid in offensu vario, compage soluta,
Turbaretur, eos montes, præruptaque curru,
Sive levi potius scandebat culmina cimba,
Consimilis cervo quam bellua juncta trahebat,
Ocyor at multo, multoque ferocior illo,
Perque nives, glaciemque, per horrida saxa volabat.
Indigenæ, rude vulgus, iners, nullisque juvare
Consiliis, operisque potens, cum sæpe viderent
Circum alienigenas fundi, atque, ut sacra ferentes,
Lente onus id vectare Viros, intus latitare
Numina credebant, Divum et procedere magnam
Matrem inter Gallos; namque illos stulta premebat
Relligio, exanimesque Deos, et inania signa
Thure coli, votisque jubens, et sanguine fuso.

494. The supplementary dissertations with which we are concerned extend from page 359 to page 426 of the second volume.

495. The first dissertation is entitled *De inæqualitate gravitatis per superficiem telluris, et figura ipsius telluris ex æquilibrio:* it occupies pages 359...380.

This may be described as an abridgement of the matter on the same topics given by Boscovich in the treatise we have already examined. Boscovich says on his page 361, referring to the former treatise:

......Ego rem totam ad solius finitæ Geometriæ vires redegi in memorato opusculo,......Singula fuse persequi, et accurate demonstrare non sinit ipsa horum supplementorum brevitas; quamobrem indicabo tantummodo methodum, quam adhibui, et theoremata præcipua, ac formulas inde erutas; ubi tamen occurrent quædam et perpolita magis, et promota ulterius, quam ibi.

I shall notice some miscellaneous matters of interest which present themselves.

496. In his Article 203, on page 359, Boscovich asserts more positively than in the former treatise, that a mass of fluid in equilibrium under no external forces must take a spherical form.

497. In his Article 209, on page 361, he is speaking about the deduction of the Figure of the Earth from the theory of gravity, and he says, "in qua perquisitione Newtonus incassum laboravit,... feliciter autem rem confecit Mac-Laurinus." This seems scarcely just to Newton, whose investigation was satisfactory as far as it went; and this is admitted by Boscovich himself elsewhere; while we do not know that Newton tried to do more and failed, as is suggested by the words *incassum laboravit.* See Art. 501.

498. In his Articles 228 and 229, on pages 366 and 367, we have a more elaborate investigation than in the corresponding part of the former treatise, which we have noticed in Art. 468. He is discussing the case in which there is a spherical nucleus surrounded by fluid; and in the present investigation, the radius of the nucleus is not assumed at first to be approximately equal to the radius of the outer surface of the fluid.

In his Article 232, on page 368, he proposes the name *fractio gravitatis,* for what we have called *Clairaut's fraction:* see Art. 336.

By the aid of what he had given in his Articles 228 and 229, Boscovich is now able to supply an investigation of Clairaut's theorem, which is rather more general than that in the former treatise: see his Article 237, on page 369.

499. His Article 238, on page 370, is important. He quotes the words from the second edition of Newton's *Principia* to which we have drawn attention in Art. 30, namely, "Hæc ita...adhuc major." It would however have been right to remark that the words were omitted in the third edition of the *Principia*. Boscovich adopts the same opinion as Clairaut, with respect to the origin of Newton's error; but states it I think more clearly; see Art. 37. Boscovich says:

......et hunc quidem Newtoni errorem Clerautius deprehendit, ac protulit. Censuit fortasse Newtonus conjectura quadam usus, et re ad geometricam trutinam nequaquam redacta, in quavis hypothesi, ut in casu homogeneitatis, vires in æquatore, et in polo, esse reciprocas distantiis, quas vidit magis augeri in polo, si massa nuclei fiat major, ob excessum gravitatis in illam massam adjectam pro loco viciniore ipsi in polo.

500. His Article 244, on pages 371 and 372, is important. He is correct as to a matter in which there is at least the appearance of error in the former treatise: see Art. 466. At the end of his Article, Boscovich indicates that he is about to investigate a certain theorem more generally than in his former treatise: the theorem is that the increase of gravity in proceeding from the equator to the poles varies as the square of the sine of the latitude.

On his pages 375 and 376, he gives tabular results as to the value of gravity at different places which are fuller than in the corresponding part of the former treatise, namely pages 479 and 480.

501. On his page 378, Boscovich expounds Newton's method of determining the Figure of the Earth; he says in his Article 264:

Clerautius in opere de figura Telluris miratur, Newtonum vidisse figuram Telluri debitam hac methodo, velut trans nebulam quandam; at mihi quidem videntur prona omnia in hac ejus methodo......Nihil in toto hoc progressu mihi videtur alienum a sagaci quidem, sed et solida, et usitata Newtoni perquirendi ratione.

But I do not find any such remark made by Clairaut as is here attributed to him; perhaps Boscovich was really thinking of a sentence with respect to Newton, which occurs in the Essay on the Tides, by Daniel Bernoulli, Chapter II., Article VIII.:

Quant à son raisonnement, il n'y a peut-être que lui, qui pût y voir clair; car ce grand homme voyoit à travers d'un voile, ce qu'un autre ne distingue qu'à peine avec un microscope.

502. The next dissertation is entitled "*De deviationibus pendulorum ex asperitate superficiei terrestris, et methodo definiendi massam terræ:* it occupies pages 380...384.

503. On his page 381, Boscovich refers to a figure which is not to be found in the book; so the reader must draw it for himself.

In the section we are now considering, Boscovich advocates the plan for determining the mass of the Earth which he had proposed in the former treatise: see Art. 476. He also suggests a modification of it. He would have constructed at royal expense in certain valleys immense reservoirs, so that they could be filled with water by the mountain streams, and again emptied at pleasure; then the position of an adjacent pendulum is to be observed before and after the reservoir was filled with water. As the form and dimensions of the reservoir would be exactly known the deviation which the mass of water would produce in the pendulum could be calculated, assuming the ratio of the density of the water to the mean density of the Earth: and then by comparison with observation this ratio would be determined.

Boscovich manifestly held very decided opinions as to the duty of governments in encouraging science.

504. The next dissertation is entitled *De veterum conatibus pro magnitudine terræ determinanda:* it occupies pages 385...389.

Boscovich refers to a separate dissertation which he had published entitled, *De Veterum argumentis pro Telluris sphæricitate:* this I have not seen.

The principal matter to notice here, is the detail of an investigation to which we alluded in Art. 475; he admits that there was a slight error in the result he formerly gave: his method is sound but laborious.

By comparing his result with that which I obtained in Art. 475, the following formula is deduced.

$$\frac{\pi}{2}-\frac{1}{6}-\frac{1}{80}-\frac{1}{336}-\frac{5}{4608}-\ldots=2\log 2,$$

that is $$\frac{\pi}{2}-u_1-u_2-u_3-u_4-\ldots=2\log 2,$$

where $$u_n=2(-1)^{n-1}\frac{\frac{1}{2}\left(\frac{1}{2}-1\right)\ldots\left(\frac{1}{2}-n+1\right)}{2n(2n+1)\lfloor n}$$

This may also be established thus:

We have $$\int_1^\infty\left\{\sin^{-1}\frac{1}{r}+\frac{\sqrt{(r^2-1)}}{r^2}-\frac{2}{r}\right\}dr=2\log 2-\frac{\pi}{2};$$

for the indefinite integral of the expression under the integral sign is $$r\sin^{-1}\frac{1}{r}+2\log\frac{r+\sqrt{(r^2-1)}}{r}-\frac{\sqrt{(r^2-1)}}{r};$$

from which the definite integral follows.

Again $$\sin^{-1}\frac{1}{r}+\frac{\sqrt{(r^2-1)}}{r^2}=\frac{2}{r^2}\int_0^1\sqrt{(r^2-x^2)}\,dx,$$

so that the definite integral $$=\int_1^\infty\left[2\int_0^1\left\{\frac{\sqrt{(r^2-x^2)}}{r^2}-\frac{1}{r}\right\}dx\right]dr$$

$$=2\int_1^\infty dr\left[\int_0^1\left\{-\frac{1}{2}\frac{x^2}{r^3}+\frac{\frac{1}{2}\left(\frac{1}{2}-1\right)}{\lfloor 2}\frac{x^4}{r^5}-\frac{\frac{1}{2}\left(\frac{1}{2}-1\right)\left(\frac{1}{2}-2\right)}{\lfloor 3}\frac{x^6}{r^7}+\ldots\right\}dx.\right.$$

Integrate with respect to r first; thus we obtain

$$2\int_0^1\left\{-\frac{1}{2}\frac{x^2}{2}+\frac{\frac{1}{2}\left(\frac{1}{2}-1\right)}{\lfloor 2}\frac{x^4}{4}-\frac{\frac{1}{2}\left(\frac{1}{2}-1\right)\left(\frac{1}{2}-2\right)}{\lfloor 3}\frac{x^6}{6}+\ldots\right\}dx,$$

that is $$2\left\{-\frac{1}{2}\,\frac{1}{2.3}+\frac{\frac{1}{2}\left(\frac{1}{2}-1\right)}{\lfloor 2}\,\frac{1}{4.5}-\frac{\frac{1}{2}\left(\frac{1}{2}-1\right)\left(\frac{1}{2}-2\right)}{\lfloor 3}\,\frac{1}{6.7}+\ldots\right\}.$$

Thus the required formula is established.

505. The next dissertation is entitled *De primis recentiorum conatibus pro determinanda magnitudine telluris:* it occupies pages 390...393.

In his Article 304 on page 391, after shewing that a certain process which seems theoretically advantageous fails by reason of practical difficulties, he concludes with this reflection:

ut quæ methodi directæ videntur primo fonte omnium aptissimæ ob theoriæ simplicitatem, plerumque fato quodam conditionis humanæ fiant maxime omnium ineptæ, et per ambages sæpe indirectas ægrè demum eo, quo tenditur, liceat evadere.

In his Article 307 on page 392, he points out the changes successively made in the French degree of the meridian originally measured by Picard, and concludes with this reflection:

Inde autem vel in hoc solo Piccarti gradu facile constat, per quas ambages, et inter quos errorum scopulos ad veritatem emergat humana mens.

506. The next dissertation is entitled *De dimensione graduum meridiani, et paralleli:* it occupies pages 393...400. This gives a good sketch of the process of measuring an arc of meridian or of longitude.

507. The next dissertation is entitled *De figura, et magnitudine terræ ex plurium graduum. comparatione:* it occupies pages 400...405.

In his Article 337 on page 402, Boscovich works out one case to which he had only alluded on page 490 of his former treatise; namely having given the length of a degree of meridian at one latitude and the length of a degree of longitude at another, to determine the axes of the Earth.

But he seems to attach the greatest importance to some approximate formulæ for the length of a degree of meridian or of longitude to which he had drawn attention in the last two pages of his former treatise. These formulæ all depend on the following approximate expression for the radius of curvature at any point of an ellipse, $\frac{a^2}{b} - \frac{3ae^2}{2}\cos^2\lambda$, where λ is the latitude, and a, b, e have their usual meaning. He says as to his formulæ in his Article 344, on page 403:

Ego quidem vix crediderim posse simpliciore, et magis uniformi methodo solvi hæc quatuor problemata....

508. The next dissertation is entitled *De recentissimis graduum dimensionibus, et figura, ac magnitudine terræ inde derivanda:* it occupies pages 406...426.

Boscovich takes the same five arcs as in his former treatise; see Art. 481. These furnish as before ten binary combinations, and therefore ten values of the ellipticity: see Art. 482. He gives the result in a Table on page 408, which may be compared with that on page 501 of the former work. He has used a slightly different formula for computing the ellipticity, so that in the later Table each denominator should exceed by 2 the corresponding denominator in the former Table. The ellipticities deduced from the ninth and tenth combinations are however quite wrong in the later Table.

509. Boscovich lays great stress on the discrepancies between the various measures of degrees; he attributes them mainly to deviations of the pendulum, produced by inequalities in the surface and the crust of the Earth. He in fact holds, as in his former treatise, that very little was really known as to the true figure of the Earth: see Art. 484. He expresses his opinions with some emphasis, and indeed it seems to me that he has allowed his feelings to disturb his attention or his judgment, for there are various misprints and some difficulties in the dissertation.

In his Article 360, on pages 409 and 410, he alludes to Bouguer's hypothesis, that the increment of the length of a degree of the meridian in passing from the equator to the pole varies as the fourth power of the sine of the latitude; but he has omitted Bouguer's name, so that the hypothesis seems to be ascribed to Clairaut or Maupertuis.

In his Article 365, on page 411, he refers to an objection he had formerly expressed when Maupertuis was supposed to have settled the exact Figure of the Earth; and for this he says, "tanquam audacissimus, et ineptus traductus sum." He goes on to speak of "illam ipsam tantam compressionem, quam in eo opusculo Maupertuisius vulgaverat,"...; but this is not accurate, for Maupertuis did not explicitly assign any value to the compression in his book, though he gave the length of his own degree, and also what he then considered to be the correct length of

Picard's degree: see Maupertuis's *Figure de la Terre*...page 126. But we have seen in Art. 177, that Clairaut once suggested incidentally a very large value of the ellipticity as obtained from the operations at the polar circle.

In his Article 371, on page 413, Boscovich says:

Multo est major utique hæc ipsa Telluris asperitas, utut tam exigua respectu totius diametri, et multis partibus major, quam, quæ totam etiam possit quadringentarum hexapedarum inæqualitatem parere, quam inter Quitensem, et Laponicum gradum observationes exhibent;...

This is not intelligible. The *difference* between the lengths of a degree of the meridian in Lapland and Peru is according to Boscovich's own Table 671 toises, not 400. But perhaps by *inæqualitas* he means not the *difference* of the two lengths, but the deviation from some theoretical standard: if so, he should have explained what the standard was, and how the deviation was estimated.

510. On his page 414, Boscovich criticises some statements made by Maupertuis in a work on Geography; and on his page 416 he animadverts on the Article relating to the Figure of the Earth in the *Encyclopédie:* the objections amount to this, that sufficient attention was not paid to the irregularity of the Earth's surface and crust.

Boscovich gives us on his page 416 the following depressing view of the course of human investigations:

At et hic quidem notare, et admirari licet humanæ gentis conditionem ubique uniformem, quæ per crebras positiones falsas, erroresque atque errorum correctiones multiplices, post erroneas observationes, erroneas etiam ratiocinationes multas ægre demum per longam observationum, et contrariarum opinionum seriem enitatur ad veritatem.

511. The most important part of this dissertation is that contained in pages 420...425. Boscovich here explains a method of his own invention for combining discordant observations so as to evolve an advantageous result. As applied to the present subject it may be stated thus: to determine the generating ellipse of the Earth's surface from the measured lengths of degrees of the meridian, under the two conditions that the sum

of the negative errors shall be numerically equal to the sum of the positive errors, and that each sum shall have the least possible value. Boscovich's exposition of his method takes a geometrical form: it is simple, clear, and instructive. Laplace gave Boscovich's method, divested of its geometrical form, in the Paris *Mémoires* for 1789; and subsequently in the *Mécanique Céleste*, Livre III. § 40. Boscovich exemplifies his method by applying it to the five arcs he had adopted; see Arts. 481 and 508: these furnish $\frac{1}{248}$ for the ellipticity. The residual errors for the length of a degree in toises for the Equator, Cape of Good Hope, Italy, France, and Lapland, are respectively 0, – 79 2, 93 8, 75 9, and – 90 5. In the French translation of Boscovich's former treatise, besides this example another is given, which involves nine measured arcs.

512. The poem of Stay, with the commentary of Boscovich, constitute a good elementary exposition of the principal results which had been obtained relative to our subject. It may be doubted whether the system on which the book is constructed is the most economical of the student's attention; for in fact various points are often treated three times, first in the verses, next in the notes, and finally in the supplementary dissertations. But probably some readers, for whom the dissertations would be too elaborate, might find the more popular parts of the work entertaining and instructive.

513. It will be convenient to notice here, though a little out of date, the French translation of the *De Litteraria Expeditione* of Maire and Boscovich: this was published at Paris in 1770 under the title of *Voyage Astronomique et Géographique, dans l'Etat de l'Eglise....* This is in quarto, containing a Title and Introductory matter on xvi pages, and the text on 526 pages; there are also four Plates and a Map. Some notes are added to the translation, and also a copious Index: the map, notes, and index render the translation more useful than the original. The name of the translator is not given; but in the life of Boscovich in the *Biographie Universelle*, the *De Litteraria Expeditione...* is said to be

"traduit en français, sous le nom de l'abbé Chatelain, par le P. Hugon, jesuite." See also La Lande's *Bibliographie Astronomique*, page 515.

514. In the part of the translation with which we are concerned there are some matters which may be briefly noticed.

On pages 449...453 there is a long note of a controversial character relating to D'Alembert; we shall mention it hereafter in connexion with D'Alembert's *Opuscules Mathématiques*, Vol. VI.

On pages 478...483 there are notes giving the results obtained by measurements in Hungary, Piedmont, and North America, which had been executed since the publication of the original work.

On pages 501...512 we have an important note. This gives us first an account of Boscovich's method of treating discordant observations, which is a translation of the exposition by Boscovich himself, published in his commentary on Stay's poem: see Art. 511. Then the method is also applied to the case of nine measured arcs, namely, the five formerly taken by Boscovich, together with four others. Also some remarks are made as to the density of a supposed spherical nucleus in the Earth.

A curious note occurs on Article 11 of Boscovich's treatise. Boscovich is speaking of relative motion, and he says that if the space in which the Earth is situated has a motion equal and opposite to that of the Earth, then the Earth itself is at rest; the note then adds:

Voici de quoi rassurer ceux qui appréhendent que le double mouvement de la terre, dans les systêmes de *Copernic* et de *Newton*, ne soit opposé au sens littéral de l'Ecriture sainte. Rien ne les empêche de supposer la terre immobile, sans rien déranger à l'économie de ces systêmes.

A note on pages 36 and 37 of the translation informs us that various measurements of degrees were undertaken at the suggestion of Boscovich; namely, those in Austria and Hungary by Liesganig, that in Piedmont by Beccaria, and that in North America by Mason and Dixon. The connexion of Boscovich with the last is thus stated:

Enfin dans son voyage en Angleterre, il a représenté à la Société Royale l'avantage qu'il y auroit de faire mesurer un dégré en Amérique, avec d'autant plus à raison, que depuis que l'Astronomie est perfectionnée, l'Angleterre n'avoit rien fait pour connoître la figure de la Terre.

The operations in England, in India, and at the Cape of Good Hope, since the time of Boscovich, have removed the reproach which is here cast on us. Perhaps we may hereafter have measurements made in Canada, Australia, and New Zealand.

A note on page 15 records the name of a person who corrected the error of Keill and Cassini; see Arts. 76 and 81:

M. des Roubais, Ingénieur chargé de poser les signaux, donna dans un Journal de Hollande, la démonstration, que les dégrés décroissans vers le pôle, faisoient la terre allongée.

See La Lande's *Bibliographie Astronomique*, page 372.

CHAPTER XV.

MISCELLANEOUS INVESTIGATIONS BETWEEN THE YEARS 1741 AND 1760.

515. THE present Chapter will contain an account of various miscellaneous investigations between the years 1741 and 1760.

I shall not in future record the titles of memoirs relating to observations of pendulums; as those which present themselves after the period at which we have arrived are given in well-known works. See La Lande's *Astronomie*, third edition, Vol. III. pages 43 and 44; Reuss's *Repertorium*...Vol. V. pages 79 and 80; and the Article on the *Figure of the Earth*, in the *Encyclopædia Metropolitana.*

516. A work was published at London, in 1741, entitled *Mercator's sailing, applied to the true figure of the Earth. With an introduction concerning the discovery and determination of that figure. By Patrick Murdoch, M.A., Rector of Stradishall, in Suffolk.*

This is a quarto volume, containing xxxii + 38 pages, and three plates of figures.

The title points out that the work consists of two parts; we are principally concerned with the first part: on this a few remarks may be made.

517. The most distinctive part of the book is the treatment of the hypothesis that the Earth is not homogeneous, but has a central nucleus denser than the surrounding fluid. Murdoch maintains that if this central nucleus is spherical, the ellipticity of the external fluid surface will be *less* than on the homogeneous

hypothesis; but if the central nucleus is an oblatum similar to the external fluid surface the ellipticity will be *greater* than on the homogeneous hypothesis. To shew this, he first gives some general reasoning on his page xxi; then he briefly sketches a mathematical investigation, and states the formulæ to which it leads on his pages xxii and xxiii; and from his formulæ he deduces numerical results on his page xxiv.

But this distinctive part of the book is unsatisfactory. In the first place, no attempt is made to shew that the mass is in relative equilibrium; but assuming it to be in that state, an equation is obtained by considering equatorial and polar columns. In the next place, since there is supposed to be a hard nucleus the columns cannot be produced to meet at the centre, and so Murdoch has to make an arbitrary supposition. This supposition expressed in modern language is that the pressure of the fluid on the nucleus is the same at the points where the equatorial and polar columns meet the nucleus. Since his results are based on these unsatisfactory principles, they cannot be accepted.

I have, however, verified his formulæ, and find that on his assumptions they are correct. I have not gone over the calculations by which his numerical results on his page xxiv are obtained.

518. Let us take one example of his numerical results from another place, namely his page xxvi. He says:

...For in one of these Examples, where the redundant Matter was a Sphere with the Radius $\frac{1}{4}$ of the Semidiameter of the Equator, if we compute its accelerating Force at the Pole, we shall find it about $\frac{38}{100}$ of the whole; and consequently the whole Density of the concentric Sphere would be to that of the ambient Matter as 42 to 1. Proportions which will not, I presume, be thought very *natural;* whereas, if the redundant Mass is a Spheroid similar to the Earth, their like Diameters being as 1 and 4, its accelerating Force at the Pole will be only $\frac{48}{1000}$, and the whole Density of the Spheroid to that of the ambient Matter, in little more than the Ratio of 1307 to 1000.

On this passage, I remark that the great discrepancy between the two results, when so slight a change is made in the hypothesis as the transition from a spherical nucleus to an oblate nucleus, should have shaken Murdoch's faith in his whole process.

But at the same time I do not see how his numerical results are obtained.

Let σ denote the density of the *redundant* matter, and ρ the density of the *ambient* matter. Let a denote the major semiaxis of the Earth, b the minor semiaxis, and ϵ the ellipticity. Then assuming the correctness of his fraction $\frac{38}{100}$, which depends on his preceding formulæ, we have

$$\left(\frac{a}{4}\right)^3 \frac{\sigma}{b^2} = \frac{38}{100}\left\{\left(\frac{a}{4}\right)^3 \frac{\sigma}{b^2} + b\left(1+\frac{4}{5}\epsilon\right)\rho\right\},$$

so that

$$\frac{a^3}{b^2}\frac{62}{64}\sigma = 38b\left(1+\frac{4}{5}\epsilon\right)\rho,$$

and

$$\frac{\sigma}{\rho} = \frac{38\times 64}{62}\,\frac{1+\frac{4}{5}\epsilon}{1+3\epsilon} = \frac{38\times 64}{62}\left(1-\frac{11}{5}\epsilon\right)$$

If we put $\epsilon = 0$, we get $\frac{\sigma}{\rho}$ less than 40; if we put, as Murdoch does elsewhere implicitly $\epsilon = \frac{1}{91}$, we get $\frac{\sigma}{\rho}$ less than 39. I presume that he means to say we get $\frac{\sigma+\rho}{\rho} = 42$.

In the second example proceeding in a similar way, I find approximately

$$\frac{\sigma}{\rho} = \frac{48\times 64}{952}\,\frac{1+\frac{4}{5}\epsilon}{1+2\epsilon} = \frac{48\times 64}{952}\left(1-\frac{6}{5}\epsilon\right),$$

so that $\frac{\sigma}{\rho}$ is about 3, and $\frac{\sigma+\rho}{\rho}$ about 4: this differs altogether from Murdoch's result.

519. On page xxvi. the passage is quoted from the second edition of Newton's Principia which corresponds to that from the first edition which we have quoted in Art. 37.

Murdoch considers Newton's language to indicate that he intended his nucleus to be not spherical, but oblate; and Murdoch thinks that D. Gregory in his Prop. 52, Lib. III. overlooked this. But I do not believe that Newton really intended to discriminate between these two forms for a nucleus.

On page xxvii. there is a reference to the Principia, Lib. I. Prop. 91, Cor. 2; but this passage has no bearing on the matter which Murdoch is discussing.

On page xxix. there is a note on the erroneous notion which Cassini held as to the figure of the Earth, in these words: "He has, I am told, of late ingenuously owned his Mistake."

On his page xxxi. Murdoch is speaking of the operations of Maupertuis, and exhibits that inaccuracy which by some fatality seems to cling to all the derived accounts of this measurement: see Art. 199.

Murdoch says "... after proper Allowances for the Refraction of Light, the Precession of the Equinoxes and Mr Bradley's Equation...". But in fact no allowance was made for refraction, as Murdoch himself admits on his page xxx. By Mr. Bradley's equation is meant what we call Aberration. Besides Precession and Aberration there was a correction for Nutation.

520. The part of Murdoch's work which is called *Mercator's sailing applied to the true figure of the Earth*, does not really fall within our scope; and so we shall not give any great attention to it. We may say generally, that the object is to construct maps of the Earth's surface, assuming the form to be an oblatum, like the maps on what is called *Mercator's projection* for a spherical form: or it is practically equivalent to this.

Murdoch is unfortunate in the value he adopts for the ellipticity; in modern notation he takes $e^2 = \cdot 022$, so that $\frac{e^2}{2}$ which is approximately the ellipticity, is about $\frac{1}{91}$. I presume that he deduced this value from the degree in Lapland, combined with Picard's degree, taking the latter at the amount assigned to it by Maupertuis in his *Figure de la Terre*...page 126. This amount for Picard's degree was soon afterwards found to be too small. In consequence of the very large value assigned to the ellipticity, maps constructed according to Murdoch's tables would in general be more erroneous than maps constructed on the hypothesis of the spherical form of the Earth.

521. Murdoch's work was translated into French under the title *Nouvelles Tables Loxodromiques...par M. Murdoch. Traduit de l'Anglois. Par M. De Bre'mond*...Paris, 1742.

This is an octavo volume consisting of xvi + 158 pages, besides the *Privilege du Roy* on four pages. There are four plates of figures.

The translator dedicates the book to Le Comte de Maurepas, the French minister who was very much concerned with the expeditions sent to Peru and to Lapland.

The following sentence from page vii. is of interest:

Malgré ce qu'un autre Auteur Anglois prétend qu'a pensé Strabon sur l'applatissement de la Terre, celui-ci a l'équité d'avouer que tous les Philosophes et les Géographes n'attribuoient point à la Terre d'autre Figure que celle d'un Globe parfait, avant la fameuse Expérience faite à Cayenne en 1672. par M. Richer Astronome François.

On page 19, there is a note on the passage of Strabo; and it is maintained that the passage does not shew Polybius to have been acquainted with the true figure of the Earth.

It ought to have been stated that this note is due to the translator, and not to Murdoch himself.

I have noticed the passage in Strabo already: see Art. 152.

522. The pages 27...46 consist of an important addition sent by the author to the translator. The essence of this addition is to be found on page 43, namely formulæ which give the attraction at the pole and at the equator, both for an oblatum, and an oblongum. These formulæ are not demonstrated, but differential expressions are investigated which will lead to the formulæ by integration. The formulæ are correct.

Let M denote the ratio of the attraction at the equator of an ellipsoid of revolution to the attraction of a concentric sphere touching the ellipsoid at that point; let N denote the ratio of the attraction at the pole of the ellipsoid of revolution to the attraction of a concentric sphere touching the ellipsoid at that point. Then whether the ellipsoid of revolution be an oblatum or an oblongum, we shall have

$$\frac{N}{2} + M = \frac{3}{2}.$$

This formula is given, though not quite with this notation, on page 44: by attending to the formula we can discover the meaning of the first seven lines of page 45, to which the printer has not done justice.

523. The addition which Murdoch sent to his translator appeared in the same year as Maclaurin's *Fluxions:* but, as we have seen, Maclaurin had been substantially in possession of the results respecting the attraction of an ellipsoid of revolution at the time he wrote his Prize Essay on the Tides. Thus Maclaurin's claim to be the first who completely solved the problem of the attraction of an ellipsoid of revolution on a particle within the body, or on its surface, remains untouched.

524. Another addition sent by the author to the translator is given on pages 104...108. This relates to the part of the work which treats on the construction of maps. The addition is due to a suggestion made by Maclaurin to Murdoch, and it effects a great improvement in the mathematical investigation. See Maclaurin's *Fluxions,* Arts. 895...899.

525. The translation is not very well executed. Some passages are unintelligible, where the original is quite clear; as an example may be mentioned a passage about Antipodes, on page 129 of the translation and page 18 of the original.

The following is a curious specimen of a misprint. On page 142 of the translation, we have *b*El. 4. By turning to the original, page 29, we find it should be 6 El. 4; here El. stands for Elements, and so what is meant is, *Euclid,* VI. 4.

526. In the Paris *Mémoires* for 1742, published in 1745, there is an article, entitled *Sur la Figure de la Terre,* which occupies pages 86...104 of the historical part of the volume: the article is by Mairan, as appears from page 92.

We have here a notice of Clairaut's *Figure de la Terre,* preceded by a sketch of the history of the subject; there is, however, nothing of importance for us.

It is remarked that even before the observations made by Richer at Cayenne, on the length of the seconds pendulum, it had been suspected at the Academy that the length ought to become shorter as the equator is approached; to support this remark the fourth Article of Picard's work is cited.

Mairan uses the word *Pesanteur*, not in the sense adopted by Maupertuis and Clairaut, but for the Earth's action apart from centrifugal force: see Art. 299. Mairan also uses the words *gravitation, gravité, attraction;* but without any apparent aim at precision: see his pages 98 and 103.

527. The results obtained in the geodetical operations which had been carried on during some years in France, were published at Paris in 1744, by Cassini de Thury under the title of *La Meridienne de l'Observatoire Royal de Paris, vérifiée dans toute l'étendue du Royaume par de nouvelles observations.* The volume is in quarto: it contains Half-title, Title, Table, then pages 292 + ccxxxv; followed by an alphabetical list of places, and the *Privilege du Roy:* there are xiv Plates.

On pages 42...51 of the historical part of the Paris *Mémoires* for 1744, published in 1748, we have an account of this work. On pages 237...244 of the Paris *Mémoires* for 1758, published in 1763, we have some corrections by La Caille of the results obtained in the work.

The *Discours Preliminaire* with which the volume commences gives a brief account of the operations. The formal admission is made by Cassini, that the length of a degree of the meridian increases from the equator; and that the Earth is therefore oblate: see his page 25. Thus the error which he had maintained after his father and grandfather is abandoned.

The present work supersedes the *De la Grandeur et de la Figure de la Terre,* and has in its turn been superseded by the *Base du Système Métrique...*

528. The volume of the Paris *Mémoires* for 1745, which was published in 1749, contains a controversy between Clairaut and Buffon, which we must notice.

Clairaut, in investigating the Lunar Theory, obtained for the motion of the apse line a result about half as great as that assigned by observation. In order to explain the difficulty, he proposed to change the law of attraction, by adding another term to the ordinary expression, which varies inversely as the square of the distance. But he soon discovered and admitted his error as to the Lunar Theory: see page 577 of the volume.

In the controversy, Buffon attempted to shew that it was necessarily impossible for the law of gravity to be expressed by the aggregate of two terms, one varying inversely as the square of the distance, and the other varying inversely as the fourth power of the distance: but his reasons are quite inconclusive. Clairaut maintained justly that there was nothing absurd in such a supposition. The controversy consists of six papers, three by each disputant; but it does not seem that all which was spoken or written, was printed.

Clairaut refers to the discrepancy between theory and observation relative to the figure of the Earth, as throwing suspicion on the ordinary law of attraction; but he admits that he had not attempted to discuss the problem on his hypothetical law: see his pages 531 and 547.

529. We have next to notice a memoir entitled *Eustachii Zanotti De figura Telluris.* This memoir was published in the *De Bononiensi Scientiarum et Artium Instituto atque Academia Commentarii,* Vol. II. Part 2, Bononiæ, 1746. The memoir occupies pages 210...227 of the volume.

Assuming that the Earth is an ellipsoid of revolution, Zanottus shews how the dimensions of the ellipse may be found from the measured lengths of two arcs, either of the meridian, or of a normal section at right angles to the meridian. There is nothing remarkable about the geometrical processes. Zanottus employs the theorem which had been demonstrated by Clairaut respecting the radius of curvature of the section at right angles to the meridian, and he refers to Clairaut's memoir: see Art. 161.

There is an account of the memoir on pages 442...451 of Part 1, of Vol. II. of these Bologna Transactions, which is dated 1745. This is a very lively and interesting notice. The liberality of the French king is commended for undertaking the expense of the Arctic and Equatorial expeditions. Zanottus thought that it would be an honour to the Italians, if they contributed something towards the solution of the problem, before Godin returned from America, and finally settled the question. Accordingly, Zanottus proposed to execute a measure of an arc at right angles to the meridian of Bologna; he explained and enforced his plan in a meeting of the Academy, but without success. We read

...Invitavit; rogavit; obsecravit. Multos etiam commovit; laboris socios sibi adjunxit; sed Ludovicus Magnus in corona non adfuit. Tamen, etsi rem non perfecit, spem retinuit, et voluisse non pœnituit. Quod dicimus, ut qui italorum ingenio nihil tribuunt, voluntati certe, si quid voluntas apud ipsos mereri potest, dent aliquid. Quamquam et ingenio tribuent fortasse non nihil, si Cassinum meminerint fuisse nostrum.

530. The volume of the Paris *Mémoires* for 1747, published in 1752, contains a memoir by La Condamine on an invariable measure of length. An abstract of this memoir in viii pages is found in some copies of the work XII of Art. 352.

The volume of the Paris *Mémoires* for 1748, published in 1752, contains a memoir by Cassini de Thury on the junction of the Meridian of Paris with that which had been traced by Snell in Holland: see Art. 105.

531. A problem occurs connected with our subject on pages 175, 176 of the *Mémoires de Mathématique...par divers Sçavans...* Vol. I. Paris, 1750. The problem is entitled *Supposant la loi d'attraction en raison inverse du quarré de la distance, trouver la*

nature du solide de la plus grande attraction. Par M. de Saint-Jacques.

The author's name is elsewhere increased by the addition of the words *de Silvabelle.*

The problem is well solved in two ways. In one solution, the early method of treating problems in the Calculus of Variations is used. In the other solution, a simpler method is adopted. Both ways of solution have been since reproduced. See my *History... of the Calculus of Variations,* pages 361 and 484.

For some account of Saint-Jacques de Silvabelle, see De Zach's work, *L'attraction des montagnes*...page 588.

532. The *Philosophical Transactions,* Vol. XLVIII. part I. for the year 1753, published in 1754, contains an article on our subject under the following title: *An account of a Book intitled, P. D. Pauli Frisii Mediolanensis, &c. Disquisitio mathematica in causam physicam figuræ et magnitudinis Telluris nostræ; printed at Milan, in 1752. Inscribed to the Count de Sylva, and consisting of Ten Sheets and a half in Quarto: By Mr. J. Short, F.R.S.*

This article occupies pages 5...17, of the volume.

I have never seen this dissertation by Frisi; but I presume, it was incorporated by Frisi in his *Cosmographiæ...Pars altera...* which was published in 1775, of which we shall give an account hereafter.

Short speaks in high terms of Frisi, thus:

This does not, however, in the least detract from the merit of F. Frisi; who discovers throughout this work much acuteness and skill, joined with all the candour and ingenuity, that become a philosopher. And as he has not yet exceeded his 23d year, it may be expected, that the sciences will one day be greatly indebted to him; especially as we find him actually engaged in composing a complete body of physico-mathematical learning.

533. The most important part of the article consists of the defence of a passage in Newton, which Frisi had misunderstood and asserted to be erroneous. The passage is that of which we give an account in Art. 22. Newton uses twice in the sentence

the phrase *in eadem ratione.* Then as Short says: "In which the expression *eadem ratione* occurring a second time has misled F. Frisi and others, to think this last ratio to be likewise that of the axes, or of 101 to 100." In fact, however, in the second case, the ratio is not that of 101 to 100; but that of 126 to $125\frac{1}{2}$. The context shews clearly, that Newton is quite correct in what he means to say. Frisi, however, was not convinced that the error lay with himself, instead of with Newton: see page 123 of the book cited in Art. 532. As we have seen in Art. 137, Maupertuis also appears to have been misled by Newton's words.

534. The defence of Newton seems to have been supplied by Murdoch, as appears from the following passage in Short's article:

I sent F. Frisi's book to my ingenious and learned friend the reverend Mr. Murdock, Fellow of this Society, who has fully consider'd the question concerning the figure of the earth; and who, after having perused the book, and discover'd the above mistake of F. Frisi, sent me the above theorem, and its demonstration. He likewise sent me the following theorems, which, he says, he had communicated to M. de Bremond, in the year 1740, when he was translating his treatise on sailing: But M. de Bremond dying soon after, those, who had the care of publishing the translation, printed it incorrect in several places; particularly the theorems for the prolate spheroid: On which account, he says, if they are thought worth preserving, they may be inserted in the *Philosophical Transactions.*

Accordingly expressions follow, which amount to giving the values of the attraction of an oblatum or an oblongum, at the pole or at the equator. But it was unnecessary to publish them now, because Maclaurin had completely solved the problem of the attraction of an ellipsoid of revolution on a particle at the surface. Moreover all that is here given is also in my copy of Bremond's translation, pages 43 and 44, and printed quite correctly: so that the above statement seems unjustifiable: it is however possible that the original page was cancelled, and a reprint substituted.

I may say that I do not understand how a numerical result is obtained, which is ascribed to Frisi on the fourth line of page 10. And on page 17, after "whose tangent is $\sqrt{(m^2-1)}$" some words follow which I do not understand, but which seem to me unnecessary. Murdoch spells his name so himself; but others sometimes spell it Murdock.

535. We pass to another article which is connected with Short's, and is published in the same volume of the *Philosophical Transactions.* This is entitled *A Translation and Explanation of some Articles of the Book intitled, Theorie de la Figure de la Terre; by Mons. Clairaut, of the Royal Academy of Sciences at Paris, and* F.R.S.

This article occupies pages 73...85 of the volume. I am not certain, but presume, that the paper is by Clairaut himself; it is written in the first person throughout, but it is not ascribed to Clairaut in Maty's *General Index to the Philosophical Transactions.*

536. Frisi considered that as to the Figure of the Earth, Boscovich underestimated the observations, while Clairaut, Bouguer, and others, underestimated the theory of Newton. Short, in his account of Frisi's dissertation, quoted the opinions. The present paper begins thus:

Mr. Short, in his account of Father Frisius's *Disquisitio mathematica in causam physicam figurae et magnitudinis telluris nostrae,* having reported that philosopher's sentiments on my reflections upon the same matter, without taking the trouble to examine whether they were founded upon the truth or not, I find myself under the necessity of laying before the Royal Society the passages of my book, which, having been misunderstood by F. Frisius, have occasioned the misconstruction, which he has made of my sentiments, either upon the trust I give to the actual operation made for discovering the figure of the earth, or Sir Isaac Newton's theoretical inquiries about the same subject.

The expressions of Father Frisius, referr'd to by Mr. Short, are as follow:

Quia tamen plerique omnes hucusque, aut nihil pro figura telluris determinanda ex iis observationibus deduci posse cum geometra celeberrimo Ruggero Boscovik autumârunt, aut exinde cum Ill. Clairaut, Bouguer, aliisque, contra incomparabilem virum ac prope divinum Isaacum Newton insurgentes, admirabilem ipsius theoriam facto minus respondentem dixerunt, assignatamque in prop. 19. lib. 3. *Princip. Mathem.* terrestrium axium proportionem à vera absonam omnino esse, alios mihi observationibus parum, alios nimis tribuere visum est, omnes ferme oppositis erroribus peccâsse, ubi res neque aurificis lance, neque molitoris, ut aiunt, statera librandæ sunt.

537. Clairaut makes various conjectures as to what was the precise meaning of the charge brought against him by Frisi. Clairaut shews quite clearly that he had not given undue importance to observations, and had not undervalued Newton. The paper contains translations of the sections 51, 68, 69, and a portion of 70, from the second part of Clairaut's treatise.

With respect to the matter we discussed in Art. 533, Clairaut says:

After F. Frisius has examined himself the 19 problem of the third book of the *Principia*,......the truth of which is incontestable, he finds, by his own mistake, a disagreement with the result of that proposition, and charges that illustrious author, without the least apology, with an error, which, says he, (quite from the purpose) is the sixth, that has been found in the same work, and also gives an enumeration of the five others, altho' they are not at all concerned in the question.

538. The volume of the Paris *Mémoires* for 1751, published in 1755, contains a memoir by Bouguer, entitled *Remarques sur les observations de la parallaxe de la Lune, qu'on pourroit faire en même temps en plusieurs endroits, avec la méthode d'évaluer les changemens que cause à ces parallaxes la Figure de la Terre.* The memoir occupies pages 64...86 of the volume. There is an account of it on pages 152...158 of the historical part of the volume.

The memoir contains some interesting mathematical results, connected with the curve which Bouguer called the *gravicentrique* in his *Figure de la Terre:* see Art. 363.

Bouguer maintains very sound opinions on the subject he discusses. If we were uncertain as to whether the figure of the Earth is oblate or oblong, then observations of the Moon's parallax might remove the uncertainty. But at the actual epoch this point was settled; the only question was to fix the amount of the ellipticity of the oblate figure, and the observations could not, practically, be of sufficient accuracy for this purpose. But if we assume a value of the ellipticity, the corrections which have to be made to the observations of the parallax in consequence of the figure of the Earth may be calculated: and the results will be important and useful.

539. La Caille was sent from France in 1750 to the Cape of Good Hope for the purpose of making astronomical observations. He proposed to determine the positions of the southern stars; the parallax of the Moon, of Mars, and of Venus; and the latitude and longitude of the Cape.

La Caille resided in the country from April 1751 until March 1753. Besides the duties which he had specially undertaken, he found time to measure an arc of the meridian. The amplitude was rather more than 1° 13′; and he obtained 57037 toises for the length of the degree of the meridian which has its middle point in latitude 33° 18′½ S. La Caille also determined the length of the seconds pendulum.

The details of the operations connected with our subject are given in the Paris *Mémoires* for 1751, published in 1755. The volume contains two memoirs embodying observations made by La Caille: the pages 425...438 are devoted to the lengths of the degree and of the pendulum.

There is also a short account of the voyage on pages 519...536 of the volume. La Caille touched at Rio Janeiro on his outward passage, and there he met Godin who was returning from his long sojourn in South America. La Caille states on his page 524, that the southern hemisphere has more stars than the northern; this statement is confirmed by actual enumeration: see *Monthly Notices of the Royal Astronomical Society*, Vol. XXXI. page 30.

There is a notice of La Caille's voyage and work on pages 158...169 of the historical portion of the Paris *Mémoires* for 1751. In one point this contradicts La Caille; for it says that he determined the position of Rio Janeiro, while he says himself that it was unnecessary for him to do this as he had been anticipated by Godin.

The positions of the stars in the southern hemisphere, between the Pole and the Tropic of Capricorn, as determined by La Caille, are given in pages 539...592 of the Paris *Mémoires* for 1752.

La Caille found it expedient to construct fourteen new constellations; but at the same time he suppressed the constellation of *Charles's Oak*, which he considers that Halley had fabricated by pillaging fine stars from the neighbourhood. He says on his page 591:

On n'y trouvera pas la constellation nouvelle que M. Halley a insérée dans son Planisphère en 1677, sous le nom de *Robur Carolinum*, parce que j'ai rendu au Navire les belles étoiles que cet Astronome, âgé alors de vingt-un ans, en a détachées pour faire sa cour au roi d'Angleterre. Quelque louable qu'ait été ce motif, je ne puis approuver la façon dont M. Halley s'y est pris pour faire passer sa constellation;...

540. A volume was published in Paris in 1763, entitled *Journal historique du Voyage fait au Cap de Bonne-Espérance...* 12mo., pages xxxvi + 380, besides the *Approbation* on four pages. There is a planisphere of the stars between the South Pole and the Tropic of Capricorn, which is reduced from that published in the Paris *Mémoires* for 1752; and a map of the country in the vicinity of the Cape, which is reduced from that published in the Paris *Mémoires* for 1751: the triangles of the survey are marked on the map. As to the authorship of the book see La Lande's *Bibliographie Astronomique*, page 482.

On page 25 of this book we find the work *La Meridienne de Paris vérifiée*, ascribed to La Caille, though his name is not on the title-page. Delambre also considers that the entire operation belonged to La Caille: see the *Base du Système Métrique*...Vol. III. *Avertissement*, page 13.

Among the memoirs ascribed to La Caille in the *Journal historique...*, on pages 71 and 72, we have one *sur la précision de la mesure de M. Picard*, and one *sur la base de Ville-Juive*: the titles, however, do not seem given with great accuracy. The former we identify by aid of a note on page 102, with the memoir published in the Berlin *Mémoires* for 1754; the latter is probably the memoir published in the Paris *Mémoires* for 1758: we shall notice these memoirs in Arts. 546 and 553.

The earliest entry in La Caille's journal which suggests the measurement of an arc of the meridian is dated September, 1751: see page 144 of the book.

541. The length of a degree of the meridian assigned by La Caille was always perplexing to theoretical investigators, being apparently much greater than it should have been; at the same time, the reputation of La Caille for accuracy ensured respect for his result: see for example, Delambre's opinion, *Base du Système Métrique*...Vol. III. page 544; and Airy's in the article on the *Figure of the Earth*, in the *Encyclopædia Metropolitana*, page 207. Consult also pages 463...465 of De Zach's work, *L'Attraction des montagnes*.... I do not know whether De Zach ever published his promised memoir on this arc.

542. A very extensive geodetical operation has been executed in South Africa in recent times, and the results published in two quarto volumes entitled *Verification and extension of La Caille's Arc of Meridian at the Cape of Good Hope, by Sir Thomas Maclear, Astronomer Royal at the Cape of Good Hope*, 1866. See also the *Proceedings of the Royal Society*, Vol. xviii. page 109.

These volumes have no Index, and no general summary of contents to guide the reader; so that it is difficult to ensure perfect accuracy in noticing their contents.

The amplitude of Sir T. Maclear's arc exceeds $4\frac{1}{2}^{\circ}$, and the length agrees closely with the value which it should have in order to correspond with the average of the arcs measured in the Northern hemisphere: see Vol. I. page 609.

The amplitude of La Caille's arc was redetermined; the result does not differ from La Caille's by so much as half a second. Sir T. Maclear's observations were made with the zenith sector, which Bradley had used in his discovery of Aberration and Nutation; the object glass however was not the same: see Vol. I. page 80. We read in Vol. I. page 111 with respect to the redetermination of La Caille's amplitude:

Although this work does not clear up the anomaly of LA CAILLE'S arc, yet it redounds to the credit of that justly distinguished astronomer, that with his means, and in his day, his result from 16 stars is almost identical with that from 1133 observations on 40 stars made with a powerful and celebrated instrument.

The remarkable accuracy of La Caille's *amplitude* would seem naturally to have suggested a recomputation of the *length* of his arc; this will not be found in the volumes, and so we are left uncertain whether La Caille made some mistakes in his geodetical work, or whether the amplitude owing to deviations of the pendulum really was greater than corresponds to the terrestrial arc. There are indications that some investigation on this point was contemplated; see Vol. I. pages 232, 403, 452, 456: it is much to be regretted that this interesting question was not settled. We learn the nature of La Caille's northern station from Vol. I. pages 39, and 403; we are told of a mountain about half a mile distant which is not less than 2500 feet high.

Reference is made to the attraction of Table Mountain, Vol. I. pages 3 and 83; but the subject does not seem to have been followed up afterwards.

A letter however from Sir T. Maclear will be found in the *Astronomische Nachrichten,* Number 574, September 3rd, 1846, in which he does give some comparison between La Caille's geodetical work, and his own: the opinion is there expressed that "The chief cause of the failure of the measurement of 1752 rests with the circumstances of the terminal points."

543. In the Paris *Mémoires* for 1752, published in 1756, there is a memoir entitled *Premier Mémoire sur la Parallaxe de la Lune...Par M. Le François de la Lande.* The memoir occupies pages 78...114 of the volume.

The memoir discusses the observations of the Moon made simultaneously, by La Lande at Berlin, and La Caille, at the Cape of Good Hope. It touches on our subject in pages 100...114; here we find some theory as to the evolute of the meridian, which is borrowed from Bouguer's *Figure de la Terre:* see Art. 363.

La Lande notices three hypotheses as to the form and dimensions of the Earth.

First, he supposes the Earth to be an oblatum in which the excentricity is $\frac{1}{179}$.

Secondly, he takes Bouguer's hypothesis, that the increment of the length of a degree of the meridian varies as the fourth power of the sine of the latitude.

Thirdly, he returns to the oblatum, but applies *arbitrary* corrections to the three measured degrees of meridian then received; namely he adds 77 toises to the length of the degree as found from the arc between Paris and Amiens, and subtracts 77 toises from the Lapland degree, and he adds 26 toises to the Peruvian degree. All that he says in justification of this process is on his page 110:

Il me paroit d'abord naturel de supposer dans les mesures faites au Pérou, une erreur qui ne soit que le tiers de celle que je supposerai dans le degré de Lapponie et dans celui de Paris à Amiens, puisque dans ces deux derniers on n'a mesuré qu'une amplitude d'un degré, tandis qu'au Pérou l'arc se trouve de trois degrés, et mesuré avec différens instrumens.

By these changes, La Lande obtains for the ratio of the axes of his ellipse $\frac{1}{1\ 043}$, which he says is nearly $\frac{232}{233}$, and so does not differ much from Newton's value, namely $\frac{229}{230}$. La Lande's fraction should be $\frac{1}{1\ 0043}$

There is an account of the memoir on pages 103...110 of the historical portion of the volume of *Mémoires*. We have only to notice an important error on page 108; here it is stated that La Lande had to apply some rather large corrections to the lengths of the degrees of the meridian to make them fit Bouguer's hypothesis; whereas it really was to make them fit with the figure of an oblatum.

La Lande seems to have viewed his arbitrary corrections with some satisfaction, for he refers to them about 40 years later: see his *Astronomie*, 1792, Vol. III. page 32.

544. La Lande's second memoir on the Parallax of the Moon, is in the Paris *Mémoires* for 1753, published in 1757. Here, La Lande continues to use Bouguer's hypothesis; and he also

takes another modification of the elliptic hypothesis founded on the arcs in Lapland and Peru, from which he gets $\frac{1}{185}$ for the ellipticity There is an account of this memoir on pages 225...228 of the historical portion of the volume; reference is again made to the first memoir, without the error which occurs in the historical portion of the volume for 1752.

545. A memoir by Euler appears in the Berlin *Mémoires* for 1753, published in 1755, entitled *Elémens de la Trigonométrie sphéroïdique tirés de la Méthode des plus grands et plus petits.* The memoir occupies pages 258...293 of the volume.

The memoir may be said to consist of two parts.

In the first part, Euler takes the lengths of a degree of the meridian as determined in Peru, South Africa, France, and Lapland. He assumes that there are errors in all the measures, and by means of arbitrary corrections he adjusts the lengths to coincide with the ellipticity $\frac{1}{230}$ obtained by Newton from theory. Euler's corrections increase the Peruvian degree by 15 toises, and the French degree by 125; they diminish the African degree by 43 toises, and the Lapland degree by the same amount, supposing here no allowance to be made for refraction. I presume that this is the memoir which Boscovich had in view, though the numbers are rather less extravagant than Boscovich stated: see Art. 483.

In the second part of the memoir, Euler gives approximate investigations respecting the shortest line between two points on the surface of an ellipsoid of revolution. He suggests a method of using such a line for determining the Figure of the Earth: the angles which the line at its extreme points makes with the meridians are to be observed. But at the end of the memoir, Euler admits that the method could not practically be applied.

546. In the Berlin *Mémoires* for 1754, published in 1756, we have a memoir by La Caille, entitled *Eclaircissemens sur les erreurs qu'on peut attribuer à la mesure du degré en France, entre Paris et Amiens.* The memoir occupies pages 337...346 of the volume.

La Caille strenuously defends the French measurement from the charge of serious error which Euler had in fact brought against it in the Berlin *Mémoires* for 1753: see Art. 545. La Caille is willing to stake his reputation on the statement that there cannot be an error of from 12 to 15 toises in the distance which had been determined between Paris and Amiens.

A few explanatory sentences by Euler are given on the last page of the memoir. There is an allusion to the memoir in the *Base du Système Métrique*, Vol. III. page 543.

547. In the fourth volume of the *Commentarii Soc. Reg. Gottingensis* 1754, we have a memoir entitled *Succinctam attractionis historiam, cum epicrisi, recitavit Sam. Christ. Hollmannus.* The memoir occupies pages 215...244 of the volume.

This memoir is not mathematical, and so does not fall within our range. The author holds that the word *attraction* is ambiguous, that Newton himself did not always use it in the same sense, and that it ought to be abandoned. He says on his last page:

...illi, qui hac attractionis voce illudantur, intelligere et explicare sibi posse videantur, quæ neque ipsi intelligant, neque explicare aliis valeant;...

548. I have alluded in Art. 301, to the memoir by Euler on the equilibrium of fluids, which appeared in the volume of the Berlin *Mémoires* for 1755, published in 1757. This memoir is of essential importance in the history of Hydrostatics; but it is not necessary in connexion with our subject to give an account of it.

549. In the Paris *Mémoires* for 1755, published in 1761, we have a memoir by La Caille, entitled *Sur la précision des Mesures géodésiques faites en* 1740, *pour déterminer la distance de Paris à Amiens; à l'occasion d'un Mémoire de M. Euler inséré dans le neuvième tome de l'Académie de Berlin.* The memoir occupies pages 53...59 of the volume.

This memoir resembles that which we have noticed in Art. 546, but is not identical with it. La Caille strenuously defends the accuracy of the operations which had been mainly performed by himself. He is convinced that there is no distance on the Earth more correctly determined than that between Paris and Amiens, in which there could not be 10 toises of error.

La Caille's confidence has been justified since: see *Base du Système Métrique*...Vol. III. page 162.

550. In the *Philosophical Transactions*, Vol. 49, Part II. which is for 1756, published in 1757, we have an *Extract of a Letter of Mons. la Condamine, F.R.S. to Dr. Maty, F.R.S. translated from the French.* It occurs on pages 622...624.

This is a fragment of no great importance; among other matters, it touches on our subject. La Condamine says that La Caille's measure, and that of Maire and Boscovich, do not agree with the elliptical curve of the meridian, or with the circularity of the parallels. He thinks that the Earth has immense cavities, and that it is of very unequal density; consequently its figure is a little irregular.

551. We have stated that a base near that of Picard was measured five times in 1740, and that the conclusion was drawn that there had been an error in Picard's original measure: see Art. 236. The subject was however again brought into discussion, apparently owing to an opinion expressed by Le Monnier in favour of Picard's result. The Paris Academy accordingly appointed two companies, each of four members, to test the operations. One company consisted of Bouguer, Camus, Cassini de Thury, and Pingré; the other company consisted of Godin, Clairaut, Le Monnier, and La Caille. Each company worked independently; and the proceedings were reported in two volumes published in 1757. I have not seen these volumes. The report of the first named company is however reprinted in the Paris *Mémoires* for 1754, published in 1759: it occupies pages 172...186 of the volume, and there is an account of it on pages 103...107 of the historical portion of the volume.

The result was a decisive confirmation of the accuracy of the operations of 1740, and consequently of the error of those originally made by Picard. See La Lande's *Bibliographie Astronomique*, page 462.

We may observe that the Toise of the North and the Toise of Peru were both employed in the course of the operations; the former appeared to be very slightly shorter than the latter.

552. In the *Philosophical Transactions*, Vol. 50, Part II., which is for the year 1758, and was published in 1759, there is a memoir by Charles Walmesley entitled *Of the Irregularities in the motion of a Satellite arising from the spheroidical Figure of its Primary Planet.* The memoir occupies pages 809...835 of the volume.

All that we have to notice in this memoir is the investigation of the attraction of an ellipsoid of revolution on a distant particle. The ellipsoid is supposed to differ but little from a sphere, and the investigation is approximate. The attraction of a sphere is known, so that we have only to find the attraction of the difference between the ellipsoid and the sphere described on its axis as diameter. By cutting this sphere by planes at right angles to the axis, we divide it into circular rings. Accordingly Walmesley first finds the approximate value of the attraction of the perimeter of a circle on a distant particle, and then applies his result to each element of the sphere.

Let a be the equatorial radius, and b the polar radius; let ξ and ζ denote the corresponding coordinates of the distant particle: put $R^2 = \xi^2 + \zeta^2$. Then Walmesley obtains the following expressions for the component attractions of the shell parallel to the directions of ζ and ξ respectively:

$$\frac{2\pi a (a-b) b\zeta}{R^3} \left\{\frac{4}{3} - \frac{4}{5}\frac{b^2}{R^2} + \frac{2\xi^2 b^2}{R^4}\right\},$$

and

$$\frac{2\pi a (a-b) b\xi}{R^3} \left\{\frac{4}{3} + \frac{2}{5}\frac{b^2}{R^2} - \frac{2\zeta^2 b^2}{R^4}\right\}.$$

Walmesley adds a corollary on his page 815 which deserves to be noticed. I adapt his words to my own notation. He says then that the former force is to the latter as

$$\zeta\left\{\frac{4}{3} - \frac{4}{5}\frac{b^2}{R^2} + \frac{2\xi^2 b^2}{R^4}\right\} \text{ is to } \xi\left\{\frac{4}{3} + \frac{2}{5}\frac{b^2}{R^2} - \frac{2\zeta^2 b^2}{R^4}\right\};$$

He adds that if the former is represented by ζ, the latter must be represented by $\xi - \frac{3\xi b^2}{5R^2}$; and so the resultant of the two does not pass through the centre of the ellipsoid, but crosses the plane of the equator at a point distant $\frac{3\xi b^2}{5R^2}$ *towards* the attracted particle.

To obtain this result we must find the value of

$$\frac{\xi\left\{\frac{4}{3} + \frac{2}{5}\frac{b^2}{R^2} - \frac{2\zeta^2 b^2}{R^4}\right\}}{\frac{4}{3} - \frac{4}{5}\frac{b^2}{R^2} + \frac{2\xi^2 b^2}{R^4}};$$

this is $$\xi\left\{1 + \frac{3}{10}\frac{b^2}{R^2} - \frac{3\zeta^2 b^2}{2R^4}\right\}\left\{1 - \frac{3}{5}\frac{b^2}{R^2} + \frac{3\xi^2 b^2}{2R^4}\right\}^{-1}$$

that is approximately

$$\xi\left\{1 + \frac{3}{10}\frac{b^2}{R^2} + \frac{3}{5}\frac{b^2}{R^2} - \frac{3(\xi^2 + \zeta^2)b^2}{2R^4}\right\},$$

that is $$\xi\left\{1 + \frac{9}{10}\frac{b^2}{R^2} - \frac{3}{2}\frac{b^2}{R^2}\right\}, \text{ that is } \xi\left\{1 - \frac{3b^2}{5R^2}\right\}$$

553. In the volume of the Paris *Mémoires* for 1758, published in 1763, there is a memoir by La Caille, entitled *Mémoire sur la vraie longueur des Degrés du Méridien en France.* The memoir occupies pages 237...244 of the volume.

The astronomical observations in the work entitled *La Meridienne de Paris vérifiée*, 1744, had not been corrected for what we now call *Nutation.* This irregularity had been discovered before that work appeared, but the theory had not been published; and it was supposed that the error produced during the interval of sixteen months, over which the operations extended, might be neglected. La Caille now applies the proper corrections to the amplitudes, and to the deduced lengths of degrees of the meridian.

554. The volume of the Paris *Mémoires* for 1758, published in 1763, contains a memoir entitled *Mémoire sur les Degrés de*

l'ellipticité des Sphéroïdes, par rapport à l'intensité de l'attraction. Par M. le Chevalier D'Arcy. The memoir occupies pages 318...320 of the volume.

We may say, in modern language, that this short memoir draws attention to the principle of the *conservation of areas,* as holding in the case of a mass set in rotation, and acted on by no forces except the mutual attractions of its particles. The writer calls the principle the *conservation of action,* and claims it for his own. See Walton's *Mechanical Problems,* 1855, page 479.

Laplace gives an application of the principle in the *Mécanique Céleste,* Livre III. § 21 : see Art. 286.

555. The Academy of Toulouse proposed the *Figure of the Earth* as the subject of an Essay, with a double prize, for the year 1750. The prize was obtained by Clairaut. The volume containing the Essay appears to have been published at Toulouse in 1759. I have not seen this volume.

There is an account of the Essay in the *Journal des Sçavans* for October 1759; this account occupies pages 281...301 of the Amsterdam edition of the volume of the Journal. The account is obscure and uninteresting, like most of the attempts to translate mathematical investigations into ordinary language. Hence I do not submit with much confidence the following brief notice of what Clairaut's Essay seems to have contained.

Clairaut considered that both the ellipticity of the Earth, and Clairaut's fraction, were found by observation to be greater than they would have been for a homogeneous fluid. Hence Clairaut's theorem does not hold for the Earth; and so it becomes necessary to devise some hypothesis which differs from those on which that theorem may be established.

Clairaut first examines an hypothesis which he attributes to Bouguer; namely, that the parts of the Earth in the vicinity of the axis of rotation are denser than the rest of the Earth. Clairaut comes to the conclusion that this is inadmissible. He finds that if the density in the vicinity of the axis differs from the density of the rest of the earth, it will not be possible to obtain an ellipticity and a Clairaut's fraction which shall *both* be greater than for a homogeneous fluid. We are not referred to the place where Bouguer has maintained this hypothesis.

Then Clairaut proposes his own new hypothesis. He assumes a solid nucleus. The generating curve is to differ slightly from an ellipse; every ordinate exceeding the corresponding ordinate of the ellipse by a small quantity which varies as the cube of the cosine of the latitude. Thus, in addition to the attraction of an ellipsoid of revolution, he has to consider the attraction of a certain shell which is also a figure of revolution.

By investigating the problem, and following the hints which may be extracted from the *Journal des Sçavans*, it will be seen that Clairaut's processes, though tedious, would not have involved any very serious difficulty.

556. La Lande mentions this Essay: see his *Bibliographie Astronomique*, page 464. He ascribes the account in the *Journal des Sçavans* to Clairaut himself, "... où Clairaut en donna lui-même l'extrait."

But from the commendation bestowed on the Essay in the account of it, I think that La Lande must be wrong. It is difficult, for example, to believe that Clairaut could have praised himself in these words:

Toutes ces transformations que nous indiquons, et que M. Clairaut emploie avec tant d'art et de succès, doivent être regardées comme le sceau du Géometre supérieur qu'il imprime à tous ses Ouvrages.

The Essay can be regarded only as a mathematical exercise; and it does not seem ever to have attracted attention. It is not mentioned in the translation of Newton's *Principia*, which was prepared by Madame du Chastellet under the guidance of Clairaut; nor in Poisson's reprint of Clairaut's *Figure de la Terre:* in this reprint some account of the Essay might with advantage have been given.

557. A problem in the Integral Calculus is mentioned with approbation as the foundation of many of Clairaut's investigations: see page 296 of the account in the *Journal des Sçavans*. I will

endeavour to reconstruct this problem from the obscure traces which are given.

The equation to the generating curve of the nucleus which Clairaut adopts will be of the form

$$y = \frac{b}{a}\sqrt{(a^2 - x^2)} + \lambda (a^2 - x^2)^{\frac{3}{2}} \dots\dots\dots\dots\dots\dots(1).$$

Let this curve revolve round the axis of x, so that the equation to the nucleus is

$$\sqrt{(y^2 + z^2)} = \frac{b}{a}\sqrt{(a^2 - x^2)} + \lambda (a^2 - x^2)^{\frac{3}{2}} \dots\dots\dots(2).$$

We suppose λ so small that its square may be neglected.

The area of a section of the solid made by a plane at right angles to the axis of x, and at the distance x from the origin, will be πy^2, that is by (1) approximately

$$\pi \frac{b^2}{a^2}(a^2 - x^2) + 2\pi\lambda \frac{b}{a}(a^2 - x^2)^2.$$

It is required to shew that the area of a section made by a plane at right angles to the axis of y, and at the distance y from the origin, can be put in an analogous form.

We have from (2)

$$z^2 = \frac{b^2}{a^2}(a^2 - x^2) - y^2 + \frac{2\lambda b}{a}(a^2 - x^2)^2$$

$$= b^2 + 2\lambda b a^3 - y^2 - \left(\frac{b^2}{a^2} + 4\lambda b a\right) x^2 + \frac{2\lambda b}{a} x^4$$

$$= B^2 - y^2 - \frac{B^2}{A^2} x^2 + \frac{2\lambda b}{a} x^4, \text{ say.}$$

Thus $$z = \sqrt{\left(B^2 - y^2 - \frac{B^2}{A^2} x^2\right)} + \frac{\frac{\lambda b}{a} x^4}{\sqrt{\left(B^2 - y^2 - \frac{B^2}{A^2} x^2\right)}}$$

The area of the section to our order of approximation will be $4\int_0^c z\,dx$, where c stands for $\frac{A}{B}\sqrt{(B^2 - y^2)}$.

Hence by assuming $x = \frac{A}{B}\sqrt{(B^2 - y^2)}\sin\theta$ we easily find that the area is

$$\pi\frac{A}{B}(B^2 - y^2) + \frac{3\pi}{4}\frac{\lambda b}{a}\cdot\left(\frac{A}{B}\right)^5(B^2 - y^2)^2.$$

This expression is of the required form.

The use of the problem to Clairaut consists in this: as the formulæ for the attraction of an ellipsoid of revolution may be considered known, as soon as he has determined the attraction of the nucleus for a point on the axis of x, he can readily infer the attraction for a point on the axis of y.

558. A celebrated French lady translated Newton's *Principia* and added a commentary; the work was published after her death under the title of *Principes Mathématiques de la Philosophie Naturelle, par feue Madame la Marquise du Chastellet.* 2 vols. 4to. Paris, 1759.

Besides *Chastellet* we have the variations Chastelet and Châtelet: see pages iv and v of the first volume of the work. The work has an *Avertissement de l'Editeur,* and a *Préface Historique* by Voltaire.

From these it appears that Madame du Chastellet was a pupil of Clairaut's; and the commentary was constructed out of the materials which she obtained from him. The translation occupies the first volume and part of the second; the commentary occupies the remainder of the second volume. We will notice those pages of the commentary which bear on our subject.

559. Pages 56...67 give an analysis of Newton's method of treating the Figure of the Earth.

On page 62 the cause of a mistake made by Newton is assigned as in Clairaut's *Figure de la Terre,* page 256; though here apparently with more confidence: see Art. 37.

On page 66 the criticism on Newton's conjecture with respect to Jupiter is given as in Clairaut's *Figure de la Terre,* page 224; though here apparently with more confidence: see Art. 31.

The pages 155...183 constitute an analytical treatise on Attractions in three sections. First we have spherical bodies, then bodies of other forms, and lastly an ellipsoid of revolution with the attracted particle on the prolongation of the axis. The investigations are simple and satisfactory.

One of these investigations relates to the attraction of a sphere on an internal particle, when the force varies inversely as the fourth power of the distance. To avoid the infinite expressions which might occur, it is assumed that if the particle be at the centre of such a sphere, the resultant attraction must be zero.

The pages 193...259 form a section entitled *De la Figure de la Terre,* in two parts. The first part on pages 193...221 is an abridgement of the theory of Hydrostatics which constitutes the first half of Clairaut's volume. The second part, on pages 221...259, is on the Figure of the Earth; this is almost a reproduction of Chapters II. and III. of the second half of Clairaut's volume.

Two simple examples constitute all the novelty which the commentary furnishes; we will mention these.

On page 238 the general formula for the value of attraction given by Clairaut on his page 247 is applied to the case in which the strata are all similar, so that the ellipticity is constant, and the density varies as the distance from the centre: see Art. 336. The result is found then to be independent of s, so that the attraction is approximately constant at all points of the surface: see Case III. of Art. 266. The result in our notation will be found to be $\pi r_1^2 \lambda (1 + \frac{2}{3} \epsilon_1)$, where λ is the density at the unit of distance from the centre.

The other example is discussed on pages 238, 239 and 246. Using the notation of Art. 336, suppose that the density and the ellipticity are given by these formulæ

$$\rho = \lambda r_1 - pr, \quad \epsilon = \epsilon_1 \frac{r}{r_1},$$

where λ and p are constants. Then the commentary finds the expression for gravity and the value of ϵ_1. See Arts. 336 and 327.

560. According to the *Préface Historique,* page ix, great care was bestowed on the Commentary. When the lady had written a

chapter, Clairaut examined and corrected it; and subsequently the fair copy was revised by a third party. Nevertheless there are numerous errors or misprints, some of which are very serious. Thus, for example, the formulæ which occur in the investigation of the attraction of an oblongum on pages 182 and 183 are much disfigured; the former part of page 197 is unintelligible, owing to the omission of important matter; the binomial expressions on pages 218 and 219 are extremely inaccurate; the reference to a supposed property of the ellipse on page 242 is absurd; and the numerical application of Clairaut's Theorem on page 257 is quite wrong.

Playfair mentions Madame du Chastellet in his *Dissertation... of the Progress of Mathematical and Physical Science...*; see page 655 of the *Encyclopædia Britannica*, eighth edition, Vol. I. In reference to her writings on the dispute as to the measure of force, he says:

...from the fluctuation of her opinions, it seems as if she had not yet entirely exchanged the caprice of fashion for the austerity of science ......

Voltaire however finds merit in a similar fluctuation. He says in the *Préface Historique*, page VI.:

...Ainsi, après avoir eu le courage d'embellir Léibnitz, elle eut celui de l'abandonner :...

Playfair speaks highly of the translation of Newton and the Commentary. I do not agree with his estimate of the Commentary. The title is really inappropriate. Instead of any explanation of Newton, we have merely other investigations drawn from well known works exhibiting more recent solutions of the problems which Newton discussed.

561. The first volume of the series published by the Turin Academy which is usually called the *Miscellanea Taurinensia* is dated 1759. On pages 142...145 Lagrange supplies a note to a memoir by another person. The note relates to the attraction of an indefinitely thin spherical shell, and undertakes to explain the paradox which had disturbed D'Alembert: see Art. 434.

Lagrange's explanation is rather a hint than a strict mathematical investigation; but the idea is sound and valuable. When the attracted particle is *very* near the shell, an infinitesimal part of the shell close to the particle produces a finite portion of the whole attraction, in fact a half. When the attracted particle forms an element of the shell, this part of the attraction vanishes; and when the attracted particle is inside the shell it becomes negative.

Lagrange's idea for the spherical shell is really the same as Coulomb afterwards used for a shell of any form in the Paris *Mémoires* for 1788; Laplace developed it in an investigation which occurs in Poisson's memoir on the distribution of electricity, in the Paris *Mémoires* for 1811.

562. The second volume of the *Miscellanea Taurinensia*, is for 1760 and 1761; the date of publication is not recorded.

This volume contains the memoir by Lagrange on Maxima and Minima, which is famous in the early history of the Calculus of Variations. In a memoir immediately following, and connected with this, Lagrange treats of various problems in Dynamics; and among others, he considers the motion of fluids.

Lagrange, on his page 282, makes a remark respecting a passage in D'Alembert's...*Résistance des Fluides* to which I have already alluded: see Art. 397.

Lagrange makes a remark that surfaces of equal density will be level surfaces provided a certain condition holds. He says on his page 284:

Cependant un grand Géométre a crû que il n'étoit pas toujours nécessaire que les surfaces des différentes couches fussent de niveau, et il a donné un autre Principe pour connôitre la figure de ces surfaces.

Lagrange commences a mathematical investigation; and in effect he says that if we proceed according to D'Alembert's manner, as given in Art. 401, we shall find that the surfaces of equal density are level surfaces. D'Alembert as we have stated, subsequently admitted his error: see Arts. 368 and 400.

Lagrange criticises other opinions of D'Alembert on pages 275 and 323 of the second volume of the *Miscellanea Taurinensia:* but these do not belong to our subject.

CHAPTER XVI.

D'ALEMBERT.

563. WE shall now resume our examination of the labours of D'Alembert in our subject. With a few unimportant exceptions, the present Chapter will be devoted to memoirs published by D'Alembert in various volumes of his *Opuscules Mathématiques.*

564. The article entitled *Figure de la Terre* in the *Encyclopédie,* was by D'Alembert; the date of the volume in which it was published is 1756.

The article occupies pages 749...761 of the volume; it gives an interesting account of the measurements and of the theoretical investigations on the subject up to the date of publication.

D'Alembert awards high praise to Maclaurin, and to Clairaut; and refers with obvious satisfaction to his own researches. He notices especially the Articles 166...169 of his *Essai sur la Résistance des Fluides;* these he appreciates at a value far beyond their worth: see Arts. 404...406. He also refers to his *Recherches ...Systême du Monde,* and it may be admitted that these volumes are not without merit as regards our subject.

D'Alembert discusses at some length in a popular manner the question as to whether the Earth can be assumed to be a figure of revolution.

In the *Encyclopédie Méthodique* the account which D'Alembert gave of the *theory* of the subject is reproduced; but his account of the measurements is omitted, and a shorter article respecting them by La Lande is supplied.

In the *Encyclopédie Méthodique* there is a reference to the fifth volume of the *Opuscules Mathématiques* which of course was not in the original *Encyclopédie.*

The following sentence of the original article is worth notice:

...ceux qui les premiers mesurerent les degrés dans l'étendue de la France, préoccupés peut-être de cette idée, que la *Terre* applatie donnoit les degrés vers le nord plus petits que ceux du midi, trouverent en effet que dans toute l'étendue de la France en latitude, les degrés alloient en diminuant vers le nord.

In the original article speaking of what his *Recherches...* contained D'Alembert says: il pourroit trés-bien être en équilibre sans avoir la figure elliptique. This is not so strong as the preface to Vol. III. of the *Recherches*...page xxxvi.

565. The article entitled *Gravitation* in the *Encyclopédie* was by D'Alembert; the date of the volume in which it was published is 1757.

The only part of the article which concerns us consists of some observations respecting the paradox which D'Alembert considered that he had discovered as to the attraction of an infinitesimally thin spherical shell: see his *Recherches...Système du Monde*, Vol. III. page 199; and Arts. 434 and 561.

D'Alembert shews analytically, that if a particle be outside the shell the resultant attraction on it is the same as if the mass of the shell were collected at the centre; this result is no more than Newton had given in his *Principia:* see Art. 4.

D'Alembert's observations to which we are here referring are omitted in the article *Gravitation* of the *Encyclopédie Méthodique;* but the substance of them is reproduced as we shall see in the first volume of the *Opuscules Mathématiques.*

566. The first volume of D'Alembert's *Opuscules Mathématiques* was published in 1761. On pages 246...264 we have a memoir entitled *Remarques sur quelques questions concernant l'attraction.*

567. According to D'Alembert's formula on page 42 of his *Réflexions...des Vents,* which is reproduced in our Art. 376, relative equilibrium might subsist in the case of a solid oblatum covered with fluid, such that the external surface of the fluid was an

oblongum; the whole rotating on a common axis. This result, D'Alembert says, had been attacked by *un Géometre Italien, qui a du nom dans les Mathématiques;* pages 246...252 constitute a reply to this attack.

The Italian Geometer was doubtless Boscovich; see page 463 of his *De Litteraria Expeditione...*, and Art. 470. The objection urged against D'Alembert's result amounts to this in modern language: that the relative equilibrium would not be *stable.* D'Alembert says, very justly, he might reply that in such researches no mathematician had as yet attempted to consider whether this condition was satisfied. However, he makes some remarks on the point. He contents himself with shewing when the tangential force at the external surface of the fluid would act towards the pole, and when towards the equator. He arrives at the conclusion that the relative equilibrium would be stable in the case to which objection had been taken, provided the density of the fluid were less than five-thirds of the density of the solid. D'Alembert's investigation is not adequate to solve the problem of the motion of the fluid when disturbed from its position of relative equilibrium: but his defence is at least as good as the attack of the Italian Geometer.

The subject will appear before us again in the sixth volume of the *Opuscules Mathématiques.*

568. On his pages 252...257, D'Alembert corrects an error into which he had fallen in his *Réflexions...des Vents,* pages 155...157. We have already paid attention to this correction: see our Art. 381.

569. On his pages 257...264, D'Alembert recurs to what he considered a paradox, as to the attraction of an infinitesimally thin spherical shell: see Art. 565. D'Alembert reproduces the substance of some remarks originally published in the *Encyclopédie.* He takes objection to Lagrange's explanation, and he says he gives one of his own: see Art. 561. What D'Alembert really does is to translate Lagrange's idea from popular language to mathematical language; and then to ascribe the entire merit to himself. He shews that the infinitesimal part of the shell to which Lagrange

refers may be taken to be the part determined by tangents from the external particle.

It may be observed that neither Lagrange nor D'Alembert uses a symbol to express the infinitesimal thickness of the shell. If we consider the shell to be very thin, though not infinitesimally thin, and suppose the attracted particle to pass gradually from the outside of the shell to the inside of the shell, all the so-called paradox disappears: for the attraction changes gradually and not discontinuously.

570. The fifth volume of D'Alembert's *Opuscules Mathématiques* was published in 1768. On pages 1...40 we have a memoir entitled *Sur l'équilibre des Fluides;* and the pages 23...40 of this memoir constitute an *Appendice sur la Figure de la Terre.*

571. The memoir begins by corrections of errors in preceding investigations.

D'Alembert had supposed that he had obtained in his *Essai sur la Résistance des Fluides* more general results than his predecessors in the theory of the equilibrium of fluids and the figure of the Earth. He now admits that the supposition was unfounded. The quantity denoted by K on page 210 of that work he now allows should be zero; and so his result coincides with Clairaut's: see Art. 405.

In like manner he admits that the same simplification ought to be made in various equations which he had given in the third part of his *Recherches......Système du Monde,* beginning with page 229. I have already, in my account of this work, noticed the correction: see Art. 444.

572. D'Alembert returns to the subject he had introduced on page 203 of his *Essai sur la Résistance des Fluides:* see Art. 400. He maintains, and rightly, that in a fluid in equilibrium the surfaces of equal density are not necessarily level surfaces. He admits, however, that for such forces as occur in nature, the surfaces of equal density are level surfaces; his original error on this point was corrected by Lagrange: see page 2 of the *Opuscules Mathématiques,* Vol. v.; also Arts. 405 and 562.

573. D'Alembert gives a form, at once simpler and more general, to the equations which he had used in the *Essai sur la Résistance des Fluides:* see page 6 of the *Opuscules Mathématiques,* Vol. V.; also Art. 402.

574. D'Alembert occupies his pages 10...22 with remarks on the conditions of fluid equilibrium. The remarks are sound, must have been valuable at the time, and may even now be read with profit. D'Alembert objects with justice to Clairaut's explanation of a paradox in the subject; see Art. 312.

The main principle which D'Alembert asserts, expressed in modern language, is in effect this: Consider only forces in one plane; then we have for the equilibrium of a fluid the equations

$$\frac{dp}{dx} = \rho X, \quad \frac{dp}{dy} = \rho Y.$$

Take the simple case of homogeneous fluid; then it is *not* sufficient for equilibrium that $Xdx + Ydy$ should be a perfect differential. If we suppose that $Xdx + Ydy$ is the differential of $\phi(x, y)$ then $\phi(x, y)$ must have only *one* value for given values of x and y. Thus, for example, $\phi(x, y)$ must not be such a function as $\tan^{-1}\frac{y}{x}$.

Again, suppose we use polar coordinates, and find that $p = F(r, \theta)$; then when $r = 0$ we have p apparently a function of θ only. But unless this apparent function of θ reduces to a constant, the pressure would not be the same in all directions about the origin; which is contrary to the nature of a fluid.

In the two preceding paragraphs we have translated D'Alembert's ideas into modern language; he himself does not speak of pressure, nor does he use the symbol p.

575. D'Alembert devotes his pages 23...40 to an Appendix on the Figure of the Earth. His object is to enquire if the oblatum is the only form of relative equilibrium for a rotating mass of homogeneous fluid.

He says it follows from what he has proved in his *Recherches sur les Vents,* Art. 28, that if the fluid mass is originally spherical, and is then put into rotation, so that the ratio of the centrifugal force to gravity is small, the form of relative equilibrium must be an oblatum. It is almost needless to remark that D'Alembert's statement is not demonstrated: the motion of such a fluid mass is too difficult for his rough approximative analysis to master.

However, he now proceeds to discuss the problem without assuming that the mass is originally spherical. He arrives at the conclusion that if the fluid is in the form of a figure of revolution, and is nearly spherical, there cannot be relative equilibrium for any other form than an oblatum. Unfortunately, his demonstration is unsound. The theorem, however, is now admitted to be true. Legendre indeed gave a demonstration, which does not assume the figure to be nearly spherical; but the demonstration is not quite free from objection. Laplace, assuming the figure to be nearly spherical, but not assuming it to be of revolution, demonstrated the theorem: he omits to mention the condition that the figure is nearly spherical when he refers to the subject in the *Méoanique Céleste*, Vol. v., page 10.

576. We will now indicate the nature of D'Alembert's method, and the point at which it fails.

Let P be the pole of the body, Q any point on the surface. We shall require the attraction at Q resolved in the direction which is in the meridian plane of Q, and is at right angles to the radius from the centre to Q.

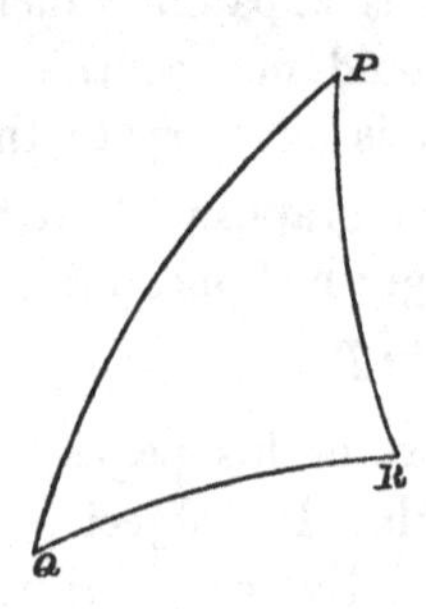

Let R be any point on the surface; let $PQ = \beta$, $QR = u$, $PR = z$; and let $PQR = \pi - \psi$, so that ψ is the angle between RQ and PQ produced. Let the polar radius be denoted by 1, and the radius at R by $1+\alpha F(z)$, where α is a small quantity. Then proceeding as in Art. 424, we find that the element of the required attraction, estimated *from* the pole, to the order we have to regard

$$=\frac{du\,d\psi\sin u\cos\psi\cos\frac{1}{2}u}{4\sin^2\frac{1}{2}u}\,\alpha F(z)$$

$$=\frac{du\,d\psi\sin^2 u\cos\psi}{8\sin^3\frac{1}{2}u}\,\alpha F(z)$$

$$=\frac{du\,d\psi\sin^2 u\cos\psi}{2^{\frac{3}{2}}(1-\cos u)^{\frac{3}{2}}}\,\alpha F(z).$$

This agrees with D'Alembert's formula at the top of his page 26; his Δ is our ψ.

The transverse attraction which we require would be obtained by integrating the above expression between the limits 0 and 2π for ψ, and 0 and π for u. Let T denote this transverse attraction.

Let V denote the attraction at Q resolved along the radius; and χ the angle between this radius and the tangent to QP at Q. Then $V\cos\chi$ is the resolved part of V along the tangent to QP at Q. Hence, supposing the body to be fluid, or at least the outer stratum to be fluid, we must have for equilibrium

$$V\cos\chi=T\ldots\ldots\ldots\ldots\ldots\ldots\ldots\ldots\ldots\ (1).$$

If the body rotates, then to secure relative equilibrium, we must supply in this equation a term corresponding to the resolved centrifugal force.

We must now give some specific form to $F(z)$ before we can carry the investigation further. Assume, with D'Alembert, that

$$F(z)=A+B\cos z+C\cos^2 z+\ldots\ldots+M\cos^m z.$$

We shall then have to our order of approximation

$$\cos\chi=-\sin\beta\,(B+2C\cos\beta+\ldots\ldots+mM\cos^{m-1}\beta)\,\alpha\,;$$

and it will be sufficient in (1) to put $\frac{4\pi}{3}$ for V. Thus (1) becomes

$$-\frac{4\pi\alpha}{3}\sin\beta\,(B+2C\cos\beta+\ldots\ldots\,mM\cos^{m-1}\beta)=T\ldots\ldots\ldots(2).$$

Now $\cos z = \cos\beta \cos u - \sin\beta \sin u \cos\psi$. Hence, corresponding to the term $M\cos^m z$ in $F(z)$ we have in T the term

$$\frac{aM}{2^{\frac{3}{2}}}\int_0^{\pi}\int_0^{2\pi}\frac{\sin^2 u \cos\psi\,(\cos\beta\cos u - \sin\beta\sin u\cos\psi)^m}{(1-\cos u)^{\frac{3}{2}}}\,du\,d\psi.$$

When we integrate with respect to ψ all the terms which involve odd powers of $\cos\psi$ vanish; so that we are left with

$$-\frac{aM}{2^{\frac{3}{2}}}\int_0^{\pi}\int_0^{2\pi}\frac{\sin^2 u\, Z\, du\, d\psi}{(1-\cos u)^{\frac{3}{2}}}$$

where $Z = m\cos^{m-1}\beta\sin\beta\cos^{m-1}u\sin u\cos^2\psi$

$$+\frac{m(m-1)(m-2)}{\underline{|3}}\cos^{m-3}\beta\sin^3\beta\cos^{m-3}u\sin^3 u\cos^4\psi$$

$+ \ldots\ldots$

Here every term involves some odd power of $\sin\beta$. Now suppose we put $(1-\cos^2\beta)\sin\beta$ for $\sin^3\beta$, and $(1-\cos^2\beta)^2\sin\beta$ for $\sin^5\beta$, and so on. Then Z takes the form

$$\sin\beta\,(N_1\cos^{m-1}\beta + N_3\cos^{m-3}\beta + N_5\cos^{m-5}\beta + \ldots),$$

where $N_1, N_3, N_5, \ldots$ are functions of u and ψ.

In like manner the other terms in $F(z)$ will give rise to corresponding terms in T, involving the product of $\sin\beta$ into various powers of $\cos\beta$; but these powers of $\cos\beta$ will all be less than the $(m-1)^{\text{th}}$ power.

Hence equating the coefficients of like terms in (2), we see that besides other relations we must have

$$-\frac{4\pi m}{3} = -\frac{1}{2^{\frac{3}{2}}}\int_0^{\pi}\int_0^{2\pi}\frac{N_1\sin^2 u\,du\,d\psi}{(1-\cos u)^{\frac{3}{2}}} \quad \ldots\ldots\ldots\ldots(3).$$

D'Alembert then has to shew that equation (3) cannot be satisfied if m be a positive integer greater than 2. His demonstration, however, fails completely, because he has given a wrong value to the quantity which we denote by N_1: his error begins with his Article 50, on his page 26.

Take the second term which we have expressed in the value of Z; and put $\sin\beta(1-\cos^2\beta)$ for $\sin^3\beta$: then we have as part of N_1

$$-\frac{m(m-1)(m-2)}{\lfloor 3}\cos^{m-3}u\sin^3u\cos^4\psi.$$

Instead of keeping this, D'Alembert puts $\sin u(1-\cos^2 u)$ for $\sin^3 u$, and then omits $\sin u$, retaining only $-\sin u\cos^2 u$, so that instead of what we have just given, he has

$$\frac{m(m-1)(m-2)}{\lfloor 3}\cos^{m-1}u\sin u\cos^4\psi.$$

He treats the other terms of Z in the same unwarrantable manner; and the consequence is that his value of N_1 is altogether wrong. The error renders all the rest of his argument worthless.

Laplace, as we shall see, alludes in his first and second memoirs to D'Alembert's demonstration, but says nothing about its unsoundness. Legendre, who may be considered to have been the first to solve the problem here involved, does not even allude to D'Alembert's demonstration.

577. It is important to notice what D'Alembert's process would have established if it had been sound. It would have shewn that $F(z)$ cannot be a *finite* series of powers of $\cos z$, in which the highest power is greater than 2. But it would not have shewn that $F(z)$ cannot be an *infinite* series of powers of $\cos z$.

578. D'Alembert gives on his page 29 the value of the definite integral $\int_1^{-1}\frac{x^{m-1}\,dx}{(1-x)^{\frac{1}{2}}}$ when for m we put 1, 2, 3, 4, 5, 6, or 7. There is no objection to his method. We may, if we please, transform the integral to $2(-1)^m\int_0^{\sqrt{2}}(y^2-1)^{m-1}dy$: thus it is easy to verify his values.

After leaving this subject, D'Alembert on his pages 36...40, makes a few other remarks; they are not of great importance, but they are correct, except those contained in his Art. 78, which are erroneous.

579. The sixth volume of D'Alembert's *Opuscules Mathématiques* was published in 1773; a large part of it is devoted to our subject.

580. A memoir entitled *Sur la Figure de la Terre,* occupies pages 47...67. It begins thus:

Feu M. Maclaurin est le premier qui ait démontré rigoureusement qu'une masse fluide homogene, tournant autour d'elle-même, devoit prendre la figure d'une ellipse dans l'hypothèse de l'attraction en raison inverse du quarré des distances. Mais personne, que je sache, n'avoit encore remarqué que dans ce cas le problême est susceptible de deux solutions, c'est-à-dire, qu'il y a deux figures possibles à donner au spheroïde, et dans lesquelles l'équilibre aura lieu. Cette considération est l'objet des Recherches suivantes.

We have already remarked that Thomas Simpson had implicitly shewn the possibility of this double solution: see Art. 285. However, D'Alembert now gives an explicit investigation, which, in substance, was afterwards incorporated by Laplace in the *Mécanique Céleste,* and thus constitutes a permanent part of the subject: see the *Mécanique Céleste,* Livre III., Chapitre III.

On his page 47, D'Alembert makes the undemonstrated assertion, that if a spherical mass of homogeneous fluid be put in rotation it *will* take the form of an oblatum: see Art. 575.

581. We will, in giving an account of D'Alembert's process, adopt to a great extent Laplace's notation.

Suppose ω the angular velocity of rotation, ρ the density of the fluid; put q for $\dfrac{\omega^2}{\dfrac{4\pi\rho}{3}}$. Suppose the major axis of the Earth to be $\sqrt{(\lambda^2+1)}$ times the minor axis; then the excentricity of the ellipse is $\dfrac{\lambda}{\sqrt{(\lambda^2+1)}}$. Denote the minor axis by $2c$. Then by the formulæ of Art. 261, or by those of any elementary work on Statics we find that the attraction on a particle at the pole is

$$\frac{4\pi\rho c(1+\lambda^2)}{\lambda^2}\left\{1-\frac{\tan^{-1}\lambda}{\lambda}\right\};$$

and the attraction on a particle at the equator is

$$\frac{2\pi\rho c\sqrt{(\lambda^2+1)}}{\lambda^2}\left\{(1+\lambda^2)\frac{\tan^{-1}\lambda}{\lambda}-1\right\}:$$

call the latter X, and the former Y.

The centrifugal force at the equator

$$=c\sqrt{(\lambda^2+1)}\,\omega^2=\frac{4\pi\rho cq}{3}\sqrt{(\lambda^2+1)}.$$

Then, as in Art. 262, the condition for relative equilibrium is

$$\frac{X-\frac{4\pi\rho cq}{3}\sqrt{(\lambda^2+1)}}{Y}=\frac{1}{\sqrt{(\lambda^2+1)}},$$

which reduces to

$$\frac{2q}{3}=\frac{(\lambda^2+3)\tan^{-1}\lambda-3\lambda}{\lambda^3}\;\ldots\ldots\ldots\ldots\ldots\ldots\ldots(1).$$

This is the standard equation on the subject: see the *Mécanique Céleste*, Livre III. § 18. D'Alembert has the same equation: see his page 50. He uses k for Laplace's λ, and ω for Laplace's q. Neither D'Alembert nor Laplace uses the symbol $\tan^{-1}$, which is of more recent origin. It will be observed that if a sphere of the same density as the fluid were to rotate with the same angular velocity, q would be the ratio of the centrifugal force to the attraction at the equator.

D'Alembert shews that the equation (1) will give two values of λ for a given value of q, provided q be not too great. Denote by $\phi(\lambda)$ the right-hand member of equation (1); then considering λ as an abscissa, and $\phi(\lambda)$ as the corresponding ordinate, he in fact traces the curve which thus arises. We have $\phi(\lambda)$ zero when λ is zero, and also when λ is infinite; when λ is very small $\phi(\lambda)$ is approximately equal to $\frac{4\lambda^2}{15}$. Since $\phi(\lambda)$ vanishes when

λ vanishes, and when λ is infinite, there must be some maximum value of λ; this maximum is determined by putting $\phi'(\lambda)=0$; this leads to

$$\tan^{-1}\lambda = \frac{9\lambda + 7\lambda^3}{(1+\lambda^2)(9+\lambda^2)} \text{ .... } (2).$$

It is evident from what has been said that this equation must have a root. We may also establish the existence of a root in the following way. When λ is very small the left-hand member is approximately

$$\lambda - \frac{\lambda^3}{3} + \frac{\lambda^5}{5} - \ldots,$$

and the right-hand member is approximately

$$\lambda - \frac{\lambda^3}{3} + \frac{7\lambda^5}{27} - \ldots;$$

thus, when λ is very small, the right-hand member is the larger. When λ is infinite the left-hand member is the larger. Hence for some intermediate value the two members will be equal. See D'Alembert's pages 51 and 52.

582. Suppose that q has a given value; let λ_1 denote the smaller of the two values which equation (1) furnishes. By comparing the weights of a polar and an equatorial column of fluid, without assuming that there is equilibrium, D'Alembert finds that if λ is a little less than λ_1 the weight of the polar column predominates, and that if λ is a little greater than λ_1 the weight of the equatorial column preponderates. Then he argues thus: Let the fluid be in relative equilibrium with the value λ_1. Suppose the oblatum a little elongated; this amounts to diminishing λ; then the weight of the polar column preponderates, and pushes out the equatorial column: thus there is a tendency to restore the equilibrium figure. Again, suppose that we start from the equilibrium figure, and compress it a little; this amounts to increasing λ; then the weight of the equatorial column preponderates, and pushes out the polar column: thus there is a tendency to restore the equilibrium figure. Hence in modern language the relative equilibrium is *stable;* D'Alembert uses the word *ferme.*

In like manner he concludes that the relative equilibrium corresponding to the larger of the two values which equation (1) furnishes is unstable.

His discussion on these points will be found on his pages 55...57: it cannot be considered adequate for such a difficult matter. I do not find that the later writers Laplace, Poisson, and Pontécoulant have followed D'Alembert in determining the stability or instability.

If the angular velocity is such as corresponds to a single solution, so that (1) and (2) are simultaneously satisfied, D'Alembert arrives at what he considers a singular result. This result expressed in modern language is that the relative equilibrium is stable with respect to an elongation of the oblatum, and unstable with respect to a compression of the oblatum: see his page 57.

583. On his page 58, D'Alembert says:

Ceci me porteroit à croire, pour le dire en passant, que dans les Théories données jusqu'ici sur la Figure de la Terre, on a peut-être trop cherché à faire accorder entr'eux les deux principes, celui de la perpendicularité de la pesanteur à la surface, et celui de l'équilibre des colomnes. Car ce dernier n'est nécessaire que quand la Terre est fluide, et n'est jamais suffisant, soit que la Terre soit solide ou fluide; au lieu que le premier est nécessaire dans les deux cas, et suffit si la Terre est solide.

By the principle of columns he probably means the balancing of columns *at the centre.* Boscovich had shewn that if at *every* point *every* pair of rectilinear columns balances, then also Huygens's principle of equilibrium is satisfied: see Boscovich's *De Litteraria Expeditione*...page 424; and Art. 463.

584. When the angular velocity is very small, one of the forms of relative equilibrium determined by equation (1) is very nearly spherical, and the other is very much compressed; D'Alembert calls this a *singulier paradoxe:* see his page 58.

Let us suppose that q is very small; one value of λ is very large as we have said. Thus (1) becomes approximately

$$2q = \frac{3\lambda^2 \frac{\pi}{2}}{\lambda^3} = \frac{3\pi}{2\lambda};$$

therefore
$$\lambda = \frac{3\pi}{4q} \quad \ldots\ldots\ldots\ldots\ldots\ldots\ldots\ldots\ldots\ldots \quad (3).$$

Let r be the radius of a sphere having the same volume as the oblatum; then with the notation of Art. 581,
$$(\lambda^2+1)\, c^3 = r^3 \quad \ldots\ldots\ldots\ldots\ldots\ldots\ldots\ldots \quad (4).$$

D'Alembert shews that the velocity of a point at the equator is very small when λ is very great; that is, the smallness of the angular velocity more than counterbalances the largeness of the radius.

For the square of this velocity
$$= (\lambda^2+1)\, c^2\omega^2 = (\lambda^2+1)\, c^2 \frac{4\pi\rho}{3} q$$
$$= r^2 (\lambda^2+1)^{\frac{1}{3}} \frac{4\pi\rho}{3} q \text{ by (4)} = r^2\lambda^{\frac{2}{3}} \frac{4\pi\rho}{3} q \text{ approximately}$$
$$= r^2 \left(\frac{3\pi}{4q}\right)^{\frac{2}{3}} \frac{4\pi\rho}{3} q \text{ by (3)} \quad = \frac{2r^2\pi^{\frac{5}{3}}q^{\frac{1}{3}}\rho}{\sqrt[3]{6}}:$$

this is small since q is small.

D'Alembert also compares the centrifugal force at the equator in this case with the centrifugal force at the equator of the sphere of equal volume. The ratio of the former to the latter
$$= \frac{\sqrt{(\lambda^2+1)}\, c\omega^2}{r\omega^2} = (\lambda^2+1)^{\frac{1}{6}}:$$

this is large since λ is large. See his page 59; there are misprints towards the bottom of the page.

585. D'Alembert was aware that his investigations did not shew that there could *not be more* than two forms of relative equilibrium corresponding to a given angular velocity. He expressly leaves this point to be discussed by other Geometers: see his page 61. Laplace was the first who demonstrated that there could not be more than two forms of relative equilibrium: see D'Alembert's *Opuscules Mathématiques*, Vol. VIII. page 292, and Laplace's *Théorie...de la Figure des Planetes*, page 124.

586. The proposition which D'Alembert thus left to be demonstrated amounts to this, that $\phi'(\lambda)$ vanishes *only once* as λ

changes from zero to infinity, besides when $\lambda = 0$; D'Alembert draws his curve consistently with this proposition, though he did not demonstrate it. The proposition is known to be true as it is indirectly involved in Laplace's investigations; but it may be useful to give a direct demonstration.

Put $\tan\theta$ for λ; then $\phi(\lambda)$

$$= \frac{(3+\tan^2\theta)\,\theta - 3\tan\theta}{\tan^3\theta} = \frac{(1+2\cos^2\theta)\,\theta - 3\sin\theta\cos\theta}{\sin^3\theta}\cos\theta$$

$$= \frac{\theta\,(1+2\cos^2\theta)\cos\theta}{\sin^3\theta} - 3\cot^2\theta.$$

The differential coefficient of this with respect to θ is

$$\frac{\cos\theta(8+\cos 2\theta)}{\sin^3\theta} - \frac{(5+4\cos 2\theta)\,\theta}{\sin^4\theta} = \frac{\sin 2\theta\,(8+\cos 2\theta) - 2\theta\,(5+4\cos 2\theta)}{2\sin^4\theta}.$$

Put $F(\theta)$ for the numerator.

When θ is very small, we shall find that $F(\theta) = \frac{16\theta^5}{15}$. This is easily obtained by expansion, for

$$F(\theta) = 8\sin 2\theta + \frac{1}{2}\sin 4\theta - 2\theta\,(5+4\cos 2\theta).$$

Or we may proceed thus: we know that when λ is very small $\phi(\lambda) = \frac{4\lambda^2}{15}$, so that $\phi'(\lambda)$ then $= \frac{8\lambda}{15}$; hence when θ is very small we must have $\frac{F(\theta)}{2(\sin\theta)^4} = \frac{8\theta}{15}$, and therefore $F(\theta) = \frac{16\theta^5}{15}$.

When $\theta = \frac{\pi}{2}$ we see that $F(\theta)$ is negative.

If then $F(\theta)$ vanishes for more than one value of θ, besides $\theta = 0$, between $\theta = 0$ and $\theta = \frac{\pi}{2}$, it must vanish for three values: and then $F'(\theta)$ must vanish for two values of θ besides $\theta = 0$. But

$$F'(\theta) = 8\cos 2\theta + 2\cos 4\theta - 10 + 16\theta\sin 2\theta;$$

$$F''(\theta) = -8\sin 4\theta + 32\theta\cos 2\theta = 16\cos 2\theta\,(2\theta - \sin 2\theta).$$

Thus $F''(\theta)$ is positive from $\theta = 0$ to $\theta = \frac{\pi}{4}$, and then negative from $\theta = \frac{\pi}{4}$ to $\theta = \frac{\pi}{2}$; therefore $F'(\theta)$ increases continually from $\theta = 0$ to $\theta = \frac{\pi}{4}$, and diminishes continually from $\theta = \frac{\pi}{4}$ to $\theta = \frac{\pi}{2}$: hence $F'(\theta)$ cannot vanish more than once besides $\theta = 0$, as θ changes from 0 to $\frac{\pi}{2}$.

587. We may put equation (1) in the form

$$\frac{2\lambda^3 q}{3} = (\lambda^2 + 3)\tan^{-1}\lambda - 3\lambda.$$

If we suppose $\lambda = 0$, both sides vanish whatever may be the value of q. But $\lambda = 0$ is not a solution of (1); we have in fact introduced this solution by multiplying both sides of (1) by λ^3.

D'Alembert devotes his page 62 to this matter; which would now be considered too obvious to need remark.

588. D'Alembert gives some extension to his investigation on his pages 63...67 by supposing extraneous forces to act; but this extension is of little importance. D'Alembert afterwards returns to the subject and discusses it in an elaborate manner: see Art. 596.

At the top of his page 64, D'Alembert seems to say he has four forces; but his first force is in fact resolved into his second and third, and is not in addition to them.

589. The next memoir in the sixth volume of D'Alembert's *Opuscules Mathématiques* is entitled *Eclaircissemens sur deux endroits de mes Ouvrages, qui ont rapport à la Figure de la Terre*; this occupies pages 68...76: it is followed by some *Remarques sur l'Article précédent* on pages 77...84.

The passages in his previous works to which D'Alembert here alludes occur on page 42 of the *Réflexions...des Vents*, and on pages 246...252 of the first volume of the *Opuscules Mathématiques*: see Arts. 376, 378, 514, and 567.

590. We have already learned from Art. 567, that Boscovich criticised D'Alembert, and that D'Alembert defended himself. Boscovich's work was translated into French, and a long note inserted on pages 449...453 which renewed the attack on D'Alembert: and now D'Alembert replies.

The matters in controversy admit of being stated briefly though neither of the disputants defines them very clearly.

The translator ascribes great merit to Boscovich for introducing the notion of what we should call the *stability* of the equilibrium: D'Alembert replies that the notion is really due to Daniel Bernoulli. Next as to mathematical results we may say that both disputants accepted the formula of Art. 376; and also both allowed that the equilibrium would be stable if ρ were less than $\frac{5}{3}\sigma$. Then D'Alembert asserts that we may have ρ less than $\frac{5}{3}\sigma$, and ϵ' positive, and yet have ϵ negative; and the formula of Art. 376 shews that his statement is correct. The French translator denies this, and so is wrong; he seems to have assumed that $1-\frac{\rho}{\sigma}$ must be positive, which is not necessary.

The following passage of the translator's note relates to the opinion which D'Alembert held of Boscovich.

> M. d'*Alembert* se contente ici de dire *qu'il a du nom dans les mathématiques:* dans un autre opuscule postérieur, il parle du P. *Boscovich* avec éloge, en disant qu'il mérite la réputation dont il jouit; mais pour ajouter qu'il a été tellement persécuté par les Supérieurs de son Ordre, que toute l'autorité du Souverain Pontife a à peine suffi pour le délivrer de leurs poursuites. Cependant on sait très bien que le R. P. *Boscovich* a toujours été considéré et respecté dans sa Compagnie comme un de ses plus dignes membres, et comme un homme du premier mérite à tous égards.

On page 71 of the memoir by D'Alembert which we are now considering he uses the words *habile Mathématicien,* I presume with reference to Boscovich. It has been asserted in recent times that D'Alembert and Lagrange had but a low opinion of Boscovich; see Arago's *Œuvres complètes,* Vol. II. page 140.

591. D'Alembert states on his page 75 his objection to the formula which Clairaut gave on his page 226. I have discussed the point in Art. 328. D'Alembert admits on his page 82 that Clairaut's more general formula on page 217 would supply all that was needed.

D'Alembert quotes in his own favour, with respect to his controversy with Boscovich's translator, a passage from a letter to himself, written as he says, by one of the greatest geometers of Europe: see his page 83.

592. The next memoir in the sixth volume of D'Alembert's *Opuscules Mathématiques* is entitled *Sur l'effet de la pesanteur au sommet et au pied des Montagnes* and more briefly *Sur l'attraction des Montagnes;* this occupies pages 85...92: it is followed by an *Addition à l'Article précédent* on pages 93...98.

593. A certain observer had reported that on the summit of a mountain in the Alps, 1085 toises high, a seconds pendulum had gained 28 minutes in two months; so that gravity appeared to be greater at the summit of the mountain than at its base. D'Alembert proposes to shew how the fact may be explained, assuming the observation to be accurate.

D'Alembert investigates the attractions of mountains of various shapes. The investigations are simple and satisfactory. In one case he supposes the mountain to be cylindrical, its height being small compared with the radius; he obtains a result which was first given by Bouguer, and has since passed into the elementary books: see Art. 363.

D'Alembert also investigates the influence exerted on a pendulum when it is placed in a valley between two mountains.

If ρ be the mean density of the Earth, and ρ' that of the mountain, D'Alembert finds that supposing we accept the observation on the Alps as trustworthy we must have $\rho' = \frac{8\rho}{3}$. This we should now consider to be quite inadmissible, and so we should have no faith in the observation. But at the date of the memoir the state of knowledge was different; and D'Alembert says on his pages 90, 91:

...cette hypothèse n'a rien de forcé; puisqu'on peut très bien supposer que la densité moyenne de la Terre est moindre que la densité des couches qui sont à sa surface.

The words are hardly fair; for the formula would make the mean density of the Earth scarcely one-third of that of the mountain.

D'Alembert refers on his page 92 to Bouguer's work on the Figure of the Earth, pages 357 and following. D'Alembert says:

On y trouve une Théorie de l'Attraction des Montagnes, mais beaucoup moins générale que celle qui a été l'objet de ce Mémoire.

594. On his page 93 D'Alembert refers to new observations with which he had become acquainted long after he had finished the preceding memoir. These observations seemed to shew that in a certain district of the Alps, attraction in ascending the mountains varied directly (not inversely) as the square of the distance from the centre of the Earth. He traces the consequence of this hypothesis.

Let h be the height of the mountain, ρ' its mean density, ρ the mean density of the Earth, r its radius. Then by the investigation referred to in Arts. 363 and 593 it appears that the attraction at the top of the mountain is $\frac{4\pi\rho r^3}{3(r+h)^2} + 2\pi\rho' h$, that is approximately $\frac{4\pi\rho r}{3} + 2\pi h\left(\rho' - \frac{4\rho}{3}\right)$. If the attraction varies directly as the square of the distance from the Earth's centre this must be equal to $\frac{4\pi\rho r}{3}\left(\frac{r+h}{r}\right)^2$, that is approximately to $\frac{4\pi\rho r}{3}\left(1 + \frac{2h}{r}\right)$.

Hence we have

$$\frac{4\pi\rho r}{3} \cdot \frac{2h}{r} = 2\pi h\left(\rho' - \frac{4\rho}{3}\right);$$

this leads to

$$\rho' = \frac{8\rho}{3}.$$

The coincidence of this result with that in Art. 593 is certainly curious; because it is a theoretical inference from observations which do not seem to have been influenced by theory. However there can be, I presume, no doubt that the observations must have been erroneous. Frisi alludes to the matter; see his *Cosmographia*, Vol. II. page 142: he seems to treat the observations as fictitious. He says:

Notitiis enim conquisitis undique accepi alpina illa experimenta... omnino esse supposita, et circa differentiam attractionum in vertice, et ad pedes montium Bouguerii tantum experimenta superesse quæ in investigationibus figuræ terrestris locum aliquem semper habere debeant.

See also La Lande's *Bibliographie Astronomique*, page 532.

595. The next memoir in the sixth volume of D'Alembert's *Opuscules Mathématiques* is entitled *Suite des Recherches sur la Figure de la Terre;* this occupies pages 99...133; it is followed by some *Remarques sur le Mémoire précédent* on pages 134...160.

596. The problem discussed is one which D'Alembert briefly noticed on pages 63...67 of the volume: a homogeneous mass of fluid in the form of an ellipsoid of revolution rotates with uniform angular velocity round its axis of figure, and is supposed to be in relative equilibrium under its own attraction and the attraction of a distant body situated on the prolongation of the axis of figure; then the condition for this relative equilibrium is found and discussed. Although the problem cannot be considered to be of any physical importance yet the analytical processes are both interesting and instructive.

Let M denote the mass of the distant body, h its distance from the centre of the ellipsoid; the axis of revolution of the ellipsoid when produced passes through M: take this for the axis of x.

Then the distant body exerts an action $\frac{M}{h^2}$ at the centre of the ellipsoid; and then in the usual way we find that what we may call the *disturbing* action of the distant body at a point (x, y) is equivalent to $\frac{2Mx}{h^3}$ and $\frac{My}{h^3}$ parallel to the axes of x and y respectively; the former in the direction in which x increases, the latter contrary to the direction in which y increases. D'Alembert says nothing about the force $\frac{M}{h^2}$; we must in fact imagine it to be counteracted by an equal force applied at every point.

Let us suppose that the equatorial axis of the ellipsoid is m times the polar axis; and let $k = \sqrt{(m^2 - 1)}$.

Suppose the density of the ellipsoid to be unity: then taking it to be an oblatum the attractions at (x, y) parallel to the axes of x and y respectively are by Art. 581

$$\frac{4\pi}{k^3}(k^2+1)(k - \tan^{-1}k)\,x \text{ and } \frac{2\pi}{k^3}\{(k^2+1)\tan^{-1}k - k\}\,y.$$

We have also the centrifugal force $\omega^2 y$ parallel to the axis of y, where ω is the angular velocity.

Hence putting X and Y for the whole forces at (x, y) parallel to the axes of x and y respectively, and estimating these forces inwards, we have

$$X = \frac{4\pi}{k^3}(k^2+1)(k - \tan^{-1}k)\,x - \frac{2Mx}{h^3},$$

$$Y = \frac{2\pi}{k^3}\{(k^2+1)\tan^{-1}k - k\}\,y + \frac{My}{h^3} - \omega^2 y.$$

Now we may apply Huygens's principle to obtain the condition of relative equilibrium. Thus X and Y must be positive, supposing x and y to be positive; and $Xdx + Ydy = 0$, must coincide with the differential equation to the ellipse which generates the ellipsoid, that is with $xdx + \frac{ydy}{k^2+1} = 0$. Hence we obtain

$$\frac{4\pi}{k^3}(k^2+1)(k - \tan^{-1}k) - \frac{2M}{h^3}$$

$$= (k^2+1)\left[\frac{2\pi}{k^3}\{(k^2+1)\tan^{-1}k - k\} + \frac{M}{h^3} - \omega^2\right];$$

and simplifying we have

$$\frac{\omega^2}{2\pi} - \frac{M}{2\pi h^3} = \frac{(3+k^2)\tan^{-1}k - 3k}{k^3} + \frac{M}{\pi h^3(k^2+1)}.$$

This is the fundamental equation of the problem; it agrees with D'Alembert's on his page 100, though with rather different notation.

We shall, as in Art. 581, put $\phi(k)$ for

$$\frac{(3+k^2)\tan^{-1}k-3k}{k^3}.$$

597. We have hitherto supposed the ellipsoid of revolution to be an oblatum. If it be an oblongum our fundamental equation still holds, only as $k=\sqrt{(m^2-1)}$, and m is now less than unity, $\phi(k)$ contains impossible quantities which must be transformed. We have

$$\phi(k)=\frac{3+k^2}{k^2}\cdot\frac{\tan^{-1}k}{k}-\frac{3}{k^2}=\frac{2+m^2}{m^2-1}\cdot\frac{\tan^{-1}\sqrt{(m^2-1)}}{\sqrt{(m^2-1)}}-\frac{3}{m^2-1}.$$

If m is less than 1, we find that $\dfrac{\tan^{-1}\sqrt{(m^2-1)}}{\sqrt{(m^2-1)}}$ transforms in the usual way into $\dfrac{1}{2\sqrt{(1-m^2)}}\log\dfrac{1+\sqrt{(1-m^2)}}{1-\sqrt{(1-m^2)}}$.

598. Our fundamental equation may be written thus

$$\frac{\omega^2}{2\pi}-\frac{M}{2\pi h^3}=\phi(k)+\frac{M}{\pi h^3(k^2+1)}=\phi\{\sqrt{(m^2-1)}\}+\frac{M}{\pi h^3 m^2}.$$

We have to consider whether a value or values of m between zero and infinity can be found to satisfy this equation. Moreover, if m is less than unity, we must consider that the proper form for $\phi\sqrt{(m^2-1)}$, free from impossible expressions, is

$$-\frac{m^2+2}{2(1-m^2)^{\frac{3}{2}}}\log\frac{1+\sqrt{(1-m^2)}}{1-\sqrt{(1-m^2)}}+\frac{3}{1-m^2};$$

we will denote this by $\psi(m)$.

That we have obtained the right equation for the case in which m is less than unity, may be verified by an independent investigation of the attraction of an *oblongum* on a particle at its surface. D'Alembert himself indicates this method of confirming the result obtained by the ordinary use of imaginary symbols: see his pages 134, 135.

599. Let us first consider the range of values of $\phi(k)$, as k increases from zero to infinity.

When k is very small $\phi(k)$ is approximately equal to $\frac{4k^2}{15}$, as may be easily shewn by expansion. And $\phi(k)$ obviously vanishes when k is infinite.

D'Alembert wishes to shew that $\phi(k)$ is always positive; see his pages 102 and 103. His demonstration is unsound. He shews that $\tan^{-1} k - \frac{k}{1+\frac{1}{3}k^2}$ is positive when k is infinitesimal; and he shews that this expression is positive when $\frac{k}{1+\frac{1}{3}k^2}$ has its greatest value, namely, when $k = \sqrt{3}$. It is easy then to see that the expression must be positive when k is greater than $\sqrt{3}$. But it does not necessarily follow that as k changes from 0 to $\sqrt{3}$ the expression is *always* positive.

We may proceed thus. Put $u = (3+k^2)\tan^{-1} k - 3k$; then

$$\frac{du}{dk} = 2k\left(\tan^{-1} k - \frac{k}{1+k^2}\right) = 2\tan\theta\,(\theta - \sin\theta\cos\theta), \quad \text{if } \tan^{-1} k = \theta.$$

Thus $\frac{du}{dk}$ is positive while k changes from zero to infinity; and so u continually increases with k and never vanishes.

Since $\phi(k)$ is always positive and vanishes both when k is zero and when k is infinite, it follows that $\phi'(k)$ must vanish, once at least, within this range of values of k. We have moreover shewn in Art. 586 that $\phi'(k)$ can vanish only once. We may observe that D'Alembert draws his diagrams consistently with the fact that $\phi'(k)$ vanishes only once, though as we have remarked he did not demonstrate this.

600. D'Alembert shews that $\psi(m)$ is always negative if m lies between 0 and 1. We have, in fact, to shew that

$$-\frac{2+m^2}{2(1-m^2)^{\frac{3}{2}}}\log\frac{1+\sqrt{(1-m^2)}}{1-\sqrt{(1-m^2)}} + \frac{3}{1-m^2}$$

is always negative. D'Alembert's method is rather laborious: see his page 104. The best way is to expand in powers of $\sqrt{(1-m^2)}$. Put t for $\sqrt{(1-m^2)}$; then we have

$$\psi(m) = -\frac{3-t^2}{2t^3}\log\frac{1+t}{1-t} + \frac{3}{t^2}.$$

Expanding the logarithm we find that

$$\psi(m) = -4\left\{\frac{t^2}{3.5} + \frac{2t^4}{5.7} + \ldots + \frac{nt^{2n}}{(2n+1)(2n+3)} + \ldots\right\}.$$

Thus as m increases from zero to unity, we have $\psi(m)$ always negative, and numerically *continually* decreasing from infinity to zero. This *continual* decrease is not mentioned by D'Alembert, though he draws his diagram consistently with it.

It will be convenient to give also the expansion of $\phi(k)$.

We have $$\phi(k) = \left(1+\frac{3}{k^2}\right)\frac{\tan^{-1}k}{k} - \frac{3}{k^2};$$

expand $\tan^{-1}k$; thus we get

$$\phi(k) = \frac{4k^2}{3.5} - \frac{8k^4}{5.7} + \ldots + (-1)^{n-1}\frac{4nk^{2n}}{(2n+1)(2n+3)} + \ldots$$

Since $k^2 = m^2 - 1 = -t^2$, we see by comparing these two expansions that the value of $\phi\{\sqrt{(m^2-1)}\}$ suffers no discontinuity as m passes through the value unity. This of course might have been held probable, but now it is demonstrated.

The series for $\psi(m)$ and $\phi(k)$ furnish us with an expansion for $\phi\{\sqrt{(m^2-1)}\}$, which will remain convergent for values of m between 0 and $\sqrt{2}$, the former extreme value being excluded.

601. Suppose we put $M=0$ in the fundamental equation of Art. 596; then we see that the equation cannot be solved by a value of m less than unity; for the left-hand member would be positive, and by Art. 600 the right-hand member would be negative. Hence a mass of rotating fluid cannot be in relative equilibrium if it is in the form of an *oblongum*, the axis of rotation coinciding with the axis of figure.

D'Alembert does not draw this inference from his formula. The theorem was first given by Laplace in his *Théorie...de la Figure des Planetes*, page 128.

602. From Arts. 599 and 600 we have the following results as to the value of $\phi\{\sqrt{(m^2-1)}\}$. When m increases from zero to infinity, $\phi\{\sqrt{(m^2-1)}\}$ begins by being negative infinity, increases algebraically, is zero when $m=1$, then becomes positive and increases to a maximum, and finally reduces to zero. In the diagram we take m as the abscissa, and $\phi\{\sqrt{(m^2-1)}\}$ as the ordinate of the curve, and we consider ordinates *positive* when they are *above* the straight line OM: D'Alembert reverses this arrangement.

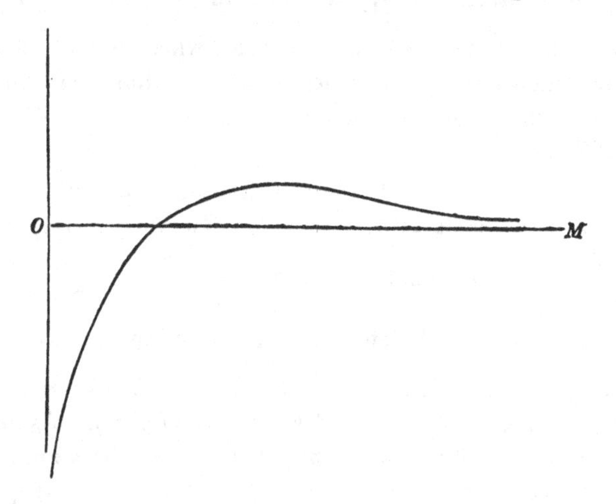

603. Next we may proceed to consider the curve, the ordinate of which is formed by adding to the corresponding ordinate of the preceding curve the term $\frac{M}{\pi h^3 m^2}$, as required by the fundamental equation of Art. 598.

Put $f(m)$ for $\phi\{\sqrt{(m^2-1)}\}+\frac{M}{\pi h^3 m^2}$, so that the fundamental equation becomes

$$\frac{\omega^2}{2\pi}-\frac{M}{2\pi h^3}=f(m).$$

When m is indefinitely small, $f(m)$ is positive and indefinitely great; when m is infinite $f(m)$ vanishes. Let y denote an ordinate corresponding to the abscissa m; then the curve determined by $y=f(m)$ may take various forms.

D'Alembert discusses the fundamental equation with great

detail, considering various cases which arise according to the values of $\frac{\omega^2}{2\pi} - \frac{M}{2\pi h^3}$ and the different forms of the curve $y=f(m)$. We will notice briefly some of the more interesting points which occur.

Let us consider some of the peculiarities of the curve $y=f(m)$.

(1) Let m_1 denote the value of m for which $\phi\{\sqrt{(m^2-1)}\}$ has its maximum value. If $\frac{M}{\pi h^3}$ is less than $\phi\{\sqrt{(m_1^2-1)}\} + \frac{M}{\pi h^3 m_1^2}$, we have $f(m)$ greater when $m=m_1$ than when $m=1$. And $f(m)$ is greater when $m=m_1$ than when $m=\infty$. Thus $f(m)$ must have some maximum value between $m=1$ and $m=\infty$. D'Alembert, pages 107 and 148.

(2) It is possible that $f(m)$ should be negative for part of the range between $m=0$ and $m=1$. For this merely requires that $\frac{M}{\pi h^3 m^2} + \psi(m)$ should be negative, or that $\frac{M}{\pi h^3} + m^2\psi(m)$ should be negative. Therefore, if $\frac{M}{\pi h^3}$ is less than the numerically greatest value of $m^2\psi(m)$, which is always negative between $m=0$ and $m=1$, there will be negative values of $f(m)$. As $m^2\psi(m)$ vanishes when $m=0$ and when $m=1$, there will be a numerically greatest value of m within this range. D'Alembert, pages 111 and 148.

(3) If, however, $\frac{M}{\pi h^3}$ is greater than the numerically greatest value of $m^2\psi(m)$ within the range from $m=0$ to $m=1$, then $f(m)$ is always positive from $m=0$ to $m=\infty$.

(4) It is possible to have such a value for $\frac{M}{\pi h^3}$ that $f(m)$ shall decrease continually from $m=0$ to $m=\infty$; that is, $f'(m)$ shall be always negative. D'Alembert, pages 117 and 120.

First, from $m=0$ to $m=1$. Here we have

$$f'(m) = \psi'(m) - \frac{2M}{\pi h^3 m^3}$$

$$= -\frac{(8+m^2)\,m}{2(1-m^2)^{\frac{5}{2}}} \log\frac{1+\sqrt{(1-m^2)}}{1-\sqrt{(1-m^2)}} + \frac{7m^2+2}{m\,(1-m^2)^2} - \frac{2M}{\pi h^3 m^3}.$$

This will be negative within the range, if algebraically

$$\frac{2M}{\pi h^3} \text{ is greater than } \frac{(7m^2+2)m^2}{(1-m^2)^2} - \frac{(8+m^2)m^4}{2(1-m^2)^{\frac{5}{2}}} \log \frac{1+\sqrt{(1-m^2)}}{1-\sqrt{(1-m^2)}}.$$

The expression on the right-hand side vanishes when $m=0$; and by èvaluation it will be found to be $\frac{8}{15}$ when $m=1$. It is always finite between these limiting values; and if $\frac{2M}{\pi h^3}$ is greater than the algebraically greatest of the values, $f'(m)$ will be negative from $m=0$ to $m=1$.

Next from $m=1$ to $m=\infty$. Here we have

$$f'(m) = \phi'(k)\frac{dk}{dm} - \frac{2M}{\pi h^3 m^3} = \left\{\frac{7k^2+9}{k^3(1+k^2)} - \frac{9+k^2}{k^4}\tan^{-1}k\right\}\frac{m}{k} - \frac{2M}{\pi h^3 m^3},$$

where $k^2 = m^2 - 1$. This will be negative between $k=0$ and $k=\infty$, if algebraically

$$\frac{2M}{\pi h^3} \text{ is greater than } \frac{(k^2+1)^2}{k}\left\{\frac{7k^2+9}{k^3(k^2+1)} - \frac{9+k^2}{k^4}\tan^{-1}k\right\}.$$

The expression on the right-hand side will be found to be $\frac{8}{15}$ when $k=0$, as it should be from above; and it is negative infinity when $k=\infty$. Hence there must be a greatest value among the positive values which it can take. If $\frac{2M}{\pi h^3}$ is greater than this value, $f'(m)$ will be negative from $m=1$ to $m=\infty$.

If then $\frac{2M}{\pi h^3}$ be greater than the greatest of the two values which have thus presented themselves, $f'(m)$ will be negative from $m=0$ to $m=\infty$.

604. The numerical result $\frac{8}{15}$ which occurs in the preceding Article may be easily verified. In fact, it is the value of $\psi'(m)$ when $m=1$, or of $\frac{d}{dm}\phi(k)$ which is required. Take the latter; then we have $\phi'(k)\frac{dk}{dm}$, that is, $\phi'(k)\frac{m}{k}$, that is, by Art. 600,

$\frac{m}{k}\left(\frac{8k}{15}-\frac{32k^3}{35}+\ldots\right)$; and when $m=1$ so that $k=0$, this becomes $\frac{8}{15}$. The same result will follow by the aid of Art. 600 from the value of $\psi'(m)$.

605. D'Alembert shews that the problem may in certain cases have two or three solutions for given values of ω, M, and h. He makes some remarks as to what we should now call the *stability* of the relative equilibrium, like the remarks on pages 56 and 57 of the volume which we have noticed in Art. 582. See his pages 112...115, 126...128, 153.

606. In the fundamental equation of Art. 598 put $m=1$; then since $\phi\{\sqrt{(m^2-1)}\}=0$ when $m=1$, we have

$$\frac{\omega^2}{2\pi}=\frac{3M}{2\pi h^3}.$$

Hence this relation must hold in order that a *sphere* may be a possible form of relative equilibrium.

607. When we have obtained a solution of the fundamental equation, it will still be necessary to advert to the condition stated in Art. 596, that X and Y must be positive if x and y are, before we can say that relative equilibrium exists. It will be sufficient to ensure that one of them is positive, because if the fundamental equation is satisfied, we know that X and Y are of the *same* sign, supposing x and y to be. D'Alembert pays proper attention to this point: see his pages 105, 116, 117, 122, 123.

Let us, for instance, consider the value of Y. Hence we see that we must have $\frac{(k^2+1)\tan^{-1}k-k}{k^3}$ greater than $\frac{\omega^2}{2\pi}-\frac{M}{2\pi h^3}$. Denote the former expression by v; then it will be found that

$$\frac{dv}{dk}=\frac{3k-(3+k^2)\tan^{-1}k}{k^4}.$$

By Art. 599 we see that $\frac{dv}{dk}$ is always negative for real values of k; and so for such values v is greatest when $k=0$: and then $v=\frac{2}{3}$.

When we put $\sqrt{(m^2-1)}$ for k, and suppose m less than 1, we get

$$\frac{dv}{dm}=\frac{dv}{dk}\frac{dk}{dm}=\frac{m}{k^4}\left\{3-(3+k^2)\frac{\tan^{-1}k}{k}\right\}$$

$$=\frac{m}{(m^2-1)^2}\left\{3-\frac{m^2+2}{2\sqrt{(1-m^2)}}\log\frac{1+\sqrt{(1-m^2)}}{1-\sqrt{(1-m^2)}}\right\}.$$

By Art. 600 we know that this is always negative if m lies between 0 and 1; and so for such values v is greatest when $m=0$.

But $v=-\dfrac{m^2}{2(1-m^2)^{\frac{3}{2}}}\log\dfrac{1+\sqrt{(1-m^2)}}{1-\sqrt{(1-m^2)}}+\dfrac{1}{1-m^2}$, so that when $m=0$ we have $v=1$.

Thus as m varies from zero to infinity, v continually diminishes from unity to zero. See D'Alembert's pages 116, 117, 151, 152.

The fact that v continually diminishes as m increases may also be shewn by putting the value of $\dfrac{dv}{dm}$ thus:

$$\frac{dv}{dm}=\frac{m}{1-m^2}\phi\{\sqrt{(m^2-1)}\};$$

this is always negative, for the factor $\phi\{\sqrt{(m^2-1)}\}$ is negative when m is less than 1, and the factor $\dfrac{m}{1-m^2}$ is negative when m is greater than 1.

It follows from this discussion that there can be no relative equilibrium if $\dfrac{\omega^2}{2\pi}-\dfrac{M}{2\pi h^3}$ is algebraically greater than unity. See D'Alembert's page 117.

608. Now let us consider the value of X. Hence we see that we must have $\dfrac{(k-\tan^{-1}k)(k^2+1)}{k^3}$ greater than $\dfrac{M}{2\pi h^3}$. This leads us to investigate the greatest value of the former expression. It will be found that this expression $=1-\dfrac{(k^2+1)\tan^{-1}k-k}{k^3}=1-v$; and as v continually diminishes from unity to zero, this expression continually increases from zero to unity. It follows that there can be no relative equilibrium if $\dfrac{M}{2\pi h^3}$ is greater than unity. See D'Alembert's page 124.

609. D'Alembert suggests another mode of obtaining solutions of the problem: see his pages 128...132. Let m be an abscissa and y an ordinate as before; and let $k = \sqrt{(m^2 - 1)}$. Then draw the curves

$$y = \frac{2(k - \tan^{-1} k)}{k^3} - \frac{M}{\pi h^3 (k^2 + 1)},$$

and
$$y = \frac{(k^2 + 1)\tan^{-1} k}{k^3} - \frac{1}{k^2} + \frac{M}{2\pi h^3} - \frac{\omega^2}{2\pi}.$$

At a point of intersection of these curves the corresponding value of m will satisfy the fundamental equation; and if the value of y at the point of intersection is positive, the resultant force at the surface tends *inwards*; therefore with the value of m thus obtained relative equilibrium will subsist.

It is sufficient by Art. 608 to confine ourselves to the case in which $\frac{M}{2\pi h^3}$ is less than unity.

In drawing the curves the results obtained in Art. 607 will be found useful. Thus, for instance, the equation to the first curve may be written

$$y = \frac{1}{m^2}\left(2 - \frac{M}{\pi h^3} - 2v\right);$$

and we know that v diminishes continually from unity to zero as m increases from zero to infinity. Hence y begins by being negative infinity, vanishes and changes sign once and only once, and is zero when m is infinite.

When $m = 1$ we have $y = 2 - \frac{M}{\pi h^3} - \frac{4}{3} = \frac{2}{3} - \frac{M}{\pi h^3}$; this is positive or negative according as $\frac{M}{2\pi h^3}$ is less or greater than $\frac{1}{3}$.

610. Instead of the two curves of the preceding Article, D'Alembert suggests in his pages 158...160, that we may take the two curves

$$y = \frac{2(k^2 + 1)(k - \tan^{-1} k)}{k^3} - \frac{M}{\pi h^3},$$

and
$$y = (k^2 + 1)\left\{\frac{(k^2 + 1)\tan^{-1} k}{k^3} - \frac{1}{k^2} + \frac{M}{2\pi h^3} - \frac{\omega^2}{2\pi}\right\}.$$

611. D'Alembert discusses at some length two analytical matters which present themselves.

On pages 134...142 he treats of difficulties which may occur in the use of the symbol $\sqrt{(-1)}$. For example, suppose we require the product of $\sqrt{(-a)}$ into $\sqrt{(-b)}$. On one hand we may take for it $\sqrt{(-a \times -b)}$, that is, $\sqrt{(ab)}$. On the other hand we may take for it $\sqrt{(a)} \times \sqrt{(-1)} \times \sqrt{(b)} \times \sqrt{(-1)}$, that is, $\sqrt{(ab)} \times \sqrt{(-1)} \times \sqrt{(-1)}$, that is, $-\sqrt{(ab)}$.

On pages 142...145 he shews in various ways that $x \log x$ is zero when x is; and so also is $x^p \log x^q$ where p and q are positive and finite.

612. The next memoir in the sixth volume of D'Alembert's *Opuscules Mathématiques* is also entitled *Suite des Recherches sur la Figure de la Terre;* this is a continuation of the preceding memoir; it occupies pages 161...197: it is followed by some *Remarques sur le Mémoire précédent* on pages 198...210.

613. In the preceding memoir D'Alembert had considered the relative equilibrium of a mass of rotating fluid in the form of an ellipsoid of revolution acted on by the disturbing force of a distant body, situated on the axis of rotation produced. In the present memoir he generalises the problem by giving any situation to the distant body, and by taking for the fluid mass the form of an ellipsoid, not necessarily of revolution.

614. We shall use notation more symmetrical than D'Alembert's.

Suppose then that the fluid is in the form of an ellipsoid. Take the axes of x, y, z to coincide with the axes of the ellipsoid; let $2a, 2b, 2c$ be the corresponding lengths of the axes. Let there be a distant body of mass M; and let its co-ordinates be l, m, n respectively: put $R^2 = l^2 + m^2 + n^2$.

Suppose the fluid to rotate with angular velocity ω round an axis, the direction cosines of which are λ, μ, ν. We have to form the conditions for relative equilibrium.

Now here we must observe that the distant body must, in fact, be supposed to share in this rotation of the fluid mass. D'Alembert never notices this fact, though it is really involved in his process. In the particular case of the preceding memoir, in which the distant body is supposed to be *on* the axis of rotation, we may practically regard the distant body as fixed; but we cannot in the present memoir. A particular case of the present memoir, as we shall see, was afterwards discussed by Laplace; in this case the Moon is taken to be the fluid mass, and the Earth to be the distant body. See Laplace's *Théorie...de la Figure des Planetes,* pages 113...116.

615. Let P be any point of the fluid; let x, y, z be the coordinates of P. The attraction of the fluid ellipsoid parallel to the axes of x, y, z respectively will be Ax, By, Cz respectively where A, B, C are certain constants. D'Alembert in effect briefly states that this can be easily shewn in the way in which Maclaurin treated the attraction of an ellipsoid of revolution; this is true, and it is to be noted that we have here, for the first time, the important extension of Maclaurin's result from an ellipsoid of revolution to the general ellipsoid. See D'Alembert's page 165. But as we shall hereafter point out, Frisi had previously gone some way in this direction: see his *De Gravitate,* pages 157 and 159.

616. The attraction of the distant body at P parallel to the axis of x is $\dfrac{M(l-x)}{\{(l-x)^2+(m-y)^2+(n-z)^2\}^{\frac{3}{2}}}$; the *disturbing* part of this is approximately

$$-\frac{Mx}{R^3}+\frac{3Ml(lx+my+nz)}{R^5},$$

say $-\dfrac{Mx}{R^3}+\dfrac{3Mlu}{R^5}$ where u is put for $lx+my+nz$.

It is only the disturbing part of the action of M which D'Alembert regards; he makes no allusion to the other part, that is, $\dfrac{Ml}{R^3}$ in this case. See Art. 596.

Let O denote the centre of the ellipsoid; let Q denote the foot of the perpendicular from P on the axis of rotation; then the so-called centrifugal force is $\omega^2 PQ$, and we require the resolved part of this. We have to *project* PQ on the axis of x; and by a known theorem of projections we may take the difference of the projections of OP and OQ for the projection of PQ.

Thus we obtain $\omega^2 \left(OP . \frac{x}{OP} - OQ \cos\lambda\right)$; and this

$$= \omega^2 (x - OQ \cos\lambda) = \omega^2 (x - OP \cos POQ \cos\lambda)$$
$$= \omega^2 \{x - (x \cos\lambda + y \cos\mu + z \cos\nu) \cos\lambda\}$$
$$= \omega^2 (x - v \cos\lambda) \text{ where } v \text{ is put for } x \cos\lambda + y \cos\mu + z \cos\nu.$$

Let X denote the whole force parallel to the axis of x, estimated inwards; then

$$X = Ax + \frac{Mx}{R^3} - \frac{3Mlu}{R^5} - \omega^2 (x - v \cos\lambda).$$

Similar expressions hold for the attractions parallel to the other axes, which we will denote by Y and Z respectively.

D'Alembert's method is substantially equivalent to this though his notation is less symmetrical.

617. The conditions for relative equilibrium are

$$X \div \frac{x}{a^2} = Y \div \frac{y}{b^2} = Z \div \frac{z}{c^2}.$$

Take the equation $Xa^2y = Yb^2x$; this must be identically true, and so we may equate the coefficients of xy, x^2, y^2, xz, yz. By equating the coefficients of xy we obtain

$$a^2 \left\{A + \frac{M}{R^3} - \frac{3Ml^2}{R^5} - \omega^2 (1 - \cos^2\lambda)\right\}$$
$$= b^2 \left\{B + \frac{M}{R^3} - \frac{3Mm^2}{R^5} - \omega^2 (1 - \cos^2\mu)\right\}.$$

By equating the coefficients of x^2, and by equating the coefficients of y^2, we arrive at the same condition, namely,

$$-\frac{3Mlm}{R^5} + \omega^2 \cos\lambda \cos\mu = 0.$$

By equating the coefficients of xz we have

$$-\frac{3Mmn}{R^5} + \omega^2 \cos\mu \cos\nu = 0.$$

By equating the coefficients of yz we have

$$-\frac{3Mnl}{R^5} + \omega^2 \cos\nu \cos\lambda = 0.$$

In like manner we may take the equation $Xa^2z = Zc^2x$; by so doing we shall find that we get only one new condition.

The whole results may be written thus:

$$\frac{3Mmn}{R^5} = \omega^2 \cos\mu \cos\nu, \quad \frac{3Mnl}{R^5} = \omega^2 \cos\nu \cos\lambda, \quad \frac{3Mlm}{R^5} = \omega^2 \cos\lambda \cos\mu.$$

$$a^2\left(A + \frac{M}{R^3} - \frac{3Ml^2}{R^5} - \omega^2 \sin^2\lambda\right) = b^2\left(B + \frac{M}{R^3} - \frac{3Mm^2}{R^5} - \omega^2 \sin^2\mu\right)$$
$$= c^2\left(C + \frac{M}{R^3} - \frac{3Mn^2}{R^5} - \omega^2 \sin^2\nu\right).$$

618. As a particular case of the preceding investigation, suppose that there is no distant disturbing body; then $M=0$; thus $\cos\mu \cos\nu = 0$, $\cos\nu \cos\lambda = 0$, $\cos\lambda \cos\mu = 0$. Hence two of the three cosines $\cos\lambda$, $\cos\mu$, $\cos\nu$ must vanish; so that the rotation must be round one of the principal axes of the ellipsoid. Hence we see that the case taken in Jacobi's theorem is the only case in which an ellipsoid of fluid rotating round a diameter can remain in relative equilibrium. A statement which has been recently made to the contrary by Dahlander and by Schell is inaccurate: see the *Proceedings of the Royal Society*, Vol. XXI.

619. Return to the conditions obtained in Art. 617. Let us suppose that l, m, and n are not zero. The first and the second of these conditions give

$$\frac{m}{l} = \frac{\cos\mu}{\cos\lambda};$$

the second and the third give

$$\frac{n}{m} = \frac{\cos\nu}{\cos\mu}.$$

Hence the radius vector to the distant body coincides in direction with the axis of rotation; thus

$$\cos\lambda = \frac{l}{R}, \quad \cos\mu = \frac{m}{R}, \quad \cos\nu = \frac{n}{R},$$

and then from any of the first three conditions we get

$$\frac{3M}{R^3} = \omega^2;$$

and the other conditions reduce to

$$a^2\left(A - \frac{2M}{R^3}\right) = b^2\left(B - \frac{2M}{R^3}\right) = c^2\left(C - \frac{2M}{R^3}\right);$$

these last will be satisfied if $a = b = c$, that is if the fluid mass be spherical.

The particular case in which the radius vector to the distant body and the axis of rotation coincide in direction presents itself in D'Alembert's memoir; but he does not pay much attention to it: see his page 200.

He also notices a particular case in which it is given that two of the three a, b, c are nearly equal: see his page 209.

But he does not notice that we may have a sphere exactly if $\frac{l}{\cos\lambda} = \frac{m}{\cos\mu} = \frac{n}{\cos\nu}$ and $\omega^2 = \frac{3M}{R^3}$.

620. It will be interesting to enquire if the conditions in the preceding Article can be satisfied in any other way besides having $a = b = c$; this enquiry leads us a little beyond the point at which the theory of the attraction of ellipsoids had arrived at this date.

Let V denote the mass of the ellipsoid; then we know that

$$A = \frac{3V}{a}\int_0^1 \frac{x^2 dx}{\sqrt{\{a^2 + (b^2 - a^2)x^2\}\{a^2 + (c^2 - a^2)x^2\}}};$$

This result was given by Laplace in his *Théorie ...de la Figure des Planetes*, page 92; as we shall see D'Alembert himself first obtained it but rejected it in the seventh volume of his *Opuscules Mathématiques*.

Assume $x = \frac{a}{\sqrt{(a^2 + s)}}$; then we find that

$$A = \frac{3V}{2}\int_0^\infty \frac{ds}{(a^2 + s)D},$$

where D stands for $\sqrt{\{(a^2 + s)(b^2 + s)(c^2 + s)\}}$.

In like manner we have

$$B = \frac{3V}{2}\int_0^\infty \frac{ds}{(b^2 + s)D}, \quad C = \frac{3V}{2}\int_0^\infty \frac{ds}{(c^2 + s)D}.$$

Put ϕ^2 for $\frac{2M}{R^3}$; then the conditions we have to examine may be written

$$(a^2 - b^2)\,\phi^2 = a^2A - b^2B = \frac{3V(a^2 - b^2)}{2}\int_0^\infty \frac{sds}{(a^2+s)\,(b^2+s)\,D},$$

$$(b^2 - c^2)\,\phi^2 = b^2B - c^2C = \frac{3V(b^2 - c^2)}{2}\int_0^\infty \frac{sds}{(b^2+s)\,(c^2+s)\,D};$$

hence we see that these conditions cannot be satisfied if a, b, c are all unequal; for they would lead to two different values of ϕ^2.

But suppose two of the three, a, b, c to be equal; say a and b: then our conditions reduce to

$$\phi^2 = \frac{3V}{2}\int_0^\infty \frac{sds}{(b^2+s)\,(c^2+s)\,D};$$

and this is quite admissible if b, c, V and ϕ be properly adjusted, whether b is greater or less than c.

621. If l, m, n are all different from zero we have the case discussed in the preceding two Articles, in which the radius vector to the distant body and the axis of rotation coincide in direction. D'Alembert himself pays little attention to this case: indeed in his page 200 he seems to consider that it cannot occur. Let us now return to the general conditions of Art. 619; and suppose that l, m, n are not all different from zero. Suppose for example that $n = 0$; then it follows from the first and second conditions that either $\cos\nu = 0$, or else $\cos\lambda = 0$, and $\cos\mu = 0$: if we suppose the latter, then l or m must also $= 0$. In the former case, the axis of rotation is in the principal plane corresponding to a and b; in the latter case the axis of rotation coincides with the axis corresponding to c. In each case the axis of rotation and the radius vector to the distant body are both in one of the principal planes of the fluid mass.

622. In Arts. 619 and 620 we see that the supposed ellipsoid is either a sphere or an ellipsoid of revolution; and in Art. 621 we see that the axis of rotation and the radius vector to the distant body must be in one of the principal planes of the fluid mass. Combining these two results, we may say that in every case in which the relative equilibrium is possible the axis of rotation and the radius vector to the distant body must be in one of the principal planes of the fluid mass. D'Alembert arrives at this result,

and confirms it by some general reasoning which is not very cogent: see his pages 198...200.

623. As a particular case of Art. 617 let us suppose we have given that the axis of rotation and the radius vector to the distant body are at right angles. This may be considered to hold with respect to the moon supposed fluid, the distant body being the Earth. Since here we have not the case of Arts. 619 and 620, it follows that one or two of the three l, m, n must be zero. Suppose $n=0$; then from the first two conditions of Art. 617, we shall find either $\cos\nu=0$, or both $\cos\lambda=0$, and $\cos\mu=0$.

I. Suppose $\cos\nu=0$. Then the third condition is

$$\omega^2 \cos\lambda \cos\mu = \frac{3Mlm}{R^5}$$

Now by our hypothesis that the two directions are at right angles, this would give $\omega^2 = -\frac{3M}{R^3}$, if we suppose that lm does not vanish; this is impossible. Therefore lm vanishes. Hence we must have either $l=0$, and $\cos\mu=0$, or $m=0$, and $\cos\lambda=0$.

II. Suppose $\cos\lambda=0$, and $\cos\mu=0$.

Then the third condition shews that $lm=0$. Therefore either $l=0$, or $m=0$.

Hence we must have the axis of rotation coinciding with one of the principal axes of the body, and the radius vector to the distant body coinciding with another.

The result might have been anticipated perhaps; and we shall find that Laplace assumes it as evident: see the reference in Art. 614.

624. We have seen in Art. 622 that the axis of rotation and the radius vector to M must always be in one of the principal planes of the ellipsoid. We will suppose that $n=0$, and $\cos\nu=0$. Hence the conditions of Art. 617, reduce to

$$\frac{3Mlm}{R^5} = \omega^2 \cos\lambda \cos\mu,$$

$$a^2\left\{A+\frac{M}{R^3}-\frac{3Ml^2}{R^5}-\omega^2\sin^2\lambda\right\}$$

$$=b^2\left\{B+\frac{M}{R^3}-\frac{3Mm^2}{R^5}-\omega^2\sin^2\mu\right\}=c^2\left\{C+\frac{M}{R^3}-\omega^2\right\}.$$

And in virtue of our supposition that n and $\cos\nu$ vanish we have

$$\cos^2\lambda+\cos^2\mu=1,\quad l^2+m^2=R^2.$$

As to whether these equations are consistent nothing is said by D'Alembert; we have discussed one case of the general problem in Arts. 619 and 620, but the matter is not of sufficient importance to detain us longer.

625. D'Alembert begins on his page 174 an investigation of the attraction of an ellipsoid on any particle at the surface. This amounts to seeking the values of the A, B, C of Art. 615.

He makes some simple and useful remarks on his pages 174...176; we will give an example of them. Suppose the semiaxes of an ellipsoid to be r, $r(1+\alpha)$, and $r(1+\beta)$, where α and β are very small. Let the approximate value of the attraction be required for a particle situated at the end of the semiaxis r. We may assume that this attraction will be $\frac{4\pi r}{3}(1+p\alpha+q\beta)$, where p and q are certain constants to be determined: this assumption depends on the fact that if α and β vanish, the body becomes a sphere, and the attraction then is $\frac{4\pi r}{3}$. Next we may admit that $p=q$; because the attraction ought to remain unchanged if we interchange the second and third semiaxes. Hence the attraction becomes $\frac{4\pi r}{3}\{1+p(\alpha+\beta)\}$. Now we can determine p. For if we suppose $\alpha=\beta$ the ellipsoid becomes an ellipsoid of revolution, and the attraction of such a solid on a particle at the pole is known: hence equating this known attraction, estimated approximately, to $\frac{4\pi r}{3}(1+2p\alpha)$ we determine p. We should thus get $p=\frac{2}{5}$.

626. D'Alembert attempted to find the attraction of an ellipsoid by decomposing it into slices in various ways; but he does not succeed in effecting the integrations. We know now that the

result can be expressed by means of elliptic integrals, but not by circular arcs or logarithms. We will briefly state the methods of decomposition of the ellipsoid which he tries. The attracted particle is supposed to be at the end of the semiaxis c.

I. Suppose a plane to pass through the attracted particle, and also through the tangent to the ellipsoid at that point which is parallel to the axis $2a$. Let this plane turn round the tangent line and cut the ellipsoid into wedge-shaped slices: see D'Alembert's page 180. This decomposition is like that used by Thomas Simpson; which we have noticed in Art. 279.

II. Instead of using the tangent parallel to the axis $2a$, we may use the tangent parallel to the axis $2b$.

III. Suppose a plane to pass through the axis $2c$ and to turn round, and thus cut the ellipsoid into wedge-shaped slices: see D'Alembert's page 183. This decomposition is like that used by Maclaurin; which we have noticed in Art. 255.

IV. Or the ellipsoid may be cut into laminæ by a plane which is always at right angles to the axis $2c$: see D'Alembert's page 184.

627. For the case of an ellipsoid in which two of the axes are very nearly equal D'Alembert obtains approximate values of the attraction at the end of the principal axes: see his page 192. A mistake in the results is corrected on page 424.

The approximate results just referred to are applied by D'Alembert to the question of relative equilibrium which was proposed at the beginning of the memoir: see his pages 194...197. He finishes in a patronising tone:

> Je ne doute point que cette nouvelle Recherche ne donnât lieu à plusieurs remarques curieuses; mais je les abandonne à d'autres Géometres, la matiere n'ayant plus aucune difficulté.

628. The next memoir in the sixth volume of D'Alembert's *Opuscules Mathématiques* is also entitled *Suite des Recherches sur la Figure de la Terre;* this is a continuation of the preceding memoir; it occupies pages 211...246: it is followed by some *Remarques sur le Mémoire précédent* on pages 247...259.

629. D'Alembert now proposes to extend the problem of the preceding memoir by supposing several distant attracting bodies instead of the single distant attracting body there considered.

This extension becomes very easy with the aid of modern symmetrical notation. Let M_1, M_2, M_3... denote the masses of the various distant bodies respectively; let l_1, m_1, n_1 be the coordinates of the first body, R_1 its distance; and let similar notation hold with respect to the other bodies.

Then instead of the first equation of Art. 617, namely

$$\frac{3Mmn}{R^5} = \omega^2 \cos\mu \cos\nu,$$

we now have $\dfrac{3M_1m_1n_1}{R_1^{\,5}} + \dfrac{3M_2m_2n_2}{R_2^{\,5}} + \dfrac{3M_3m_3n_3}{R_3^{\,5}} + \ldots = \omega^2 \cos\mu \cos\nu,$

which we may write thus $3\Sigma \dfrac{Mmn}{R^5} = \omega^2 \cos\mu \cos\nu.$

similarly the other equations may be expressed.

D'Alembert himself does not proceed in this way nor adopt this notation. He uses spherical trigonometry. It may be observed that he demonstrates the expression for the cosine of an angle of a spherical triangle in terms of the sines and cosines of the sides; he starts from formulæ for a right-angled spherical triangle which he assumes: see his pages 247 and 248.

As we have remarked in Art. 614, the distant bodies must be supposed to rotate with the fluid mass; though D'Alembert does not notice this fact. And as in Arts. 596 and 616, D'Alembert says nothing about certain forces which are not what I have called *disturbing* forces.

630. The only point which appears to be of any interest in the problem is a remark which D'Alembert makes on his page 253; the remark amounts to this: if the axis of rotation and the radii vectores to the distant attracting bodies are all in one plane that plane must be a principal plane of the ellipsoid. He does not demonstrate this, but seems to rely on the principle of symmetry as in the corresponding theorem for a single distant attracting body: see Art. 622. We will examine the theorem. Suppose that the equation to the plane is $\alpha x + \beta y + \gamma z = 0$; so that

$$\alpha \cos \lambda + \beta \cos \mu + \gamma \cos \nu = 0,$$
$$\alpha l_1 + \beta m_1 + \gamma n_1 = 0,$$
$$\alpha l_2 + \beta m_2 + \gamma n_2 = 0,$$

and so on.

Take the three equations

$$3\Sigma \frac{Mmn}{R^5} = \omega^2 \cos \mu \cos \nu, \qquad 3\Sigma \frac{Mnl}{R^5} = \omega^2 \cos \nu \cos \lambda,$$
$$3\Sigma \frac{Mlm}{R^5} = \omega^2 \cos \lambda \cos \mu.$$

Substitute in the first of these for n_1, n_2, ..., and for $\cos \nu$;

thus $$3\Sigma \frac{Mm(\alpha l + \beta m)}{R^5} = \omega^2 \cos \mu \, (\alpha \cos \lambda + \beta \cos \mu);$$

therefore by means of the third equation we obtain

$$3\Sigma \frac{Mm^2}{R^5} = \omega^2 \cos^2 \mu.$$

Similarly $3\Sigma \frac{Ml^2}{R^5} = \omega^2 \cos^2 \lambda$, $3\Sigma \frac{Mn^2}{R^5} = \omega^2 \cos^2 \nu$.

Hence the first of the three equations becomes

$$\Sigma \frac{Mmn}{R^5} = \sqrt{\Sigma \frac{Mm^2}{R^5} \, . \, \Sigma \frac{Mn^2}{R^5}} \, .$$

Squaring we get $\left\{ \frac{M_1 m_1 n_1}{R_1^5} + \frac{M_2 m_2 n_2}{R_2^5} + \frac{M_3 m_3 n_3}{R_3^5} + \dots \right\}^2$

$$= \left\{ \frac{M_1 m_1^2}{R_1^5} + \frac{M_2 m_2^2}{R_2^5} + \frac{M_3 m_3^2}{R_3^5} + \dots \right\} \left\{ \frac{M_1 n_1^2}{R_1^5} + \frac{M_2 n_2^2}{R_2^5} + \frac{M_3 n_3^2}{R_3^5} + \dots \right\}.$$

This by common Algebra leads to $\frac{m_1}{n_1} = \frac{m_2}{n_2} = \frac{m_3}{n_3} = \dots$

In this way we see that all the radii vectores to the distant bodies must coincide. Thus the case reduces to that of Art. 619.

But suppose, as in fact D'Alembert does, that the plane in which the axis of rotation and the radii vectores to the distant bodies lie is perpendicular to a principal plane; let its equation be

$$\alpha x + \beta y = 0.$$

Then as before we can obtain from our three equations,

$$3\Sigma \frac{Ml^2}{R^5} = \omega^2 \cos^2 \lambda, \qquad 3\Sigma \frac{Mm^2}{R^5} = \omega^2 \cos^2 \mu;$$

but we do *not* now have also $3\Sigma \dfrac{Mn^2}{R^5} = \omega^2 \cos^2 \nu$.

The equations which correspond to the last two of Art. 617 are

$$a^2 \left\{A + \Sigma \frac{M}{R^3} - \omega^2\right\} = b^2 \left\{B + \Sigma \frac{M}{R^3} - \omega^2\right\} = c^2 \left\{C - 2\Sigma \frac{M}{R^3}\right\};$$

for $\omega^2 \sin^2 \nu = \omega^2 (\cos^2 \lambda + \cos^2 \mu) = 3\Sigma M \dfrac{l^2 + m^2}{R^5} = 3\Sigma \dfrac{M (R^2 - n^2)}{R^5}$.

If D'Alembert's remark were universally true the equations connecting a, b, and c ought to be impossible, or inconsistent with the others, if a, b, and c are unequal. But this does not seem to be the case. By the method of Art. 620, we get from them

$$\omega^2 - \Sigma \frac{M}{R^3} = \frac{3V}{2} \int_0^\infty \frac{sds}{(a^2 + s)(b^2 + s) D},$$

and
$$2c^2 \Sigma \frac{M}{R^3} = \frac{3V}{2} \int_0^\infty \frac{c^2 s + c^2 (a^2 + b^2) - b^2 a^2}{D^3} sds;$$

and these if $c^2 (a^2 + b^2) - b^2 a^2$ is positive present nothing impossible.

As an example we might suppose two distant bodies, and take

$$l_1 = 0, \quad m_1 = 0, \quad n_1 = R,$$

$$l_2 = R_2 \cos \lambda, \quad m_2 = R_2 \cos \mu, \quad n_2 = R_2 \cos \nu.$$

Then it will be found that our first three equations give $\omega^2 = \dfrac{3M_2}{R_2^3}$; and we have only to ascertain if this is consistent with the last two equations, the form of which has just been given. Thus we have to put

$$\frac{2M_2}{R_2^3} - \frac{M_1}{R_1^3} = \frac{3V}{2} \int_0^\infty \frac{(c^2 + s) sds}{D^3},$$

and
$$2c^2 \left(\frac{M_2}{R_2^3} + \frac{M_1}{R_1^3}\right) = \frac{3V}{2} \int_0^\infty \frac{c^2 s + c^2 (a^2 + b^2) - b^2 a^2}{D^3} sds.$$

It will be found that these lead to values of $\frac{M_2}{R_2^3}$ and $\frac{M_1}{R_1^3}$ which are certainly positive if $(a^2-c^2)(c^2-b^2)$ is positive; for then also $c^2(a^2+b^2)-b^2a^2$ is positive. It is manifest that this condition *may* be satisfied; and thus D'Alembert's remark is not true.

631. D'Alembert on his page 216, refers to Maclaurin's Essay on the Tides, as containing a little matter bearing on the problem discussed in this memoir; but Maclaurin had not effected much. Maclaurin did not shew that the figure of an ellipsoid would satisfy the conditions of equilibrium; nor did he show how to determine the position of the axes of the ellipsoid. D'Alembert says of his own memoir: Nous avons de plus démontré dans celui-ci que la figure du sphéroïde est elliptique... However he does not shew that the figure *is* an ellipsoid, but only that it *may* be an ellipsoid.

632. D'Alembert says on his page 217, that he will conclude with some detached reflexions bearing on the Figure of the Earth.

633. He says that among the solutions hitherto given of the problem the only one which is exact is that which supposes the spheroid to be fluid and homogeneous; the other solutions being approximations. Suppose that α is a very small quantity; and we have found that neglecting α^2 the equation of relative equilibrium is satisfied for a certain figure; we must not say that this figure *exactly* satisfies the conditions of relative equilibrium. But D'Alembert suggests that if we give to the figure a certain small change of the order α^2 the conditions of relative equilibrium may be rigorously satisfied; and he considers it a plausible supposition that there may be an *infinity* of figures in which the relative equilibrium will subsist rigorously: see his page 223. Probably few persons will agree with D'Alembert in considering this supposition plausible.

634. D'Alembert returns on his pages 225...230 and 254...259, to his favourite equation relating to the ellipticity of fluid surrounding a solid nucleus: see Arts. 376, 430, and 590.

We shall briefly notice some points that arise.

On his pages 227...229, D'Alembert criticises as inexact certain formulæ on page 247 of Clairaut's work, and thus as affording an insufficient proof of Clairaut's theorem which is founded on them.

But, as might be expected, D'Alembert is wrong and Clairaut is right. The fact amounts to this: what I have called for instance A in Art. 336, is called A by Clairaut. Now D'Alembert really supposes A to stand for an integral taken not from 0 to r_1, but from some value say r_2 up to r_1: and thus he wants to add terms to Clairaut's formulæ. Plana rightly takes the side of Clairaut: see *Astronomische Nachrichten*, Vol. XXXVIII, page 245.

On his pages 254, 255, D'Alembert gives, without any preparatory statements what is really a more exact investigation of the problem of Art. 376. He thus arrives at the result which I have given in Art. 377, in which the difference between r' and r_1 is not neglected. In this investigation however he *assumes* on the second line of his page 255 the expression for the force at right angles to the radius. In Clairaut's investigations the necessary results are *demonstrated*. D'Alembert does not observe that the theorem is included in a more general one which he had demonstrated like Clairaut: see Art. 443.

In the formulæ of Art. 376, suppose that $\epsilon = \epsilon'$; then we get $\epsilon = \frac{5}{4}\phi$ where ϕ stands for $\omega^2 \div \frac{4\pi\sigma}{3}$. This result *is independent* of ρ; *it is the same as we should get for a homogeneous fluid.* D'Alembert seems to attach special importance to this result: see pages 79, 225, 256 of the Volume. But the result is what might be expected. Suppose a homogeneous fluid rotating in relative equilibrium: solidify all but a film of fluid; the relative equilibrium will not be disturbed. If we consider the film so thin that its action on itself may be disregarded, it is kept in relative equilibrium by the attraction of the solid part. Hence if we alter the density of the fluid film, it will still be kept in relative equilibrium.

635. On his page 231 D'Alembert refers to the demonstration he had given of the proposition that an oblatum is the only form of relative equilibrium for a revolving fluid: see Art. 575. That demonstration we pronounced a failure. From what he now says, it appears to me that he overlooks the consideration brought forward in Art. 577, as to what his theorem would have established if the demonstration had been sound.

636. D'Alembert devotes his pages 232...246 to investigations relative to the attraction of an ellipsoid on an external particle. He confirms by analysis Maclaurin's proposition respecting the attraction of confocal ellipsoids of revolution on an external particle which is on the line of the axis or in the plane of the equator. But D'Alembert was unable to extend this as Maclaurin did, to the case of ellipsoids not of revolution. D'Alembert says on his pages 242 and 243.

Je soupçonne donc que M. Maclaurin s'est trompé dans l' *art.* 653 de son *Traité des Fluxions*, quand il a dit que sa méthode pour trouver l'attraction d'un sphéroïde de révolution dans le plan de l'équateur, ou dans l'axe, pouvoit s'appliquer à un solide qui ne seroit pas de révolution......Au reste, ce n'est ici qu'un doute que je propose, n'ayant pas suffisamment examiné la proposition de M. Maclaurin, qu'il se contente d'énoncer sans la démontrer.

As we have stated in Art. 260 Maclaurin really demonstrated the theorem which D'Alembert considers to have been only enunciated, and the truth of which he here doubts. Subsequently, as we shall see, D'Alembert conquered his doubts and demonstrated the theorem: he was the first person who drew attention to the theorem and demonstrated it after Maclaurin himself.

637. The next memoir in the sixth volume of D'Alembert's *Opuscules Mathématiques* is entitled *Sur les Atmospheres des Corps Célestes;* it occupies pages 339...359.

638. The first paragraph explains the object of the memoir:

Le but des Recherches suivantes est de donner sur l'Atmosphere des Planetes quelques Remarques que je crois nouvelles, et de corriger en même-temps quelques méprises où des Auteurs célébres sont tombés sur cette matiere.

D'Alembert refers on his pages 345, 347, 349 and 350 to Mairan's Treatise on the Aurora Borealis; he refers to Euler on his page 350; and to Maupertuis on his page 358. Thus, I presume, these are the celebrated authors whose mistakes he proposes to correct.

639. D'Alembert obtains his fundamental equation in an unsatisfactory manner. He assumes that the stratum of the air in

contact with the surface of the planet is a level surface; then he takes an exterior level surface; and he makes what he calls the weight of a column terminated at these surfaces constant. He ought not to assume that the surface of the planet is a level surface for the air.

Suppose ω the angular velocity, r the distance of a point in the atmosphere from the centre of the Earth, θ the angle which r makes with the polar axis, M the mass of the Earth. Then by the usual equations for relative equilibrium

$$\frac{1}{\rho}\frac{dp}{dy} = -\frac{M}{r^2}\cos\theta, \quad \frac{1}{\rho}\frac{dp}{dx} = -\frac{M}{r^2}\sin\theta + \omega^2 x.$$

Hence the equation to a level surface is

$$\frac{M}{r} + \frac{\omega^2 x^2}{2} = \text{constant}.$$

Let r_1 and r_2 be the values of r at the equator and the pole respectively in the same level surface; then

$$\frac{M}{r_1} + \frac{\omega^2 r_1^2}{2} = \frac{M}{r_2}.$$

The matter is discussed by Laplace, as we shall see hereafter; but nothing is really added to what we find in D'Alembert's memoir. D'Alembert shews that the zodiacal light cannot be caused by the atmosphere of the Sun: the remark is repeated by Laplace. See the *Mécanique Céleste*, Livre III., Chapitre VII.

640. The form of the atmosphere is determined by a curve of which the equation in polar coordinates is

$$\frac{r^3 \sin^2\theta}{c^3} - \frac{r}{b} + 1 = 0.$$

It may happen that corresponding to a given value of θ we have two positive values of r, and one negative value. The three values would all be regarded in tracing the curve according to modern notions. D'Alembert touches on the subject in his pages 347...349. We may state that his opinion briefly amounts to rejecting the negative value of r entirely. He observes, in fact, that if we put $\sqrt{(x^2+y^2)}$ for r, and clear of radicals, we obtain an equation of the sixth degree; and this gives a branch corresponding to the negative value of r just mentioned. But according to him this new branch does not belong to us. However, he is not so much regarding the curve itself as the physical problem from which it arose.

641. Hitherto we have not supposed any action on the atmosphere except that of the planet to which it belongs; but D'Alembert proceeds to consider the action of one or more other planets. As in the case of a revolving fluid, when he introduces a distant planet he first puts it on the prolongation of the axis of rotation: see his pages 354 and 355. Next he supposes the distant planet to have any position. As before too he really supposes the distant planet to preserve the same relative position, so that, in fact, the distant planet must be supposed to rotate with the planet which carries the atmosphere. See Art. 629.

642. The mode in which D'Alembert finds what we should now call the pressure at any point of the atmosphere, when there is besides the planet itself, a distant planet acting, may be noticed. See his pages 355...357.

We know that the polar equations for relative equilibrium are

$$\frac{1}{\rho}\frac{dp}{dr} = R, \qquad \frac{1}{\rho}\frac{dp}{rd\theta} = T.$$

Now, in fact, he only considers the first of these equations. The value of p found from this must give the right result, provided we remember that the so-called arbitrary constant must be, if necessary, regarded as a function of θ. But without working out the problem fully in rectangular coordinates, we easily see that the value of p must be such that θ never enters alone but always accompanied by r. Thus p cannot contain any arbitrary function of θ alone. Therefore, the first equation alone is sufficient for finding p. D'Alembert himself, however, gives no explanation of his process.

643. In the *Nouveaux Mémoires de l'Académie...* of Berlin, for 1774, published in 1776, we have extracts from two letters addressed by D'Alembert to Lagrange; see pages 308...311 of the

volume. D'Alembert had discovered that Maclaurin's theorem, about which he formerly doubted, was really true; and here he sends to Lagrange sketches of two demonstrations: see Art. 636. The demonstrations are given at full in the seventh volume of the *Opuscules Mathématiques*, to which we now proceed.

644. The seventh volume of D'Alembert's *Opuscules Mathématiques* was published in 1780; a memoir entitled *Sur l'attraction des Sphéroides Elliptiques*, occupies pages 102...207; this is followed by some *Remarques sur le Mémoire précédent* on pages 208...233. From page 208 we learn that the Remarks were written long after the memoir; and, therefore, the memoir must have been written long before 1780.

D'Alembert says that his attention had been turned to the subject again by reading the excellent memoir by Lagrange in the Berlin *Mémoires* for 1773.

645. The first part of the memoir, which occupies pages 103...116, is devoted to the proof of Maclaurin's theorem: see Art. 636. D'Alembert starts from formulæ given in the sixth volume of his *Opuscules Mathématiques;* and by three different methods arrives at the required result.

One of these methods occupies D'Alembert's Article 30; it is curious from its obscurity. When carefully examined it is found to be equivalent to a circuitous method of arriving at the expression πab for the area of an ellipse, of which a and b are the semiaxes.

D'Alembert, on his page 114, corrects an important misprint in Lagrange's memoir in the Berlin *Mémoires* for 1773.

646. The second part of the memoir, which occupies pages 116...159, is devoted to the discussion of two formulæ relating to the attraction of an ellipsoid, which were given on pages 180 and 184 of the sixth volume of the *Opuscules Mathématiques:* see Art. 626. We will briefly indicate, by modern methods and notation, the nature of these formulæ.

Suppose we wish to find the attraction of an ellipsoid, the axes of which are $2a$, $2b$, $2c$, on a particle at the end of the axis $2c$.

We use the method I. of Art. 626. Taking this point as origin, we have for the equation to the ellipsoid,

$$\frac{2z}{c}=\frac{x^2}{a^2}+\frac{y^2}{b^2}+\frac{z^2}{c^2}.$$

Put $x=r\cos\theta$, $y=r\sin\theta\cos\phi$, $z=r\sin\theta\sin\phi$; then

$$\frac{2\sin\theta\sin\phi}{c}=r\left(\frac{\cos^2\theta}{a^2}+\frac{\sin^2\theta\cos^2\phi}{b^2}+\frac{\sin^2\theta\sin^2\phi}{c^2}\right).$$

The attraction which we require is equal to

$$\iiint dr\sin\theta\,d\theta\,d\phi\,.\sin\theta\sin\phi,$$

that is, $2a^2b^2c\displaystyle\iint\frac{\sin^3\theta\sin^2\phi\,d\theta\,d\phi}{b^2c^2\cos^2\theta+a^2c^2\sin^2\theta\cos^2\phi+a^2b^2\sin^2\theta\sin^2\phi}.$

The limits for θ are 0 and π; the limits for ϕ are $-\frac{\pi}{2}$ and $\frac{\pi}{2}$. Now suppose we integrate first with respect to θ; put t for $\cos\theta$; thus we obtain the form

$$\int\frac{(1-t^2)\,dt}{a^2c^2\cos^2\phi+a^2b^2\sin^2\phi+t^2(b^2c^2-a^2c^2\cos^2\phi-a^2b^2\sin^2\phi)}.$$

There is no difficulty in integrating this; but the form of the integral is different according to the sign of

$$b^2c^2-a^2c^2\cos^2\phi-a^2b^2\sin^2\phi;$$

involving circular functions if this quantity is positive, and logarithms if this quantity is negative. It is this double form which renders the process troublesome, if we adopt this order of integration; and D'Alembert discusses the matter at great length.

The best mode would be to integrate with respect to ϕ first; this would lead to a result which we shall presently obtain in another way: but D'Alembert does not adopt this order of integration.

647. Let us now consider the problem by the method III. of Art. 626.

Suppose that instead of an ellipsoid we had an ellipsoid of revolution, in which the semiaxes are c, b, and b. Then, by

Art. 255, the attraction on a particle at the end of the semiaxis c would be

$$2cb^2 \int_0^{2\pi} \int_0^{\frac{\pi}{2}} \frac{\sin\theta \cos^2\theta \, du \, d\theta}{b^2 + (c^2 - b^2)\sin^2\theta}.$$

Then for the case of an ellipsoid, not of revolution, we must put ρ^2 instead of b^2, where

$$\rho^2 = \frac{b^2}{1 + \left(\frac{b^2}{a^2} - 1\right)\cos^2 u},$$

so that our formula becomes

$$2c \int_0^{2\pi} \int_0^{\frac{\pi}{2}} \frac{\sin\theta \cos^2\theta \, du \, d\theta}{1 + \left(\frac{c^2}{\rho^2} - 1\right)\sin^2\theta}.$$

If we integrate with respect to θ first, we shall have two forms, according as $\frac{c^2}{\rho^2}$ is greater or less than unity; and D'Alembert discusses the matter at great length.

648. But suppose we integrate the formula of the preceding Article with respect to u first. We have

$$\frac{2c \sin\theta \cos^2\theta}{1 + \left(\frac{c^2}{\rho^2} - 1\right)\sin^2\theta} = \frac{2c \sin\theta \cos^2\theta}{\cos^2\theta + \frac{c^2}{\rho^2}\sin^2\theta}$$

$$= \frac{2c \sin\theta \cos^2\theta}{\cos^2\theta + \frac{c^2}{b^2}\left\{1 + \left(\frac{b^2}{a^2} - 1\right)\cos^2 u\right\}\sin^2\theta}$$

$$= \frac{2c \sin\theta \cos^2\theta}{\cos^2\theta + \frac{c^2}{b^2}(\sin^2 u + \frac{b^2}{a^2}\cos^2 u)\sin^2\theta}$$

$$= \frac{2c \sin\theta \cos^2\theta}{(\cos^2\theta + \frac{c^2}{a^2}\sin^2\theta)\cos^2 u + (\cos^2\theta + \frac{c^2}{b^2}\sin^2\theta)\sin^2 u}$$

Integrate with respect to u between the limits 0 and $\frac{\pi}{2}$ and multiply the result by 4. Then we find that the required attraction

$$= 4\pi abc \int_0^{\frac{\pi}{2}} \frac{\sin\theta \cos^2\theta \, d\theta}{\sqrt{(a^2 \cos^2\theta + c^2 \sin^2\theta)} \sqrt{(b^2 \cos^2\theta + c^2 \sin^2\theta)}}$$

Thus the attraction is made to depend on a single definite integral. We may say that this result is the point at which modern investigations have finally arrived.

We shall presently see that D'Alembert absolutely rejected this important formula which was within his reach.

649. D'Alembert himself draws attention to the fact, that when we have to find the value of a double integral, the facility of the process may depend very much on the order in which we effect the two integrations. See his page 158. He makes this remark after he has considered a way of finding the volume of a right cone, by cutting it into hyperbolic slices, by planes parallel to the axis: this way is difficult, though the final result is necessarily very simple.

650. The third part of the memoir occupies pages 159...207; it considers different ways of calculating the attraction of elliptic spheroids, and treats also of the attraction of some other spheroids.

I will notice some of the more important points.

We know that a plane may be so moved, parallel to itself, that all the sections which it makes of an ellipsoid shall be *circular* sections. D'Alembert suggests the problem of finding the attraction of an ellipsoid at the extremity of the diameter which passes through the centres of a series of circular sections. But the integrations are too complex to be worked out. See his pages 159...164.

Let any segment of an ellipse revolve round its bounding chord; then the attraction exerted by the solid thus generated on a particle at the extremity of the chord can always be found, or at least expressed as a single definite integral without radicals. See D'Alembert's page 164.

In fact, this attraction $= 2\pi \int_0^\beta r \sin\theta \cos\theta \, d\theta$, where β is the angle between the chord and the tangent to the ellipse at the origin; and $r = \dfrac{h\cos\theta + k\sin\theta}{a\sin^2\theta + b\sin\theta\cos\theta + c\cos^2\theta}$.

A theorem which presents itself incidentally may be noticed: see D'Alembert's page 167. Take any diameter of an ellipse, and let a solid be generated by the revolution of one of the halves of the ellipse about this diameter: then the volume generated varies inversely as this diameter.

If this diameter be called $2l$, and the axes of the ellipse be $2a$ and $2b$, the volume of the solid is $\dfrac{4\pi a^2 b^2}{3l}$.

D'Alembert invites mathematicians to continue their attempts to express the attraction of an ellipsoid without the use of arcs of conic sections; he says that the attempt does not appear to him hopeless: see his page 171. We now know that he was seeking what it is impossible to obtain. Plana has drawn attention to this passage in Crelle's *Journal für...Mathematik*, Vol. XX. page 190.

D'Alembert gives some simple examples of the process for the change of the independent variables in a double integral which Lagrange had developed in the volume of the Berlin *Mémoires* for 1773. See D'Alembert's pages 176 and 177.

651. We now arrive at a very singular passage. D'Alembert in effect gives the process of our Art. 648 and rejects it as inadmissible. See his pages 177...180. His y is our θ. He says:

J'avois imaginé d'intégrer d'abord la formule de la page 183 du Tome VI. de nos *Opuscules* en faisant varier u, et ensuite y, et j'avois cru trouver un résultat qui me conduisoit à une formule algébrique d'attraction pour les sphéroïdes elliptiques. Comme cette méthode pourroit en tromper d'autres, il ne sera peut-être pas inutile de la détailler ici.

In his process there is nothing wrong in principle, but he has omitted a bracket in the third line of his Art. 147; thus his result is slightly inaccurate. He gives some invalid arguments against the method. Thus D'Alembert deliberately rejects one of the most important formulæ of the subject, which in fact quite supersedes a large part of the present memoir. This is perhaps the strangest of all his strange mistakes.

652. D'Alembert shews in his page 199 that a theorem given by Laplace in the Paris *Mémoires* for 1775 might be investigated with ease. Laplace himself found afterwards a much simpler demonstration than that which he originally gave: for this see the Paris *Mémoires* for 1776, page 261. D'Alembert says in his page 221, with respect to Laplace and his two proofs of his theorem:

Ce même Académicien, à qui j'avois communiqué ma démonstration très-simple de son théorême, en a aussi trouvé depuis une autre plus simple que la premiere, et qu'il a lue à l'Académie au mois de Juillet 1778. Il m'a appris en même-temps que M. de la Grange avoit aussi trouvé de son côté une démonstration de ce théorême général.

The following is the theorem. Let the radius vector of a spheroid be $1 + \alpha\mu$, where α is very small and μ a function of the colatitude ψ. At any point of the surface let A denote the attraction along the radius vector, and B the attraction at right angles to the radius vector in the meridian plane from the pole: then will

$$\frac{dA}{d\psi} = \frac{B}{2} - \frac{2\pi\alpha}{3}\frac{d\mu}{d\psi} \quad\text{..................... (1).}$$

We shall demonstrate the theorem when we treat on Laplace's contributions to our subject.

Suppose T the whole force along the tangent, ω the angular velocity: then to the order of approximation we regard

$$T = B - A\frac{d(1+\alpha\mu)}{d\psi} + \omega^2 \sin\psi \cos\psi :$$

in the small term we may put $\frac{4\pi}{3}$ for A; thus

$$T = B - \frac{4\pi}{3}\alpha\frac{d\mu}{d\psi} + \omega^2 \sin\psi \cos\psi \quad\text{.......... ... (2).}$$

From (1) and (2) we have

$$\frac{dA}{d\psi} = \frac{T}{2} - \frac{\omega^2}{2} \sin\psi \cos\psi \text{ .... (3).}$$

Let P denote the gravity, so that $P = A - \omega^2 \sin^2\psi$; then (3) becomes

$$\frac{d(P + \omega^2 \sin^2\psi)}{d\psi} = \frac{T}{2} - \frac{\omega^2}{2} \sin\psi \cos\psi,$$

that is $$\frac{d}{d\psi}\left(P + \frac{5\omega^2}{4} \sin^2\psi\right) = \frac{T}{2} \text{ .... (4).}$$

D'Alembert contemplates the theorem under the form (4), and puts it into words: see his pages 200 and 201.

If the body is fluid and in relative equilibrium the condition $T = 0$ must be satisfied; and thus the theorem is simplified.

653. On pages 391...392 of the volume D'Alembert suggests a process for the calculation of the attraction of a hemispherical mountain on a pendulum occupying any position close to the mountain; but it is not fully intelligible, and nothing is really effected. He makes the erroneous statement that the direction at right angles to any radius of the mountain will also be at right angles to the radius of the Earth.

654. The eighth volume of D'Alembert's *Opuscules Mathématiques* was published in 1780. A memoir entitled *Nouvelles réflexions sur les loix de l'équilibre des fluides* occupies pages 1...35; and there are some remarks relating to the memoir on pages 354..357.

655. This memoir is not very closely connected with our subject; we will briefly indicate the nature of the topics discussed. We may observe that the old and obscure language with respect to fluid equilibrium is still retained; no advantage is taken of the capital improvement effected by Euler in introducing the notion of *pressure* and its appropriate symbol p.

D'Alembert notices an objection which he says an able mathematician had brought to him. It amounts to this. Suppose for simplicity the density of the fluid to be uniform; then we have shewn in Art. 394, that

$$\frac{dY}{dx} = \frac{dX}{dy};$$

but in the demonstration small quantities of the second order have been neglected: thus we may be in doubt whether any inference from this result is *rigorously* true. D'Alembert's words adapted to the notation and diagram of Art. 394 are:

...il est certain que Xdx ne représente la force du canal PS qu'à un infiniment petit du second ordre près, puisqu'on néglige les quantités infiniment petites du premier ordre qui entrent dans X, pour en exprimer la valeur le long du canal PS; il est certain aussi qu'il en est de même de Ydy; ne peut-on pas conclure delà, m'a objecté un habile Mathématicien, que l'équation $\frac{dX}{dy} = \frac{dY}{dx}$ ne représente l'équilibre du canal rectangulaire $PQRS$, qu'à un infiniment petit du second ordre près, et qu'ainsi elle ne représente pas rigoureusement l'équilibre, qui doit exister *rigoureusement* entre les parties du fluide, et qui seroit nécessairement troublé, s'il n'étoit pas tel?

D'Alembert discusses the matter in his pages 2...8.

D'Alembert considers whether it is necessary that X and Y should be continuous; that is whether throughout the fluid X should always be the same function of x and y, and also Y the same function of x and y. He maintains correctly that this is not necessary: see his pages 9...15.

But he seems on his page 355 to lose faith in his own demonstration.

In his pages 16...20 he adverts to a supposition he had formerly made with the view of giving greater generality to the equations of fluid equilibrium: see Art. 397. In effect he now abandons that supposition.

In his pages 20...26, to use modern language, he makes some remarks on the equations of fluid equilibrium, when referred to polar coordinates; he had formerly considered this topic: see Art. 574.

His pages 26...28 he devotes to shewing that if a fluid occupies an infinite tube, and a finite portion of the fluid be put in motion, no sensible movement in the mass will be produced. It does not seem to me that the investigation is of any value.

In his pages 28...30 he professes to demonstrate the statement, commonly admitted by writers on hydrostatics, that if a fluid mass be in equilibrium any portion of it may be supposed to become solid without disturbing the equilibrium. The demonstration does not seem to me of any value.

We have *three last* remarks in conclusion. On page 30 he says: Terminons ces recherches par quelques réflexions sur la loi de la compression de l'air en raison des poids dont il est chargé. Then on page 32 he says: Nous ajouterons ici en finissant, une remarque à laquelle il est bon de faire attention dans la graduation des baromètres. And on page 33 he says: Je terminerai ces recherches par une nouvelle remarque sur la théorie de l'équilibre des fluides. This new remark however is substantially old, having been given in page 206 of the *Théorie de la Résistance des Fluides:* see Art. 448.

656. In pages 292...297 of the eighth volume of the *Opuscules Mathématiques* we have a memoir entitled *Sur la Figure de la Terre;* and some remarks on it are given on pages 389...392: these form D'Alembert's last contribution to our subject.

657. We have already observed that D'Alembert having arrived at a certain equation shewed that it would sometimes have two roots; but left for others to demonstrate the proposition that there could not be more than two roots; and this was first established by Laplace: see Arts. 581 and 585. D'Alembert says:

M. de la Place m'en a communiqué une démonstration assez simple qui m'en a fait aussi trouver une très-simple, presque sans aucun calcul.

D'Alembert's demonstration is ingenious in principle but unsound.

In equation (1) of Art. 581 put x for λ; thus we get

$$\frac{2qx^3 + 9x}{3x^2 + 9} = \tan^{-1} x.$$

Let y be an ordinate corresponding to the abscissa x, and let the curve be drawn whose equation is

$$y = \frac{2qx^3 + 9x}{3x^2 + 9}$$

Again let η be an ordinate corresponding to the abscissa x, and let the curve be drawn whose equation is

$$\eta = \tan^{-1} x.$$

We have to shew that the curves cannot intersect more than twice for positive values of x, besides the intersection at the origin.

We have
$$y = \frac{9x}{9+3x^2} + \frac{2qx^3}{9+3x^2};$$

and thus when x is very small

$$y = x - \frac{x^3}{3} + \frac{2qx^3}{9} + \dots$$

Also when x is very small

$$\eta = x - \frac{x^3}{3} + \dots$$

Thus when x is very small y is greater than η; and so near the origin the first curve is above the second.

When x is infinite y is infinite and η is finite. Thus if the curves intersect at one point, say x_1, they must intersect at another say x_2. At this second point therefore $\frac{dy}{dx}$ will be greater than $\frac{d\eta}{dx}$. To ensure that the curves never intersect again we have only to shew that $\frac{dy}{dx} - \frac{d\eta}{dx}$ is always positive when x is greater than x_2; for if this be the case, y is always greater than η if x is greater than x_2.

Now D'Alembert says that $\frac{dy}{dx} - \frac{d\eta}{dx}$ is of the form

$$\frac{Ax^6 + Bx^4 + Cx^2 + D}{(1 + x^2)(9 + 3x^2)^2};$$

and this is true. Then he asserts that every quantity of the form $Ax^6 + Bx^4 + Cx^2 + D$, which is positive for a certain value of x, will be positive if x is increased. His words are:

...toute quantité de cette forme $Ak^6 + Bk^4 + Ck^2 + D$, qui sera positive pour une certaine valeur de k, doit l'être si on augmente k; car cette quantité est toujours $= Ak^2 (k^2 + E)^2 + G$, qui augmente quand k augmente.

But this statement is untrue. In the present case $\frac{dy}{dx} - \frac{d\eta}{dx}$ is positive when x is very small, but it is not always positive: it must be negative when $x = x_1$.

The demonstration may be made sound by shewing that in the present case the values of A, B, C, D are such that

$$Ax^6 + Bx^4 + Cx^2 + D$$

is always positive when x is greater than x_2. This method is really adopted by Cousin in his *Astronomie Physique*, page 148.

But we do not require the values of A, B, C, D to establish the point; it is sufficient to observe that D is zero: this is obvious from the fact that when x is very small we must have

$$\frac{dy}{dx} - \frac{d\eta}{dx} = \frac{d}{dx}\frac{2qx^3}{9} = \frac{2qx^2}{3}.$$

Since then $D = 0$, we have

$$\frac{dy}{dx} - \frac{d\eta}{dx} = \frac{x^2 (Ax^4 + Bx^2 + C)}{(1 + x^2)(9 + 3x^2)^2}:$$

now we know that the quadratic expression $Ax^4 + Bx^2 + C$ cannot change sign more than twice; and in the present case the sign is positive when $x = 0$, negative when $x = x_1$, and positive when $x = x_2$; therefore the sign must always be positive when x is greater than x_2.

658. The memoirs of D'Alembert on our subject which we have thus analysed in our Chapters XIII. and XVI. occupy about 700 pages in their original form. But the amount of important matter which they contain is not in proportion to their great extent. Probably the researches in the third volume of the *Recherches...Système du Monde* are the most valuable. The sixth volume of the *Opuscules Mathématiques* contains much interesting matter; but this matter is rather of a speculative kind than of physical importance.

On the whole we may sum up D'Alembert's contributions to our subject thus: He shewed how to calculate the attractions of a nearly spherical body of a form more general than an ellipsoid of revolution: see Art. 452. He drew explicit attention to the fact that more than one oblatum would correspond to a given angular velocity, a fact which had indeed been implicitly noticed before: see Art. 580. He considered the action of a distant body or bodies on a mass of rotating fluid supposed in relative equilibrium: see Arts. 596...630.

On the other hand we must observe that there are numerous and striking faults. Laplace, referring more especially to the *Recherches...Système du Monde*, says: Les recherches de D'Alembert, quoique générales, manquent de la clarté si nécessaire dans les calculs compliqués. *Mécanique Céleste*, Vol v. page 8. The full import of the criticism becomes apparent when we remember that with French writers *clarté* is the supreme indispensable requisite: want of clearness with them is on the same level as want of utility with Englishmen, or want of learning with Germans. The errors of D'Alembert are certainly surprising; they seem to me to indicate that he was little in the habit of enlarging his own views by comparing them with those of others. His criticisms of Clairaut prove that he had not really mastered the greatest work which had been written on the subject he was constantly studying. His readiness to publish unsound demonstrations and absolute errors is abundantly shewn in the course of our criticism: see for instance Arts. 576, 651, and 657. On the whole the blunders revealed in the *History of the Mathematical Theory of Probability*, and in the present History, constitute an extraordinary shade on a fame so bright as that of D'Alembert.

CHAPTER XVII.

FRISI.

659. IN the present Chapter I shall give some account of three works by Paul Frisi. As I have stated in Art. 532, I have not seen the first publication by Frisi on our subject; but probably it was incorporated in his later works.

660. The first of these works is entitled *De Gravitate Libri Tres.* This was published at Milan in 1768: it is a quarto volume of 420 pages, besides 12 pages which contain the Title, Dedication, Preface, and Index; there are six plates of figures.

This work forms a treatise intended for didactic purposes, the object being to conduct a student with elementary mathematical knowledge through a course of Mechanics and Physical Astronomy: see the first page of the Preface. The two volumes published by Frisi about six years later, under the title of *Cosmographia,* may be regarded as an improved and enlarged edition of the present work.

The part of the volume with which we are concerned consists of pages 135...189; they form the first four Chapters of Frisi's Second Book.

The pages 135...145 are introductory. They contain an outline of the facts then known as to the lengths of degrees and to the lengths of the seconds pendulum.

661. The first Chapter is entitled *De Figura Terræ,* it occupies pages 146...154.

This Chapter contains approximate formulæ for the lengths of a degree of meridian and of a degree of longitude. From these formulæ, and the results furnished by observation, the ellipticity of the Earth may be deduced. Frisi maintains that all the measurements hitherto made agree reasonably well with the ratio of the axes assigned by theory for an oblatum of fluid, namely that of 231 to 230.

Some erroneous statements occur in the second Corollary on page 151. Frisi has given a formula for determining the ellipticity from the lengths of a degree of meridian in two different latitudes. Then he says that if the arc in Lapland and the arc at the Cape of Good Hope be taken the ellipticity deduced is $\frac{1}{1282}$; but in the *Cosmographia*, Vol. II., page 97, he gives the correct result, namely about $\frac{1}{252}$. The error probably arose from taking the ellipticity which Boscovich had deduced from the arcs in France and South Africa by mistake, instead of that which was deduced from the arcs in Lapland and South Africa: see the supplement by Boscovich to Stay's poem, Vol. II., page 408. Again Frisi gives $\frac{1}{80}$ as the ellipticity deduced from the arcs in Peru and South Africa; but in the *Cosmographia*, Vol. II., page 96, he gives the correct result, namely about $\frac{1}{182}$. The error probably arose from simply copying one made by Boscovich: see Art. 508.

Frisi's first Chapter closes thus:

Quam ex amplissima Pensilvaniæ planitie Clariss. Mason, et Dixon afferent mensuram gradus figuræ terrestris inquisitioni novam lucem affundent. Interim certum manet Terram sub æquatore, et polari circulo, et in meridionali Africæ, et Galliæ Narbonensis parte, atque in Anglia etiam, et Stiria, ac Pedemonte, a figuræ sphæroidicæ, et proportionis assumptæ hypothesi non magis recedere, quam ut in minimos errores observationum differentia omnis refundi possit.

The anticipation as to the light to be derived from the American arc, has scarcely been realised; for this arc has not been received with much confidence: see Bowditch's translation of the *Mécanique Céleste*, Vol. II., page 444.

662. The second Chapter is entitled *De æquilibrio particularum sese trahentium;* it occupies pages 154...164.

This Chapter contains a demonstration of the proposition that an oblatum is a form of relative equilibrium for rotating fluid; the method is that of Maclaurin and Clairaut: see Art. 318.

We have, in this Chapter, some extensions to ellipsoids in general of results which had already been established for ellipsoids of revolution: see the Corollary II. on page 157, and the Corollary II. on page 158. Thus Newton had shewn that a shell bounded by homothetical ellipsoids of revolution exerts no attraction on a particle placed within the inner surface. Frisi shews that this is true for a shell bounded by homothetical ellipsoids when the particle is *on* the inner surface. He does not expressly shew that this is true when the particle is *within* the inner surface; but it was quite in his power to infer this from what he had already given.

This seems to be the first introduction of the ellipsoid, as distinguished from the oblatum and the oblongum, into our subject. D'Alembert afterwards considered the matter in the sixth volume of his *Opuscules Mathématiques:* see Art. 615.

Frisi also alludes to the results which will follow when the fluid oblatum is disturbed by the action of one or two distant attracting bodies, like the sun and the moon. His process however is brief and not very satisfactory. This matter was afterwards discussed in detail by D'Alembert in the sixth volume of his *Opuscules Mathématiques:* see Chapter XVI.

At the end of the Chapter Frisi refers to Maclaurin, Simpson, Clairaut, and Newton. The last is styled vir longe omnium ingeniosissimus: these words are omitted in the corresponding passage of the *Cosmographia.* But in both works Frisi says:

> ...ut recte propterea dixerit Daniel Bernoullius, § 8. cap. 2. de fluxu, et refluxu maris, Newtonum trans velum etiam vidisse, quæ vix ab aliis microscopii subsidio discerni possunt.

We have given the original words of D. Bernoulli in Art. 501.

663. The third Chapter is entitled *De sphærarum, sphæroidumque attractione;* it occupies pages 164...175: but the pages

170...173, which are rather difficult, do not belong to our subject, and are removed to a more appropriate place in the *Cosmographia*.

Here we have an exact investigation of the attraction of a spherical shell on an internal particle; and an application to the case in which the particle is on the surface of the shell, or forms a component of the shell. The process, like others in the Chapter, really involves the Integral Calculus, though without its notation.

Next we have an approximate investigation of the attraction of an oblatum on a particle situated on the prolongation of the axis of revolution; the result is correct to the first power of the ellipticity.

Then we have an approximate investigation of the attraction of an oblatum on any external particle; this problem is treated in Clairaut's manner: see Clairaut's pages 236...239, or Art. 335.

Frisi refers to the criticism of Short and Murdock on his supposed discovery of an error in Newton: see Arts. 533 and 534. Frisi however does not admit the accuracy of the criticism; he says:

Nævum hujusmodi cap. 6. dissertationis de Figura Terræ a nobis jam adnotatum, in Transact. anni 1753. excusare voluerunt Clariss. Short, et Murdock, postrema Newtoni verba *in eadem ratione quam proxime* intelligentes de rationis continuitate, non de identitate cum ratione semiaxium: qui tamen sensus allati textus minime videtur esse.

664. The fourth Chapter is entitled *De æquilibrio, et lege terrestrium ponderum;* it occupies pages 175...189.

Here we have first a proposition and corollaries which belong rather to a rude theory of the tides than to our subject.

Next we have an approximate investigation of the ratio of the axes, in order that an oblatum of rotating fluid may be in relative equilibrium.

Then it is shewn that if the Earth be homogeneous, or be composed of spheroidal strata, the weight of a given body on the surface of the Earth will increase in passing from the equator to the pole; the increment varying as the square of the sine of the latitude.

For a particular case Frisi finds that we may reasonably satisfy the observations by supposing the Earth to consist of a sphere having the minor axis for diameter, and of an outer portion; each of the two portions being homogeneous, but the density of the sphere to the density of the outer portion as $1+\frac{1}{8}$ is to 1. See his pages 183...185.

On his page 186, Frisi draws attention to a point as to which he differed with Newton. He had already referred to this by anticipation on his page 174, where he says Omnino falsum est illud, quod in Prop. 38. Lib. 3. assumpserat Newtonus,... We will return to this point when we give an account of Frisi's *Cosmographia*.

665. On his pages 224...235 Frisi has a Chapter entitled *De variationibus Maris, quæ oriri possunt ex Sole aut Luna*. The first half of this Chapter bears rather more on our subject than the title might seem to indicate; but we will reserve our notice of it until we speak of the *Cosmographia*.

666. Frisi himself gave an account of the contents of his work before it was published; this account is contained on pages 514...530 of the Bologna *Commentarii*, Vol. V., part 2, 1767. This account adds nothing to our subject. Frisi, on page 522, draws attention to the two points at which he differs with Newton: see Arts. 663 and 664.

667. Judging from the part of Frisi's work which I have thus had to examine, I should say that it may be considered to have formed a reasonably good elementary treatise at the time of its appearance. It contains however none of the higher researches which Clairaut had given as to the Figure of the Earth, when supposed to be heterogeneous; and thus the promise held out in the Preface of conducting the reader to the summit of physical astronomy—ad summum Physicæ celestis apicem—is scarcely fulfilled.

668. We have now to notice the second work by Paul Frisi. It is in two quarto volumes. The first volume is entitled *Cosmographiæ Physicæ, et Mathematicæ Pars prior Motuum periodicorum theoriam continens*. The second volume is entitled *Cosmographiæ*

Physicæ, et Mathematicæ. Pars altera De Rotationis Motu et Phænomenis inde pendentibus.

The work was published at Milan; the first volume is not dated; the second is dated 1775. The first volume contains 266 pages, besides a page of errata, and the Title, Dedication, and Index on 6 pages. The second volume contains 276 pages, besides the Title, Dedication, and Index on 6 pages. Each volume has three plates of figures.

669. It is well known that in what is called the Jesuits' edition of Newton's Principia, there is a note by the editors in which they profess their submission to the decrees issued by the supreme Pontiffs against the motion of the Earth, although in commenting upon Newton they were obliged to adopt the same hypothesis as he did. I do not know at what date these decrees of the supreme Pontiffs were first allowed to be disregarded. Certainly in the present work Frisi has no hesitation in adopting the truth as to the Earth's motion; his language seems much more decisive than it was in his former work. We have the following words on page 28 of Vol. I. of the *Cosmographia:*

Galilæus Martem, et Venerem moveri circa Solem certissime ex eorumdem phasibus collegit. Totum vero Telluris motæ sistema novo hoc analogiæ argumento confirmatum ita in dialogis vindicavit, adornavitque, ut, qua in physicis rebus certitudine fieri poterat, ostenderit Planetas quinque primarios simul cum Terra motu periodico ab occidente in orientem revolvi circa Solem in planis transeuntibus per Solem ipsum, et parum dehiscentibus a se invicem: Lunam ab occidente pariter in orientem revolvi circa Solem,...

The context shews that the last word *Solem* is a mistake for *Terram.*

670. The *Cosmographia* may be considered as an enlarged edition of the treatise *De Gravitate,* of which we have already given an account. The part of the *Cosmographia* with which we are concerned consists of pages 83...142, and 207...219 of the second volume.

671. The pages 83...142, form the Second Book of the second volume. This Book consists of an introductory portion followed by four Chapters.

The introductory portion occupies pages 83...92. This gives an account of the various measurements of arcs on the Earth's surface up to the current date.

The following important passage relative to the errors which might arise from the use of a zenith sector, occurs on page 88:

...certo autem ostendit Clariss. Maskelinius cum in expeditione ad insulam S. Helenæ pro parallaxi Sirii, aliisque Caillii observationibus recognoscendis suscepta deprehendit iis sectoribus, quibus Maupertuisius, aliique, ad mensuram graduum usi fuerant, frictionem fili ex instrumenti centro suspensi errorem 3″, aut 4″ in partes adversas quandoque parere: ut fuse a summo ipso Astronomo cum Grenovicii essem accepi.

The substance of this passage occurs also in the *De Gravitate*, page 140; but the words from *ut fuse* to the end are not given there.

Frisi notices that the arc measured at the Cape of Good Hope by La Caille, was longer than might have been expected from the results of other measurements. He suggests that this may be owing to the circumstance that the continent of Africa supplies an excess of matter toward the end of the arc which is nearer to the equator when contrasted with the ocean near which the other end of the arc is situated. Thus the plumb line at the end of the arc which is nearer to the equator may be considered to be affected as it would be by a range of mountains at the equator; so that the amplitude of the arc would be rendered a little too short. This suggestion was also made by Cavendish; see the *Philosophical Transactions* for 1775, page 328. We have spoken of the modern remeasurement of the South African arc in Art. 542.

672. The first Chapter is entitled *De dimensione graduum, et, quæ inde colligitur, Figura Terræ;* it occupies pages 92...104.

This Chapter corresponds to the first Chapter which we examined in the former work; see Art. 661. Frisi maintains, as before, that all the measurements hitherto made agree reasonably well with the ratio of the axes assigned by theory for an oblatum of fluid, namely that of 231 to 230.

673. The second Chapter is entitled *De æquilibrio particularum omnium sese trahentium:* it occupies pages 104...113. This Chapter corresponds to the second Chapter of the former work: see Art. 662.

There is a slight mistake in the second corollary on page 112. The mistake is not of much importance, but the correct expressions involved are often useful: see Art. 596.

Suppose a body acted on by a very distant particle of mass M. Take a fixed point in the first body as origin; and let the axis of x pass through the second body. Let k represent the distance of the particle of mass M from the origin.

Then the action of M at a point (x, y) will be equivalent to $\frac{M(k-x)}{R^3}$ parallel to the axis of x, and $-\frac{My}{R^3}$ parallel to the axis of y; where $R^2 = (k-x)^2 + y^2$: both resolved attractions being estimated outwards.

Expand and neglect powers of $\frac{x}{k}$ and $\frac{y}{k}$ above the first. Thus we obtain $\frac{M}{k^2} + \frac{2Mx}{k^3}$ parallel to the axis of x, and $-\frac{My}{k^3}$ parallel to the axis of y.

The force $\frac{M}{k^2}$ is constant. Thus in the language of the Planetary Theory we may say that we have as *disturbing forces*, $\frac{2Mx}{k^3}$ parallel to the axis of x, and $-\frac{My}{k^3}$ parallel to the axis of y.

Frisi's mistake consists in changing the coefficient 2 to 3.

We may if we please arrange these disturbing forces differently. Since $\frac{2Mx}{k^3} = \frac{3Mx}{k^3} - \frac{Mx}{k^3}$, we may say that we have $\frac{3Mx}{k^3}$ parallel to the axis of x, besides $-\frac{Mx}{k^3}$ and $-\frac{My}{k^3}$ parallel to the axes of x and y respectively: then the latter two may be combined into a single force towards the origin. Thus finally we have $\frac{3Mx}{k^3}$ parallel to the axis of x, and $\frac{M\sqrt{(x^2+y^2)}}{k^3}$ towards the origin.

Or we may if we please arrange these disturbing forces in another way. Since $-\frac{My}{k^3} = \frac{2My}{k^3} - \frac{3My}{k^3}$, we may say that we have $-\frac{3My}{k^3}$ parallel to the axis of y, besides $\frac{2Mx}{k^3}$ and $\frac{2My}{k^3}$ parallel to the axes of x and y respectively: then the latter two may be combined into a single force towards the origin. Thus finally we have $-\frac{3My}{k^3}$ parallel to the axis of y, and $\frac{2M\sqrt{(x^2+y^2)}}{k^3}$ from the origin.

674. The third Chapter is entitled *De sphærarum, spheroidumque attractione;* it occupies pages 114...123. This Chapter corresponds to the third Chapter of the former work: see Art. 663.

Frisi retains his opinion noticed in that Article as to the supposed error of Newton.

675. The fourth Chapter is entitled *De Planetarum Figura, quæ ex æquilibrii legibus colligitur;* it occupies pages 123...142. This Chapter corresponds to the fourth Chapter of the former work: see Art. 664. The $\frac{1}{5}$ of Art. 664 is now replaced by $\frac{1}{5\frac{1}{2}}$.

676. I will now explain the difference between Frisi and Newton to which I have alluded in Art. 663: Frisi refers to it on his page 135.

Let m denote the mass of the moon, M that of the Earth, k the distance between their centres.

Suppose the moon to be a homogeneous fluid; then the surface of the moon may be in the form of an oblongum with the longer axis directed to the Earth.

Let b denote the minor semiaxis of this oblongum, and $b+h$ the major. Suppose the centre of the moon brought to rest; then we may consider that besides the attraction of the Earth, there is a central disturbing force, and also the disturbing force $\frac{3Mx}{k^3}$ as in Art. 673. Then if we proceed as Newton did for determining the figure of the Earth, or in some more analytical method, we shall obtain approximately

$$\frac{h}{b} = \frac{5}{4} \cdot \frac{\frac{3Mb}{k^3}}{\frac{m}{b^2}} = \frac{15}{4} \cdot \frac{M}{m} \cdot \frac{b^3}{k^3},$$

so that

$$h = \frac{15}{4} \cdot \frac{M}{m} \cdot \frac{b^4}{k^3}.$$

There is no difference of opinion as to this result; it agrees with one obtained by D. Bernoulli in his Essay on the Tides, Chapter IV. Article 8. It is also consistent with Newton's result in the *Principia*, Book III. Proposition 36.

The simplest way of connecting the result with Newton's investigations is to adopt the last method of arranging the disturbing forces given in Art. 673; so that we have a central force and also a force $-\frac{3My}{k^3}$ parallel to the axis of y. Then in Art. 28 corresponding to a centrifugal force which may be denoted by $\omega^2 y$ we obtained an *oblatum* in which $\epsilon = \frac{5j}{4}$; and now corresponding to a disturbing force $-\frac{3My}{k^3}$ we obtain an *oblongum* with a similar value for the ellipticity.

Now proceed in a similar way to determine the figure of the Earth, supposed fluid, under the action of the moon.

Let B and $B + H$ denote respectively the semiaxes of the oblongum; then we have

$$H = \frac{15}{4} \cdot \frac{m}{M} \cdot \frac{B^4}{k^3}.$$

Hence, by division,

$$\frac{h}{H} = \left(\frac{M}{m}\right)^2 \left(\frac{b}{B}\right)^4.$$

This result is given by Frisi. But Newton in his *Principia*, Book III. Proposition 38, asserts in fact that

$$\frac{h}{H} = \frac{M}{m} \cdot \frac{b}{B}.$$

It is clear that Frisi is right; but I do not know of any commentator on Newton, who has accepted the correction. Some further information on the subject will be found in a paper published in the *Monthly Notices of the Royal Astronomical Society*, Vol. XXXII. pages 234...236.

677. In his pages 140...142 Frisi treats on the attraction of mountains; he refers to what D'Alembert had given on the subject in the sixth volume of his *Opuscules Mathématiques.* Frisi in these pages uses the notation of the Differential Calculus which does not occur in the other parts of his work that have come under our examination. Frisi throws doubts on the genuineness of those observations to which D'Alembert drew attention: see Art. 594.

678. Frisi's pages 207...219 form a Chapter entitled *De æquilibrio fluidorum nucleos sphæroidicos circumambientium.*

This Chapter presents in an improved and enlarged form the propositions bearing on our subject to which we referred in Art. 665.

Frisi first shews that the attraction at any point of the surface of a nearly spherical oblatum or oblongum consists of a central force together with a small force parallel to the major axis. Then he considers the case of a spherical nucleus surrounded by fluid, the whole rotating with uniform angular velocity. Next he supposes the nucleus to be an oblatum or an oblongum; this investigation includes the preceding as a particular case. The result at which Frisi arrives amounts to the same as that which I have stated at the end of Art. 374; his investigations are fairly satisfactory.

On his pages 215 and 216, Frisi arrives at results respecting what we should now call the *stability of the relative equilibrium,* which resemble those of D'Alembert: see Art. 567. Frisi's investigations on this matter however, as might be expected, are rather rude; they were not given in his former work, and were doubtless suggested by the sixth volume of D'Alembert's *Opuscules Mathématiques,* to which Frisi refers on his page 219. Frisi refers also on this page to Boscovich, respecting the same matter, saying "de quibus casibus plura ingeniose scripserat Boscovichius." Frisi however does not say that his own conclusion does not agree with that of Boscovich; the latter, as we have stated in Art. 470, held that the oblongum could not be a stable form, whereas Frisi holds with D'Alembert that it is so in a certain case: see Art. 590.

On his page 210 Frisi attempts to investigate the approximate expressions for the attraction on a particle outside an oblatum resolved at right angles to the straight line drawn to the centre, supposing the particle very near the plane of the equator: but I cannot consider his process satisfactory: see Art. 321.

679. Frisi's Fifth Book, entitled *De Atmosphæra Planetarum,* occupies pages 231...271 of the second volume of his *Cosmographia.* It is not really a part of our subject, and I have not examined it throughout. I will however make some remarks on certain passages.

Frisi deduces in a satisfactory manner the result which we may thus express in modern language: if we omit all consideration of centrifugal force the height of the atmosphere will be infinite. But then he seems to be frightened at his own result, and makes a remark which amounts to saying that his investigation does not hold. See his page 240.

The subject seems to have been regarded as paradoxical by some of the mathematicians of the eighteenth century. See for instance Fontana's *Ricerche sopra diversi punti...*, pages 89...105. Fontana refers to an error committed by David Gregory; Frisi without mentioning a name seems to refer to the same point in his second Corollary on page 239.

Fontana also finds fault with section 36 of a memoir by Playfair, in the *Edinburgh Transactions* for 1788; but it seems to me that Fontana misunderstands what is said: Playfair wishing to shew that the height of the atmosphere is infinite attains his end by supposing that the density is zero, for then his formula gives the inadmissible consequence that the radius of the earth is infinite. He is however not so clear as he might be on the matter, and Fontana takes him to make the hypothesis that the radius of the earth is infinite. Fontana's words are:

Il Sig. Giovanni Playfair...inferisce, che il semidiametro terrestre r è infinito, e di qui che *the athmosphere on this supposition admits of no limit*, illazione visibilmente assurda essendo contro il fatto, e la natura delle cose il dare al semidiametro della terra una lunghezza infinita.

Frisi obtains, in his pages 254 and 255, an equation to determine the limit of the atmosphere; the equation representing the generating curve of the bounding figure. This equation is

$$(3a^2 - x^2)^2 (x^2 + y^2) = 4a^6,$$

where a represents the equatorial semiaxis. This agrees with what we havè already given: see Art. 640. Frisi, in fact, determines one of the two constants there occurring, inasmuch as he supposes the surface to pass through the points where the centrifugal force becomes equal to the attraction.

Frisi says in his page 257, with respect to this equation,

...atmospheræ figuram ex limitibus altitudinis sic deduximus ut rami omnes excurrentes ad infinitum, et casus alii ramorum duplicium præcluderentur, quos D. Mairan in tractatu de Aurora Boreali enumeraverat.

680. A collection of works by Frisi was published at Milan in three volumes quarto; the first volume appeared in 1782, the second in 1783, the third was issued by the author's brothers in 1785 after his death.

We are concerned only with the third volume of these works, which is entitled *Paulli Frisii Operum Tomus tertius Cosmographiam Physicam, et Mathematicam continens.* It contains 561 pages, besides the Index on 3 pages, and the Title and Dedication on 6 pages; there are 3 plates of figures.

681. The Second Book of this volume is entitled *De Figura Terræ et Planetarum;* it occupies pages 117...184. This may be described as a republication of the Second Book of the second volume of the *Cosmographia,* with some omissions and some additions; the changes however are of little importance. We may therefore refer to the notice already given of the *Cosmographia* in Arts. 668...679; and shall only add a few remarks on some points of interest.

682. On his pages 164 and 165, Frisi discusses the problem of the solid of revolution of given volume and maximum attraction. He arrives at the following differential equation for the generating curve; $dy = \left(\frac{y}{3x} - \frac{2x}{3y}\right) dx$, which is correct. But three observations suggest themselves.

(1) Frisi makes no reference to an incorrect investigation which he had formerly given; to this we shall return in the next Chapter: see Art. 686.

(2) In enunciating the problem he implies that the generating curve is to pass through two fixed points; but he pays no attention to this condition in his solution. If I had been acquainted with this passage of Frisi's work when I published my *Researches in the Calculus of Variations* I should have noticed it then. Compare page 123 of that work.

(3) Frisi in a Corollary adverts to the solution given by Silvabelle: see Art. 531. Frisi seems to think that the results obtained by himself and Silvabelle do not agree: for he uses the words "formularum diversitatem." But Silvabelle's result is $a^2x = (x^2 + y^2)^{\frac{3}{2}}$, where a is a constant; and if we differentiate this, and eliminate a, we obtain Frisi's result. Frisi then objects to Silvabelle's solution. The objection amounts to saying in modern language that Silvabelle confounds dy with δy. Silvabelle's process however is quite sound, if we are careful to understand it properly.

683. We may notice the points in which Frisi makes special reference to Newton.

(1) The passage which I have quoted in Art. 662 now appears on page 153, thus:

Daniel Bernoullius...acute dixerat Newtono trans velum etiam apparuisse quæ vix ab aliis microscopii subsidio distingui possunt.

(2) Frisi retains his opinion as to the supposed error of Newton: see Arts. 663 and 674.

(3) Frisi, with more justice, retains his opinion as to the other point on which he differed with Newton: see Art. 676. Frisi

says, in reference to this, on his page 170, "qui Newtoni locus a nemine antea fuerat emendatus."

684. We may observe that the names of Lagrange and Laplace are now mentioned by Frisi: see his pages 153 and 166.

In Frisi's pages 190...204, we have reproduced with some additions the matter contained in pages 207...219 of the *Cosmographia:* see Art. 678.

All the three works by Frisi which we have noticed are printed on stout durable paper. Either the general public must have received them with a favour not usually bestowed on mathematical treatises, or they must have obtained the private patronage of wealthy persons; for the expenses of producing them could scarcely have been otherwise sustained.

CHAPTER XVIII.

MISCELLANEOUS INVESTIGATIONS BETWEEN THE YEARS 1761 AND 1780.

685. THE present Chapter will contain an account of various miscellaneous investigations between the years 1761 and 1780. The first three of Laplace's memoirs relating to our subject were published during this period, but it will be convenient to defer our notice of them until the next Chapter.

686. In the *Novi Commentarii*...St Petersburg, Vol. VII., which is dated 1761, there is a paper by Frisi, entitled *De Problematis quibusdam isoperimetricis:* it occupies pages 227...234 of the volume.

This paper belongs to the early history of the Calculus of Variations, and not to our subject. I advert to it however because on his last page, Frisi alludes to Silvabelle's problem and two others of the same kind, but without referring to Silvabelle: see Art. 531. In particular, Frisi states definitely his result for Silvabelle's problem. But this result is wrong; and in fact the whole page is vitiated by an error which occurs at the top. It will be found that in his notation he has neglected to allow for the change of TL into PL.

687. An academical essay was published at Tübingen in 1764, entitled *Dissertatio Physico-Mathematica de ratione ponderum sub polo et æquatore Telluris...auctor Wolffgangus Ludovicus Krafft.* This consists of 28 pages in small quarto, with a page of diagrams. It appears from the last page of the essay that the father of the author had been a professor at Tübingen.

688. The mechanical principles involved in the essay are not always sound. Thus in the first paragraph the author seems to think that the time of Jupiter's rotation on his axis is determined by Kepler's Third Law; for he says:

Pari modo in Jove, et multo quidem major, diametrorum inaequalitas a Cassini et Flamsteedio detecta est, cujus diurnam circa axem conversionem cum plus, quam duplo celeriorem nostra, ostendat sagacissimi KEPPLERI regula, quadrata temporum periodicorum esse ut cubos distantiarum a sole...

689. For another example, we may take some remarks which the author makes involving centrifugal force. Suppose a sphere of radius a to rotate with uniform angular velocity; let f denote the attraction at any point of the surface, and ϕ the centrifugal force at the equator: then at a point on the surface in latitude λ the gravity will be approximately $f-\phi\cos^2\lambda$. So far Krafft is correct. Now produce the radius vector of the place to a point at the distance x from the centre. Then he reduces the expression for gravity in the ratio of a^2 to x^2, and takes for the centrifugal force resolved along the radius vector the usual expression: thus he obtains for the gravity $\frac{a^2}{x^2}(f-\phi\cos^2\lambda)-\frac{x}{a}\phi\cos^2\lambda$. But it is obvious that he has thus introduced the centrifugal force twice, once erroneously and unnecessarily. The expression for gravity should be $\frac{a^2f}{x^2}-\frac{x}{a}\phi\cos^2\lambda$. This error occurs on pages 8 and 10. On page 10 he assigns the distance from the centre at which gravity would vanish; but the result depends on his erroneous formula, and is therefore wrong.

690. But it is curious that Krafft avoids this mistake in a problem which he discusses. The problem would be expressed thus in modern language: to find the *lines of force* outside a sphere, supposing that the sphere attracts, and that there is also a force of the nature of centrifugal force.

Take the axis of x for that which would correspond to the axis of rotation; let

$$X=\frac{fa^2x}{r^3}, \qquad Y=\frac{fa^2y}{r^3}-\omega^2y.$$

Then we have to determine a curve from the equation

$$\frac{dx}{X} = \frac{dy}{Y}.$$

Thus
$$\frac{f(ydx - xdy)\,a^2}{r^3} = \omega^2 ydx,$$

that is
$$\frac{ydx - xdy}{(x^2 + y^2)^{\frac{3}{2}}} = cydx,$$

where c is put for $\frac{\omega^2}{fa^2}$.

Krafft obtains this differential equation; to integrate it he assumes $y = \frac{x\sqrt{(1 - z^2)}}{z}$, this gives

$$cx^2dx = \frac{z^2dz}{1 - z^2}:$$

hence
$$\frac{2cx^3}{3} + \text{constant} = \log\frac{1 + z}{1 - z} - 2z,$$

that is
$$\frac{2cx^3}{3} + \text{constant} = \log\frac{\sqrt{(x^2 + y^2)} + x}{\sqrt{(x^2 + y^2)} - x} - \frac{2x}{\sqrt{(x^2 + y^2)}}.$$

Krafft gives the second term on the right-hand side incorrectly: see his page 12.

Krafft does not enunciate his problem in our modern language. According to him a tower is to be built on the Earth's surface, in such a manner that there is to be no force at any point tending to overturn it: in other words, the force at any point of what we may call the axis of the tower is to be along the tangent to the axis.

691. Krafft shews that the increment of gravity in proceeding from the equator to the pole varies as the square of the sine of the latitude very approximately; his method closely resembles that given by Boscovich, and is liable to the same remark: see Art. 467. Moreover, the theorem which Krafft uses as the foundation of his method is that to which we have drawn attention in Art. 34, and this assumes that the Earth is in the form of a fluid rotating in relative equilibrium. But Krafft has said nothing about fluidity,

so that an incautious reader might suppose him to be affirming some proposition relative to the attraction of an oblatum.

692. The main part of the essay however is the determination of the attraction of an oblatum at the pole and the equator; this is finally accomplished to the order of the square of the ellipticity: see his pages 14...20.

He first calculates the attraction of an elliptic lamina on a particle directly over the centre of the lamina. Thus his problem is the same as Euler had already considered, and Krafft investigates it in a similar way, but there is no reference made to Euler: see Art. 229.

If we put λ for $\frac{a^2-b^2}{a^2(b^2+c^2)}$, we shall find by that Article that if we neglect λ^3 and higher powers of λ, the attraction

$$=\frac{4bc}{a\sqrt{(b^2+c^2)}}\int_0^a \frac{\sqrt{(a^2-x^2)}}{c^2+x^2}\left\{1-\frac{\lambda}{2}x^2+\frac{3\lambda^2}{8}x^4\right\}dx.$$

Of the three integrals which this involves, Krafft works out the first in the laborious way which Euler adopted; he merely states the values of the other two integrals.

We have obtained the first and the second by a simple method in the Article cited. The third may be easily given. We have

$$\int_0^a \frac{x^4\sqrt{(a^2-x^2)}}{c^2+x^2}dx=\int_0^a\left(x^2-c^2+\frac{c^4}{c^2+x^2}\right)\sqrt{(a^2-x^2)}\,dx$$

$$=\int_0^a x^2\sqrt{(a^2-x^2)}\,dx-c^2\int_0^a\sqrt{(a^2-x^2)}\,dx+c^4\int_0^a\frac{\sqrt{(a^2-x^2)}}{c^2+x^2}dx$$

$$=\frac{\pi a^4}{16}-\frac{\pi c^2a^2}{4}+\frac{\pi c^4}{2}\left\{\frac{\sqrt{(a^2+c^2)}}{c}-1\right\}.$$

Thus Krafft obtains a very approximate value of the attraction of the elliptic lamina; but he speaks on his page 18 as if he had thus obtained the *accurate* value. Then knowing the attraction of the elliptic lamina, he proceeds to calculate that of an oblatum at the equator and at the pole; his investigation is rather intricate, but it is correct, and his final results are in exact agreement with those given towards the end of Art. 229.

It is strange however to see this tedious *approximate* solution of the problem of the attraction of an oblatum at its equator and pole so long after the *exact* formulæ had been obtained by Maclaurin and by Simpson.

693. Krafft says that if we take the polar axis to be to the equatorial as 100 is to 101, we find that the attraction at the pole is to the attraction at the equator as 134396 is to 134129, that is approximately as 509 is to 508. It seems to me that by his formulæ the ratio should be that of 352790 to 352091, that is approximately that of 505 to 504: moreover the approximation of 509 to 508 does not follow from his first figures.

694. Some miscellaneous observations on mathematical topics occupy pages 21...27 of the essay. They do not seem to contain anything of interest except a statement relative to the series 1, 3, 4, 7, 11, 18,... in which each term is the sum of the two preceding. This series, according to Krafft, was introduced by Daniel Bernoulli in his *Exercitationes Mathematicæ*, who gave a formula for the sum of a finite number of terms, the value of the last term being given. Let p denote the last term, then the sum is $\frac{1}{2}\{3p - 6 + \sqrt{(5p^2 \pm 20)}\}$, where in the ambiguity the upper or the lower sign is to be taken according as the number of the terms is odd or even. Krafft adds: "Sed nullo artificio detegi potest terminus generalis." But this statement is very strange; for if we put α for $\frac{1+\sqrt{5}}{2}$, and β for $\frac{1-\sqrt{5}}{2}$, it can be easily shewn that the nth term of the series is $\alpha^n + \beta^n$. The sum of n terms can then be readily obtained, and shewn to agree with the expression given above.

The series occurs on page 7 of the *Exercitationes Mathematicæ*, and there D. Bernoulli seems also to state that the general term cannot be expressed. He says:

...unicam seriem exempli loco afferam talem 1, 3, 4, 7, 11, 18, 29... in qua quilibet terminus duorum præcedentium est summa, et quam nunquàm ad terminum generalem reduci posse demonstratum habeo; hæc series, quamvis Geometris non considerata, quod sciam, attentione tamen dignissima est ob multas, quibus gaudet, proprietates:...

695. Krafft's essay does not contribute in any way to the advancement of the subject; in fact the author by his ignorance of what had been effected by Maclaurin and by Simpson, shews that his knowledge was below the level it might have reached. The problem which we have noticed in Art. 690, *may* have been new at the time, but this is very uncertain.

The address of the President, John Kies, to the author of the essay may be preserved as a specimen of the academical pleasantry of the last century:

Gravissimum est argumentum, cujus elaborationem in Te suscepisti, miratus Tuum volatum à Polo ad Aequatorem Te feliciter rediisse gratulor, calculorum quos evitare non licuit, neque multitudo neque pondera Te à Tuo tramite potuerunt avertere, et uti in series eorum infinitas incidisti, ita inde reduci Tibi seriem prosperitatis et felicitatis infinitam ex animo apprecor. Sequere porro vestigia Celeberrimi Parentis του μακαριτου, Antecessoris mei in officio academico desideratissimi. Vale et me amare perge d. 8. Octobr. 1764.

696. We next notice a memoir entitled *Pet. von Osterwald Bericht über die vorgenommene Messung einer Grundlinie von München bis Dachau....*

This memoir is contained in the *Abhandlungen der...Baierischen Akademie...*Vol. II., 1764.

It occupies pages 361...386 of the volume.

The base line to which this memoir relates may have been useful for the topography of Bavaria, but it has had I believe no influence on our subject. The base was measured twice; the first result was 43824 French feet; the second result was 10 feet 3 inches more. Five rods of fir-wood were employed, each 12 feet long. The temperature was higher on the average at the second measurement than at the first; and to this circumstance Osterwald attributes the difference of the result. He gives an account of experiments to shew that *heat contracted* and *cold expanded* his rods: but this seems very strange, and probably no one at the present day would accept such a conclusion.

697. In the *Philosophical Transactions*, Vol. LVI., which is for the year 1766, published in 1767, we have a memoir entitled *Proposal of a Method for Measuring Degrees of Longitude upon Parallels of the Æquator*, by *J. Michell*, B.D., F.R.S. The memoir occupies pages 119...125 of the volume: it was read May 8, 1766.

Michell, in fact, proposed to measure an arc perpendicular to the meridian of a given place; but he does not discuss the practical difficulties which would occur in attempting to execute the design. The memoir indeed seems to belong to an earlier generation, and to be quite out of date in 1766.

698. We have next to notice a memoir by Canterzanus, entitled *De Attractione Sphæræ*. This memoir is contained in Vol. V., part 2, of the *De Bononiensi...Academia Commentarii*, published at Bologna in 1767; the memoir occupies pages 66...70 of the volume.

There is an account of the memoir on pages 175...177 of the preliminary portion of Vol. V., part 1, of the same series; this account is by Franciscus Maria Zanottus.

Zanottus desired to give, as an appendix to a book on central forces, the theorem, that according to the ordinary law of attraction a sphere attracts an external particle, as if the mass of the sphere were collected at its centre. As in the book he had adopted short, simple, Cartesian explanations, Zanottus wished this theorem to be exhibited in like manner, or at least to be established by a strictly synthetical demonstration.

It seems curious that at this time a book should have been written using Cartesian methods—*Cartesianos calculos*. Moreover, it is difficult to see why Zanottus could not be content with Newton's demonstration; but to this he does not allude.

Zanottus refers to a demonstration by Sigorgnius, which he condemns as unsatisfactory; and justly, assuming that he has reported it faithfully.

John Bernoulli, he says, had demonstrated the theorem, using infinitesimal differences, in the manner of Leibnitz.

Gravesand also had given a demonstration, "brevem admodum, si tantum legas; si comprehendere etiam velis, non admodum." Moreover, this demonstration was not really synthetical, but only an analytical one disguised.

Finally Zanottus had recourse to Canterzanus, whom he describes as "maximo ingenio juvenem......a quo nihil non sperandum esse videbatur." Canterzanus accordingly satisfied him by a demonstration which occupies nearly five large pages. The demonstration is sound, and not devoid of elegance. We will give a brief account of it, by the aid of algebraical symbols; though this would probably have been very distasteful to Zanottus himself.

Let A denote the centre of the sphere, O any point outside the sphere; we propose to find the attraction of the sphere at O.

Let $OA = c$; let a denote the radius of the sphere. Let x lie in value between $c - a$ and $c + a$. Describe with O as centre a spherical surface of radius x, and another of radius $x + \delta x$, where δx is infinitesimal. Between these two surfaces a portion of the given sphere is contained. The attraction of the portion at O is along OA, and is ultimately equal to

$$\frac{1}{x^2} \times \pi x^2 \sin^2 \theta \delta x, \text{ that is } \pi \sin^2 \theta \delta x,$$

where θ is the angle which a straight line drawn from O to the boundary of the portion makes with OA: see Art. 8.

This is, in fact, the essence of Canterzanus's method; he obtains this result synthetically, and without difficulty.

He then has to determine the whole attraction of the sphere; this also he obtains synthetically, though with some little trouble.

In modern notation we should say that the whole attraction is $\pi \int_{c-a}^{c+a} \sin^2 \theta dx$; and we should easily effect the integration by observing that

$$\cos \theta = \frac{c^2 + x^2 - a^2}{2cx}$$

The result will be $\frac{4\pi a^3}{3c^2}$.

699. We now proceed to a memoir entitled *Eustachii Zanotti. De angulo positionis et ejus usu in determinanda Telluris figura.* This is published in the *De Bononiensi Scientiarum... Commentarii.* Vol. v. part 2, 1767. It occupies pages 256...264 of the volume.

Zanottus at Bologna observed the sun setting near a high tower of Modena. Hence he determined the angle of position of the tower by a way which he considered gave a very accurate result. Then if the difference of longitude between his observatory and this tower were known, he could calculate the difference of latitude on the assumption that the Earth is a sphere. By comparing this calculated difference with that assigned by direct observations, information would be obtained as to the Figure of the Earth. However, it is admitted, that practically the difficulty of fixing the difference of longitudes exactly renders the suggestion useless. Still Zanottus thinks that some advantages could be obtained by the use of the angle of position, when determined with the accuracy which his method would ensure.

700. A memoir by J. A. Euler entitled, *Versuch die Figur der Erden durch Beobachtungen des Monds zu bestimmen*, is contained in the *Abhandlungen der......Baierischen Akademie...* Vol. v., Munich, 1768. The memoir occupies pages 199...214 of the volume. The memoir consists of two parts. In the first part, assuming the form of the Earth, the influence exerted by this form on the meridian observations of the moon is investigated. In the second part it is proposed to determine the form of the Earth from such observations; but the author himself admits that the process is not satisfactory.

701. A memoir by Lambert entitled *Sur la Figure de l'Océan* is contained in the Berlin *Mémoires* for 1767, published in 1769. The memoir occupies pages 20...26 of the volume. The memoir is not mathematical, and does not belong to our subject, but rather to geology.

702. In Volume LVIII. of the *Philosophical Transactions* which is for 1768, published in 1769, we have an *Extract of a Letter, dated Vienna April 4, 1767, from Father Joseph Liesganig,*

Jesuit, to Dr. Bevis, F.R.S., containing a short Account of the Measurement of Three Degrees of Latitude under the Meridian of Vienna. This occupies pages 15 and 16 of the volume. It records the amplitudes and the lengths of the various parts into which Liesganig's entire arc was divided. See Art. 704.

703. In the same Volume of the *Philosophical Transactions* we have an account of the operations carried on by Charles Mason and Jeremiah Dixon, for determining the length of a Degree of latitude in the provinces of Maryland and Pennsylvania, in North America. There is an introduction by Maskelyne, then the detail of the work by Mason and Dixon, and finally some remarks and a postscript by Maskelyne. The whole occupies pages 270...328 of the volume. The peculiarity of these operations is that the whole length was actually measured with rods.

I do not know what Maskelyne means by saying on page 325 that "an error of only 1″ in the celestial measure would produce an error of no less than 67 feet in the length of the degree." Instead of 67 feet, it seems to me that we should read 100 feet, for the length of the degree is found to be more than 360000 feet: see his page 324.

Maskelyne himself had supposed that from the nature of the country no deflections of the plumb-line were to be feared; then he says on his page 328:

......But the Honourable Mr. Henry Cavendish has since considered this matter more minutely, and having mathematically investigated several rules for finding the attraction of the inequalities of the Earth, has, upon probable suppositions of the distance and height of the Allegany mountains from the degree measured, and the depth and declivity of the Atlantic ocean, computed what alteration might be produced in the length of the degree, from the attraction of the said hills, and the defect of attraction of the Atlantic; and finds the degree may have been diminished by 60 or 100 toises from these causes. He has also found, by similar calculations, that the degrees measured in Italy, and at the Cape of Good Hope, may be very sensibly affected by the attraction of hills, and defect of the attraction of the Mediterranean Sea and Indian Ocean.

Frisi, in his *Cosmographia*, Vol. II., page 91, seems to think that this diminution of 60 or 100 toises in the American arc can hardly be accepted. It is indeed difficult to see how the Atlantic ocean can have produced any appreciable effect. Bailly refers to the passage without any expression of dissent: see his *Histoire de l'Astronomie Moderne*, Vol. III., page 41.

On pages 329...335 of the same volume of the *Philosophical Transactions*, some pendulum observations are recorded, which were made by Mason and Dixon at the northern end of their arc, that is in latitude 39° 56′ 19″ North.

704. An account of the measurements of arcs of the meridian in Austria and Hungary by Liesganig, was published at Vienna in 1770, under the title of *Dimensio Graduum Meridiani Viennensis et Hungarici.* The volume is in quarto; it contains a Dedication and an Introduction, the text on 262 pages, and a leaf of Errata; there are ten plates.

The volume contains no theoretical investigations, so that it does not fall within our range. Practically the results of the operations do not seem to be esteemed of any value: see De Zach's *Correspondance Astronomique*, Vol. VII.; and the article *Figure of the Earth* in the *Encyclopædia Metropolitana*, page 170.

705. In the *Philosophical Transactions*, Vol. LXI., for 1771, published in 1772, we have a memoir entitled *An attempt to explain some of the principal Phænomena of Electricity, by Means of an elastic Fluid: By the Honourable Henry Cavendish*, F.R.S. The memoir occupies pages 584...677 of the volume. This memoir would require careful attention in a History of Electricity; but a very brief notice will suffice for our purpose, as it contributes nothing that is really new to the theory of attraction.

We have, on pages 586 and 587, the attraction of a cone on a particle at the vertex, assuming the law to be that of the inverse nth power of the distance.

We have, on page 592, the enunciation of the proposition that, on the ordinary law of attraction or repulsion, a spherical shell does not exert any action on an internal particle: for the demonstration we are referred to Newton's *Principia*, Lib. I., prop. 70.

Cavendish adds:

It follows also from his demonstration, that if the repulsion is inversely, as some higher power of the distance than the square, the particle P will be impelled towards the center; and if the repulsion is inversely as some lower power than the square, it will be impelled from the center.

Hence, if the law of attraction or of repulsion is given to be that of some *single* power of the distance it follows that a particle will not be at rest when placed inside a spherical shell, unless the law be that of the inverse square. But this does not apply if the law is merely assumed to be expressed by some *function* of the distance, as for example by $hr^m + kr^n$, where r denotes the distance, and the other letters denote constants. The general proposition was given by Laplace in the *Mécanique Céleste*, Livre II. § 12: it has since passed into the elementary treatises on Attraction.

On Cavendish's page 616, we have an investigation of the attraction of a circular lamina on an external particle symmetrically situated; the expression obtained is made to yield various easy inferences in the subsequent pages.

706. In the Paris *Mémoires* for 1772, *Seconde partie*, published in 1776, there is a memoir by La Condamine, entitled *Remarques sur la Toise-étalon du Châtelet, et sur les diverses Toises employées aux mesures des Degrés terrestres et à celle du Pendule à secondes.*

The memoir occupies pages 482...501 of the volume; see also pages 8...13 of the historical portion of the volume.

The memoir was read on the 29th of July, 1758; but was forgotten by its author, and found two years after his death.

The object of the memoir is to recommend the toise of the Equator as the standard toise; this toise is that which was used in measuring the arc of the meridian in Peru, and which is elsewhere called the toise of Peru: see Arts. 186 and 551.

A certain iron toise existed which was theoretically the standard toise; this was the *Toise-étalon du Châtelet*, so called from the place where it was kept. This standard appears to have been fixed in its place in a wall in 1668; and Picard adjusted by it the toise which he used for measuring his arc between Paris and Amiens. At the date of the composition of La Condamine's memoir, the standard was damaged and no longer trustworthy; and Picard's toise had not been preserved. La Condamine gives information respecting the toise of the Equator; and he compares with it various other toises, beginning with that used to measure the arc in Lapland, called the toise of the North.

With respect to the toise of the Equator and the toise of the North La Condamine says on his page 492:

Il est vrai que depuis que les deux toises sont revenues en France, on a cru trouver entr'elles, par une nouvelle comparaison, une légère différence, qu'on a jugée d'un vingtième ou d'un trentième de ligne (dont la toise du Nord est plus courte) en attendant une détermination plus précise. Voyez le rapport des quatre Commissaires, inséré dans les Mémoires de l'Académie de l'année 1754, p. 178; et le Journal des opérations de M. le Monnier, imprimé au Louvre en 1757, page 8, ligne 11.

We have alluded to this in Art. 551.

La Condamine thinks that the toise of the Equator and the toise of the North were originally of the same length, and that the slight difference between them arose from the shipwreck in the Gulf of Bothnia when the expedition returned from Lapland. At a later period, when these two toises were again compared, they were found to be equal: see the *Base du Système Métrique*, Vol. III., page 413.

La Condamine considers in the next place the toise which was used in the geodetical operations in France in 1739 and 1740, and in 1756; this toise is called the toise of the Observatory or of the Degrees of France. La Condamine arrives at the conclusion that this toise was practically equivalent in length to the toise of the Equator.

La Condamine also thinks that the toise used by La Caille for measuring an arc at the Cape of Good Hope, agrees with the toise of the Equator.

La Condamine refers also to a toise which had been used by De Mairan in some pendulum experiments: La Condamine considers that this toise is about a tenth of a line shorter than the toise of the Equator. But I have found a statement by D'Alembert which is not quite consistent with this; it is in the article *Figure de la Terre* in the original *Encyclopédie*, page 754:

......or la toise de M. de Mairan est aussi la même qui a servi à la mesure des degrés sous l'équateur et sous le cercle polaire, et la même qu'on a employée pour vérifier en 1740 la base de M. Picard.

I presume that the authority of La Condamine is superior to that of D'Alembert.

La Condamine briefly examines the various methods of preserving an exact record of the standard of length: he recommends that the standard should be hollowed out in a table of porphyry or of granite.

In a note at the end of the Memoir, we are told that the proposition to take the toise of the Equator as the standard toise was not adopted in 1758, owing to the opposition of De Mairan; but in 1766 the royal authority was exerted in favour of it.

The memoir is of great importance with respect to standards of length; it contains references on the subject to memoirs in the preceding volumes of the Paris Academy, and to other works. An interesting paragraph respecting a suitable universal standard of length occurs on page 500; it begins thus:

M. Mouton, Chanoine de Lyon, est le premier, que je sache, qui proposa cette mesure tirée du pendule; ce fut en 1670. *Observationes diam. Sol. Lun.* Lyon, publiées en 1670.

707. A memoir by Lagrange entitled *Sur l'attraction des sphéroïdes elliptiques* is contained in the *Nouveaux Mémoires de l'Académie*...Berlin for 1773, published in 1775. The memoir occupies pages 121...148 of the volume.

708. Lagrange refers to the investigations given by Maclaurin in his Prize Essay on the Tides. Lagrange says on his page 121:

... il faut avouer que cette partie de l'Ouvrage de M. Maclaurin est un chef-d'œuvre de Géométrie, qu'on peut comparer à tout ce qu'Archimede nous a laissé de plus beau, et de plus ingénieux.

Lagrange proposes to demonstrate by analytical methods the results which Maclaurin demonstrated by geometry.

709. Lagrange speaks thus of Simpson on his page 122:

On trouve à la vérité dans les Ouvrages de M. Thomas Simpson une solution purement analytique du probleme de M. Maclaurin, dans laquelle on ne suppose point que le sphéroïde elliptique soit à très peu près sphérique; mais d'un autre côté cette solution a le défaut de procéder par le moyen des séries, ce qui la rend non seulement longue et compliquée, mais encore peu directe et peu rigoureuse.

I suppose that the defect which Lagrange has in view when he describes Simpson's solution as deficient in rigour is the fact that the series are not always convergent: see Art. 282.

710. The general formulæ for the attraction of a body on a particle in terms of rectangular coordinates are first investigated, and it is remarked that the great difficulty is to effect the integrations which are indicated in the formulæ. This leads to the subject of the transformation of the variables in a triple integral. Lagrange gives the method which has since remained in nearly all our elementary books; although obscure and unsatisfactory. I have supplied an account of this method, and indicated its defects, in my *Integral Calculus*, where I have explained and adopted another method.

The transformation of multiple integrals is an important branch of analysis; we may consider that Lagrange was the author of it, and that the subject of attraction suggested the consideration of it to him. See Lacroix *Traité du Calcul Différentiel et du Calcul Intégral*, Vol. II., page 208.

As an example of this transformation, Lagrange gives the formula, now very familiar, by which we pass from rectangular to polar coordinates in the expression for an element of volume, namely

$$dxdydz = r^2 \sin \theta d\theta d\phi dr.$$

711. Lagrange works out the case of the attraction of an oblatum on an internal particle; the process is essentially the same as would be found in any modern treatise on attractions: see, for example, *Statics*, Chapter XIII.

712. Lagrange says on his page 139:

M. Maclaurin dans son Traité du flux et du reflux de la mer s'est contenté de chercher l'attraction d'un sphéroïde elliptique sur un point quelconque de ce sphéroïde; et les résultats de sa belle méthode synthétique s'accordent parfaitement avec ceux que nous venons de trouver par l'Analyse. M. d'Alembert vient d'étendre la solution de M. Maclaurin à des sphéroïdes où toutes les coupes seroient elliptiques, en faisant remarquer que les propositions qui servent de base à cette solution sont également vraies à l'égard de tous les sphéroïdes elliptiques, soit de révolution ou non; c'est ce que nous avons trouvé directement par notre Analyse...

The researches of D'Alembert which are here noticed are, I presume, those in the sixth volume of the *Opuscules Mathématiques.*

In the last words of the preceding extract, Lagrange alludes to the demonstration of various properties in the attraction of an ellipsoid, not necessarily of revolution, on an internal particle. Lagrange shews: that as long as we keep on the same radius drawn from the centre of the ellipsoid the attraction varies as the distance from the centre, that the attraction of a shell bounded by similar, similarly situated, and concentric ellipsoidal surfaces on a particle within the shell is zero; and that the attraction resolved parallel to an axis varies as the perpendicular distance from the plane which contains the other axes. For the case of an ellipsoid of revolution these results had been long known; the first and the second were given by Newton, and the third by Maclaurin. It had more recently been shewn that they were also true for ellipsoids not of revolution: see Arts. 615 and 662. Lagrange now establishes these propositions by analysis.

713. With respect to the absolute value of the attraction of an ellipsoid on an internal particle, Lagrange says on his page 139:

...à l'égard de la valeur absolue de l'attraction des sphéroïdes qui ne sont pas de révolution, M. d'Alembert a essayé de la déterminer par différens moyens très ingénieux, mais dont aucun ne lui a pleinement réussi ;...

714. Lagrange also investigates the attraction of an oblatum on an external particle which is situated on the prolongation of the axis of revolution. He says at the end of the investigation on his page 144:

Ce Probleme a aussi été résolu synthétiquement par M. Maclaurin dans son Traité des fluxions, et nos solutions s'accordent dans les résultats.

In the course of the investigation Lagrange allows himself to fall under the suspicion of contradicting the first principles of the Integral Calculus: see his pages 142 and 143. He says in fact that $\int Pdp$ vanishes when taken between the limits α and $-\alpha$, where P is a function of p which is always positive. The result at which he arrives is correct, but his method is unsatisfactory. Instead of integrating with respect to p between the limits α and $-\alpha$, and then with respect to q between the limits 0 and π, he ought to integrate with respect to p between the limits 0 and α, and with respect to q between the limits 0 and 2π.

His polar expression for the element of volume really assumes that $\sin p$ is always positive.

715. Lagrange alludes to the case of the attraction of an ellipsoid, not of revolution, on an external particle which is situated on the prolongation of an axis. He says on his page 145:

......mais l'intégration de la différentielle dont il s'agit étant très difficile, si même elle n'est pas impossible, nous ne nous y arrêterons pas; outre que cette matiere n'est pas proprement de l'objet auquel ce Mémoire étoit destiné, elle a d'ailleurs été déjà savamment discutée dans le sixieme Volume des Opuscules de M. d'Alembert, auquel il nous suffira par conséquent de renvoyer.

Lagrange says that we should find still greater difficulties in attempting to investigate the attraction of an ellipsoid on *any* external point. He shews what the expressions which have to be integrated become when the axes of coordinates are shifted so that one of them is made to pass through the external point.

716. The memoir does not proceed so far in the subject as Maclaurin's *Treatise of Fluxions* did; for the theorem which we have reproduced in Arts. 257 and 258 is not demonstrated by Lagrange. But as we shall see Lagrange added the demonstration in the Berlin *Mémoires* for 1775.

717. An account of a measurement of an arc of the meridian in Lombardy by Beccaria was published at Turin in 1744; I have not seen the volume. The result obtained by this operation appears however never to have been received with confidence. See the memoir by De Zach in the Turin *Mémoires* for 1811 and 1812; and De Zach's *Correspondance Astronomique,* Vol. VII., page 502; and also the article *Figure of the Earth* in the *Encyclopædia Metropolitana,* pages 170, 208, and 210.

718. A work was published in Florence in 1777, entitled *Lettere di un Italiano ad un Parigino intorno alle riflessioni del Sig. Cassini de Thury sul grado Torinese.* The work consists of 67 octavo pages: it is anonymous.

Cassini de Thury seems to have made some remarks on Beccaria's book in the *Mercure de France* for 1776; and the present work is a reply. The main purport of the reply is to shew that Beccaria's result was what might have been expected if due allowance were made for the attraction exerted by the Alps. There is however no theory nor calculation in the book, but only general considerations. I have not seen the remarks which Cassini de Thury made; but judging from the reply, they contained some inaccuracies or misprints.

719. We have referred in Art. 643 to the extracts from two letters which D'Alembert addressed to Lagrange; the letters gave rise to some investigations by Lagrange, which we shall now notice.

720. In the *Nouveaux Mémoires de l'Académie*...Berlin for 1775, published in 1777 we have on pages 273...279 an *Addition au Mémoire sur l'attraction des sphéroïdes elliptiques imprimé dans le Volume pour l'Année* 1773, *par M. de La Grange.*

This addition was read on the 9th of November 1775; it commences thus:

Les remarques contenues dans la lettre de M. d'Alembert dont j'ai eu l'honneur de faire part à l'Académie il y a huit jours, m'ont donné occasion de chercher si le Théoreme de M. Maclaurin concernant l'attraction d'un ellipsoïde sur un point quelconque placé dans le prolongement de l'un de ses trois axes ne pourroit pas se déduire des formules que j'ai données dans ce Mémoire; et je crois que les Analystes verront avec plaisir avec combien de facilité on peut parvenir par ces formules à la démonstration du Théoreme dont il s'agit.

721. Lagrange starts with a formula which he had obtained in the memoir of 1773 for the attraction of an ellipsoid on a point on the prolongation of an axis.

Let c be the polar semiaxis, a the equatorial semiaxis, of an oblatum; let the density be unity. Then for the attraction on a particle on the prolongation of the polar axis at the distance r from the centre we have by Art. 261, the expression

$$\frac{4\pi ca^2}{a^2-c^2}\left\{1-\frac{r}{\sqrt{(a^2-c^2)}}\tan^{-1}\frac{\sqrt{(a^2-c^2)}}{r}\right\}.$$

Put m for $\frac{c^2}{a^2}$; then the expression becomes

$$\frac{4\pi}{1-m}\left\{c-\frac{r\sqrt{m}}{\sqrt{(1-m)}}\tan^{-1}\frac{\sqrt{\{(1-m)c\}}}{r\sqrt{m}}\right\}.$$

Now let us suppose that instead of an oblatum we have an ellipsoid; let a and b be the semiaxes of the section which is at right angles to the distance r; and let c as before be the semiaxis in the direction of r. The attraction will be equal to the integral between 0 and 2π of

$$\frac{4\pi}{1-m}\left\{c-\frac{r\sqrt{m}}{\sqrt{(1-m)}}\tan^{-1}\frac{\sqrt{\{(1-m)c\}}}{r\sqrt{m}}\right\}\frac{d\theta}{2\pi},$$

where m now denotes $\dfrac{c^2(a^2\sin^2\theta+b^2\cos^2\theta)}{a^2b^2}$.

This is obvious; for a wedge of this ellipsoid made by two planes inclined respectively at angles θ and $\theta+d\theta$ to the plane of a and c may be considered as equivalent to the wedge of an oblatum which has for semiaxes c and a_1, where

$$\frac{a_1^2 \cos^2 \theta}{a^2} + \frac{a_1^2 \sin^2 \theta}{b^2} = 1,$$

so that
$$a_1^2 = \frac{a^2 b^2}{a^2 \sin^2 \theta + b^2 \cos^2 \theta}.$$

Thus from the known value of the attraction of an *oblatum* we have deduced the expression which must be integrated in order to determine the attraction of an *ellipsoid* in the case under consideration. The method we have used is not formally identical with Lagrange's, but it is coincident in principle.

Lagrange then by a suitable transformation of the expression just obtained, succeeds in demonstrating Maclaurin's theorem.

722. Strictly speaking, Lagrange's demonstration applies only to the case in which the attracted particle is on the prolongation of the least axis of the ellipsoid. But it will be found on examination, that the method may be applied with obvious modifications to the cases of the other axes.

Lagrange finishes thus:

C'est le théoreme que M. Maclaurin a énoncé sans démonstration dans l'Art. 653 de son *Traité des fluxions;* et que nous nous étions proposé de déduire de nos formules.

As we have already stated in Art. 260, Lagrange underrates what Maclaurin really effected.

723. A work was published in 1775, by Cassini de Thury, entitled *Relation d'un Voyage en Allemagne,... Suivie de la Description des Conquêtes de Louis* XV, *depuis* 1745 *jusqu' en* 1748.

This work is in quarto, containing xxviii + 194 pages: it may have a brief notice, though very slightly connected with our subject. Cassini de Thury travelled to Vienna, and made numerous observations of angles in the course of his journey. He calculated a large number of distances, by means of triangles; and the data and the results are recorded. There are numerous maps on which the triangles are drawn. On pages 1...6, we have an account of the measurement of a base near Munich.

On page xx. there is an allusion to an Observatory which Cassini had formerly constructed on the top of a tree 100 feet high: it seems to be the same as we noticed in Art. 226.

Cassini de Thury was present with the French army during part of the war in Flanders in 1745 and 1746. He records a set of triangles, and gives a map, extending over the range of the French conquests. He says on page 124:

> telle est l'idée que l'on doit se former de l'étendue des conquêtes du feu Roi, que j'ai tâché de représenter dans une Carte générale, qui est le seul monument qui nous en reste, si l'on compte pour rien une longue paix qui en a été la suite, et que le plus aimé des Rois préféroit à la victoire.

Cassini de Thury refers on page iii. to a work published in 1765; this I have not seen, but conjecture to be that of which the title is given on page 483 of La Lande's *Bibliographie Astronomique* with the date 1763.

At the end of the volume we have, on five pages, *Extraits des Registres de l'Academie*...: these furnish an account of the book, by Laplace, signed by Le Monnier and himself.

The work cannot be considered of any scientific importance: see the Paris *Mémoires* for 1775, pages 41...44 of the historical portion; and De Zach's *Monatliche Correspondenz*, Vol. VII., page 397.

724. In the *Philosophical Transactions*, Vol. LXV., for 1775, published in 1775, there are two memoirs by Maskelyne.

The first memoir is entitled *A Proposal for measuring the Attraction of some Hill in this Kingdom by Astronomical Observations.* This occupies pages 495...499 of the volume: it was read in 1772.

The second memoir is entitled *An Account of Observations made on the Mountain Schehallien for finding its Attraction.* This occupies pages 500...542 of the volume: it was read July 6, 1775.

Mr Charles Mason examined various hills in England and Scotland, and selected Schehallien, in Perthshire, as suitable for the proposed operations. Maskelyne's second memoir details the astronomical and geodetical proceedings, and records the observations of the stars. From a preliminary determination founded on observations of 10 stars it appeared that a deviation of 11″·6 was produced in the plumb-line by the sum of the attractions on the North and South sides of the mountain.

725. The following remarks occur in the first memoir, on page 496:

Sir ISAAC NEWTON gives us the first hint of such an attempt, in his popular Treatise of the System of the World, where he remarks, "That a mountain of an hemispherical figure, three miles high and six broad, will not, by its attraction, draw the plumb-line two minutes out of the perpendicular." It will appear, by a very easy calculation, that such a mountain would attract the plumb-line 1′ 18″ from the perpendicular.

The work to which Maskelyne here alludes is entitled *A Treatise of the system of the World. By Sir Isaac Newton.* London, 1728. This purports to be a translation of the popular exposition drawn up by Newton himself, to which he refers at the beginning of the third Book of the *Principia.* The date 1728 is after the death of Newton. The passage which Maskelyne quotes is from page 41. On the same page is a statement equivalent to that which we have noticed in Art. 125, so that Maupertuis must have taken it from this book. Since that Article was printed, I have obtained a copy of the second edition of Maupertuis's *Figure des Astres;* the statement is omitted in this edition.

The passage stands thus:

...For the attractions of homogeneous spheres near their surfaces, are as their diameters. Whence a sphere of one foot in diameter, and of a like nature to the Earth, would attract a small body plac'd near its surface, with a force about 20000000 times less, than the Earth would do if placed near its surface. But so small a force could produce no sensible effect. If two such spheres were distant by $\frac{1}{4}$ of an inch, they would not even in spaces void of resistance, come together by the force of their mutual attraction in less than a months time. And less spheres will come together at a rate yet slower, viz. in the proportion of their diameters.

Since the statement is ascribed to Newton it may be proper to give an investigation, which need not appear necessary when only the authority of Maupertuis was involved.

Let a denote the radius of each sphere, R that of the earth, $2a + c$ the original distance of the centres of the spheres. When the centres of the spheres are at the distance r the acceleration which tends to bring them nearer is $2\frac{a}{R}g\left(\frac{a}{r}\right)^2$ Thus during the motion this acceleration lies between $\frac{2ag}{R}\left(\frac{a}{2a+c}\right)^2$, and $\frac{2ag}{R}\left(\frac{a}{2a}\right)^2$ If the acceleration were constant and equal to f, the time of motion would be $\sqrt{\frac{2c}{f}}$. Hence the real time lies between $\frac{2a+c}{a}\left(\frac{cR}{ag}\right)^{\frac{1}{2}}$ and $2\left(\frac{cR}{ag}\right)^{\frac{1}{2}}$.

In the example it is said that the sphere is one foot in diameter, but this must be a mistake for one foot in radius. Thus $a = 1$, $c = \frac{1}{48}$, $g = 32$, and $R = 20000000$; therefore the time in seconds is between $\left(2 + \frac{1}{48}\right)1000\left(\frac{5}{384}\right)^{\frac{1}{2}}$ and $2000\left(\frac{5}{384}\right)^{\frac{1}{2}}$, that is less than 250 seconds.

This differs so widely from what we find in the foregoing passage, that the words must I conclude have some meaning different from that which they appear to suggest.

The preceding elementary considerations are sufficient for our purpose; but there is no difficulty in supplying an exact investigation.

Let x denote the distance of the centre of one sphere from a fixed origin at the instant denoted by t, and x' the distance of the centre of the other sphere; suppose x' greater than x, and put r for $x'-x$. Let m denote the mass of each sphere, M the mass of the earth, and T the whole time of motion. Then

$$\frac{d^2x}{dt^2}=\frac{m}{(x'-x)^2},\quad \frac{d^2x'}{dt^2}=-\frac{m}{(x'-x)^2};$$

hence by subtraction

$$\frac{d^2r}{dt^2}=-\frac{2m}{r^2};$$

therefore $$\left(\frac{dr}{dt}\right)^2=4m\left(\frac{1}{r}-\frac{1}{b}\right),$$

where b denotes the initial value of r, that is $2a+c$.

Therefore $$T=\frac{\sqrt{b}}{2\sqrt{m}}\int_{2a}^{b}\frac{\sqrt{r}\,dr}{\sqrt{(b-r)}}$$

And $$\frac{m}{a^2}=\frac{m}{M}\cdot\frac{M}{R^2}\cdot\frac{R^2}{a^2}=\frac{a}{R}g.$$

Assume $r=b\sin^2\theta$; thus

$$T=\frac{b^{\frac{3}{2}}}{a^{\frac{3}{2}}}\left(\frac{R}{g}\right)^{\frac{1}{2}}\int_{\beta}^{\frac{\pi}{2}}\sin^2\theta\,d\theta,$$

where $$\sin^2\beta=\frac{2a}{b}.$$

Therefore $$T=\frac{b^{\frac{3}{2}}}{2a^{\frac{3}{2}}}\left(\frac{R}{g}\right)^{\frac{1}{2}}\left\{\frac{\pi}{2}-\beta+\frac{\sin 2\beta}{2}\right\}$$

This is exact, and the value may be easily calculated numerically.

Suppose c small compared with b; then we may approximate thus:

Put γ for $\frac{\pi}{2}-\beta$, therefore $\sin 2\beta=\sin 2\gamma=2\gamma$ nearly.

Also $\cos\gamma=\sqrt{\frac{2a}{b}}=\left(\frac{b-c}{b}\right)^{\frac{1}{2}}$; so that approximately

$$1-\frac{\gamma^2}{2}=1-\frac{c}{2b},\text{ and }\gamma^2=\frac{c}{b}.$$

Hence $$T=\frac{b^{\frac{3}{2}}}{a^{\frac{3}{2}}}\left(\frac{R}{g}\right)^{\frac{1}{2}}\gamma=\left(\frac{cR}{ag}\right)^{\frac{1}{2}}\frac{2a+c}{a}.$$

726. Maskelyne obtained some results by calculation, which are thus stated:

By calculation...it should follow, that the sum of the contrary attractions of Whernside...on the plumb-line placed half-way up the hill, would not be less than 30″, and might amount to 46″...

By a calculation..., the sum of the contrary attractions of the plumb-line, placed alternately on the North-side of Helwellin, and the South-side of Skidda, amounts to about 20″...And although the density of the earth near the surface should be five times less than the mean density, as there is some reason to suspect, and the attractions, as here stated, should consequently be diminished in the proportion of five to one, still the sum of the contrary attractions of Whernside would be 6″ or 9″, and the sum of the contrary attractions of Helwellin and Skidda would be 4″

727. On the whole 43 stars were observed, and 337 observations taken. Maskelyne proposed at his leisure to compute the result from all the observations: see his page 530. It does not appear that this was done by him; but it was by De Zach; see *L'attraction des Montagnes*, pages 686...692. The result obtained by De Zach agrees very closely with Maskelyne's own.

Some important calculations were founded on Maskelyne's result by Hutton in 1778, and by Playfair in 1811; these we shall consider hereafter: see Art. 730.

728. We next notice a work entitled *Essai sur les Phénomènes relatifs aux disparitions périodiques de l'Anneau de Saturne. Par M. Dionis du Séjour.* Paris, 1776. This is an octavo, and contains xxxii + 444 pages, besides title-pages, and plate. A notice of the work is given in the historical portion of the Paris *Mémoires* for 1775, pages 53...55.

The work relates to the appearances presented by Saturn's ring, and barely touches on the theory with which we are concerned.

On pages 402...406 we have the equation which Maupertuis investigated for the form of the ring; for the demonstration we are referred to Maupertuis: see Art. 119.

On pages 407...411 we have a formula for the attraction of a circular lamina on a constituent particle: the formula however would not be of any use, because the expression to be integrated becomes infinite within the range of integration.

The author believes that the parts of the ring of Saturn must be animated by a centrifugal force in order to balance the effect of the attraction of Saturn: see his pages iv and 401.

This supposition has been confirmed since by the researches of Laplace, and the observations of Herschel.

729. A work was published by John Whitehurst entitled *An Inquiry into the original state and formation of the Earth...*

The first edition appeared in 1778, and the second in 1786; both are in quarto. In Hutton's *Philosophical and Mathematical Dictionary* under the head *Whitehurst*, it is stated that a third edition appeared in 1792.

The work is geological, and not mathematical, and so does not fall within our range.

I extract one sentence which reproduces an undemonstrated assertion noticed in Art. 130; it occurs on page 6 of the first edition: "...and therefore when the component parts of fluid bodies are thus assembled together, they must necessarily assume spherical forms...."

730. In the *Philosophical Transactions*, Vol. LXVIII., for 1778, part 2, published in 1779, there is a memoir by Hutton, entitled *An account of the Calculations made from the Survey and Measures taken at Schehallien, in order to ascertain the mean Density of the Earth.* The memoir occupies pages 689...788 of the volume: it was read on May 21, 1778.

The attraction of the mountain had to be calculated for each of the two stations at which Maskelyne made his astronomical observations. The form of the mountain was ascertained by a very minute survey; then it was supposed to be decomposed into slender vertical prisms, the attraction of every one of which could be calculated. There were 960 such prisms for each of the two stations.

The numerical labour was of course very great; but the memoir adds nothing to the theory of Attraction. We may notice the method adopted for facilitating the calculation, derived, as Hutton says on his page 750, "partly from some hints of the Honourable Henry Cavendish, F.R.S. and partly from some of my own, which had been communicated to the Astronomer Royal in the years 1774 and 1775..." Compare the note on page 237 of the article, *Figure of the Earth* in the *Encyclopædia Metropolitana*.

Take the horizontal plane through one of the stations for the plane of polar coordinates; let the station be the origin, and let the initial line be in the direction of the meridian.

Let there be a vertical prism standing on a base which has r and θ for the polar coordinates of a corner; and let $r\Delta r\Delta\theta$ denote the area of the base. Suppose z the height of the prism; then, taking the density as unity, the horizontal attraction of the prism is very approximately $\frac{r\Delta r\Delta\theta \, . \, z}{r\sqrt{(r^2+z^2)}}$ and the resolved part of this in the direction of the meridian is $\frac{z\Delta r\Delta\theta}{\sqrt{(r^2+z^2)}}\cos\theta$. This may be put in the form $\frac{z}{\sqrt{(r^2+z^2)}}\{\sin(\theta+\Delta\theta)-\sin\theta\}\Delta r$. Then to facilitate the calculation the values of $\Delta\theta$ were so taken as to make $\sin(\theta+\Delta\theta)-\sin\theta$ retain a constant value. Thus, to obtain the attraction of the prisms forming part of the same ring, the values of $\frac{z}{\sqrt{(r^2+z^2)}}$ must be summed, and the result obtained must be multiplied by $\{\sin(\theta+\Delta\theta)-\sin\theta\}\Delta r$.

Hutton's conclusion is that the mean density of the Earth is $\frac{9}{5}$ of that of the mountain. He conjectures that the mean density of the mountain may be $\frac{5}{2}$ times that of water; so that the mean density of the Earth is about $\frac{9}{2}$ times that of water. See Art. 17.

Hutton's memoir is reproduced in his *Tracts on Mathematical and Philosophical Subjects*, Vol. II. 1812, with the addition of a few remarks towards the end, on the character of the mountain, which were derived partly from Mr Duncan Macara, and partly from Professor Playfair.

731. In the *Philosophical Transactions* for 1811, part 2, published in 1811, there is a memoir by Playfair entitled *Account of a Lithological Survey of Schehallien, made in order to determine the specific Gravity of the Rocks which compose that Mountain.* The memoir occupies pages 347...377 of the volume: it was read June 27, 1811. It will be convenient to notice this memoir here.

Hutton, as we have seen, calculated the attraction of Schehallien, and thence deduced the mean density of the Earth, on the supposition that the mountain was homogeneous; and he assumed 2·5 for the density, that of water being unity. Playfair investigated the composition of the mountain, and modified the calculations by allowing for the actual density of the parts. Playfair found that the upper part of the mountain was composed of quartz of the mean specific gravity 2·6398; and that the lower part was composed of mica and hornblend slate of the mean specific gravity 2·83255, and limestone of the mean specific gravity 2·76607. On the whole he considered that the matter composing the mountain could be divided into two classes of rocks; namely, quartz of the mean specific gravity 2·639876, and micaceous rock, including calcareous, of the mean specific gravity 2·81039. The line separating the two classes of rocks could be accurately traced on the face of the mountain. As to the arrangements in the interior of the mountain, Playfair considered that only two suppositions could be made with any degree of probability; these amount to assuming that the two classes of rocks are separated by a vertical boundary, or by a nearly horizontal boundary. Playfair calculates the mean density of the Earth on both suppositions; on the former he obtains 4·55886, and on the latter 4·866997.

The memoir adds nothing to the theory of Attraction. Playfair availed himself of the practical method for facilitating the computation which is given in Hutton's memoir. Playfair says, on his page 364:

I have also used a theorem in these computations, which gives an accurate value of the attraction of a half cylinder of any altitude a, and any radius r, on a point in the centre of its base, and in the direction of a line bisecting the base.

Let A be equal to that attraction; then

$$A = 2a \log \frac{r + \sqrt{a^2 + r^2}}{a}$$

The investigation of this formula may be usefully supplied.

Take the origin at the attracted point; let the axis of z coincide with the axis of the cylinder, and the axis of x with the direction in which the attraction is estimated. Then the resolved attraction is

$$\iiint \frac{x\, dx\, dy\, dz}{(x^2 + y^2 + z^2)^{\frac{3}{2}}}.$$

We integrate first with respect to z, between the limits 0 and a; thus we obtain

$$\iint \frac{a\, x\, dx\, dy}{(x^2 + y^2)\,(x^2 + y^2 + a^2)^{\frac{1}{2}}}.$$

The limits for y are $-\sqrt{(r^2 - x^2)}$ and $\sqrt{(r^2 - x^2)}$; and the limits for x are 0 and r.

Assume $x = s \cos\theta$ and $y = s \sin\theta$; thus the integral transforms into

$$2a \int_0^r \int_0^{\frac{\pi}{2}} \frac{\cos\theta\, ds\, d\theta}{(s^2 + a^2)^{\frac{1}{2}}};$$

and the value is $$2a \log \frac{r + \sqrt{(a^2 + r^2)}}{a}$$

There is something wrong about the plates which ought to accompany the memoir. Playfair refers to a plan of the mountain which shews the boundary between the two classes of rocks, and also to a diagram; see his page 363: but neither of these is given. Also it appears from his page 365, that the plates which are given ought to have been coloured.

732. It will be convenient to notice here a subsequent paper connected with the Schehallien experiment. In the *Philosophical*

Transactions for 1821, part 2, published in 1821, there is a memoir by Hutton, entitled *On the mean density of the Earth.* It occupies pages 276...292 of the volume: it was read April 5, 1821.

Hutton adverts to the Schehallien operations, and to Playfair's investigation of the density of the mountain. Hutton thinks that the mean density of the Earth is very nearly five times that of water, but not greater. He prefers the Schehallien determination to that which Cavendish had obtained by experiment. At the age of 84, he had undertaken to recompute the experiments of Cavendish, and had discovered some important errors in the original computation. He suggests that observations might be made near one of the Egyptian pyramids, of the nature of those made at Schehallien.

733. We will collect here the titles of some investigations which deserve to be studied by those who are interested in the important question of the mean density of the Earth; the greater part of these investigations fall without the period over which the present history ranges. We suppose the density of water to be unity.

A famous experiment was made by Cavendish, from which he deduced that the mean density of the Earth was about 5·48. The details are given in the *Philosophical Transactions* for 1798; we shall recur to the memoir hereafter.

A memoir by Hutton in the *Philosophical Transactions* for 1821 we noticed in Art. 732. According to Hutton's calculation the result of Cavendish's own experiments is 5·31.

A paper by Carlini in the *Milan Ephemeris* for 1824 gives an account of a series of pendulum experiments made at the height of a thousand toises. The result obtained for the mean density is 4·39; but a serious error is introduced by using a wrong formula to express a certain attraction; the error was pointed out by Schmidt and by Giulio. Moreover there are other considerations which shew that the process does not seem to deserve much confidence. See the *Report on Astronomy* by the Astronomer Royal in the *Reports of the British Association*, Vol. I. page 169; and a note by Sabine in the translation of Humboldt's *Cosmos*, Vol. I. 1849, page xlvii.

The subject is discussed in Schmidt's *Lehrbuch der mathematischen und physischen Geographie,* 1830; see Vol. II. pages 469...487.

Schmidt corrects Carlini's result 4·39 to 4·837. Schmidt obtains 5 52 by calculation from Cavendish's own experiments.

In 1838, a work was published at Freiberg, entitled *Versuche über die mittlere Dichtigkeit der Erde...von F. Reich.* In his introduction he refers to an article by Muncke in the new edition of Gehler's *Physikalisches Wörterbuch* Vol. III. page 940..., as giving a valuable comparison of the results hitherto obtained. Reich himself repeated Cavendish's experiment and obtained nearly 5·44 for the mean density.

A memoir by Menabrea is published in the Turin *Memorie,* Vol. II. 1840, entitled *Calcul de la densité de la Terre.* This contains an investigation of the theory connected with Cavendish's experiment.

A memoir by Giulio is published in the Turin *Memorie,* Vol. II. 1840, entitled *Sur la détermination de la densité moyenne de la terre, déduite de l'observation du pendule faite à l'hospice du Mont Cenis par M. Carlini en Sept.* 1821.

Carlini's result 4 39 is here corrected to 4·95.

Cavendish's experiment was repeated by Baily, who made far more trials, and with greater precautions, than his predecessors. The details form Vol. XIV. of the *Memoirs of the Royal Astronomical Society,* 1843; see also the references connected with this volume in the Royal Society's *Catalogue of Scientific Papers* under the head *Baily,* No. 45. The result obtained for the mean density was about 5·67; but the results of individual experiments were found to vary considerably.

In the *Philosophical Transactions* for 1847 there is a memoir by Hearn entitled, *On the cause of the discrepancies observed by Mr. Baily with the Cavendish apparatus for determining the mean density of the Earth.*

The discrepancies are attributed to the influence of magnetism.

After the publication of Baily's result Reich again repeated the experiment: see the Leipzig *Abhandlungen*, Vol. I. 1852, and the Royal Society's *Catalogue of Scientific Papers* under the head *Reich*, No. 17. The result gives about 5·58 for the mean density.

Reich refers to Hearn's memoir, but does not agree with it.

In the *Account of the...Principal Triangulations*...1858, which forms part of the *Ordnance Survey of Great Britain*...the subject is discussed in Section X. Some account is given of preceding researches, together with the details of a new operation, like that at Schehallien, on the hill called *Arthur's Seat* at Edinburgh. The new operation gives the mean density 5·316.

Pendulum experiments were made in 1854, by the Astronomer Royal in Harton Colliery for ascertaining the mean density of the Earth: see the *Philosophical Transactions* for 1856, and the Royal Society's *Catalogue of Scientific Papers* under the head *Airy*, Nos. 100, 101, and 110. The result is the value 6·566.

The Astronomer Royal observes with respect to this result on page 342 of the *Philosophical Transactions* for 1856:

> The value thus obtained is much larger than that obtained from the Schehallien experiment, and considerably larger than the mean found by Baily from the torsion-rod experiments. It is extremely difficult to assign with precision the causes or the measures of the error of any of these determinations; and I shall content myself with expressing my opinion, that the value now presented is entitled to compete with the others, on, at least, equal terms.

Haughton, in the *Philosophical Magazine* for July, 1856, by a special method deduces the result 5·48 from the Harton Colliery experiments.

A memoir was published in Göttingen, in 1869, entitled *Ueber die Bestimmung der mittleren Dichtigkeit der Erde von Anton Schell*. This is in quarto, containing 39 pages, with three plates. It is a useful account of various researches on the subject.

734. We notice next a memoir entitled *De Figura Terræ Commentatio. Autore J. A. J. Cousin, Parisino.* This is contained in the *Acta Academiæ Electoralis Moguntinæ*...1777. Erfurti. 1778. It occupies pages 209...216 of the volume.

The memoir consists chiefly of relations between certain lines in any curve, expressed in the language of the Differential Calculus. But no diagram is supplied, at least in the only copy which I have seen, and thus part of the memoir is unintelligible. However it may be safely pronounced to be of no importance.

735. A memoir by Euler, entitled *Theoria Parallaxeos ad Figuram Terræ sphaeroidicam accommodata,* is contained in the *Acta Academiæ...Petropolitanæ* for 1779, *pars prior*, published in 1782. The memoir occupies pages 241...278 of the volume.

This memoir adds nothing to the theory of the Figure of the Earth. Euler assumes that the Earth is an oblatum, and investigates the consequent expressions for the moon's parallax. He gives tables and numerical examples, which are calculated on the supposition that the ellipticity is $\frac{1}{200}$.

736. In the *Philosophical Transactions* for 1780, published in 1780, there is a memoir by Hutton, entitled *Calculations to determine at what Point in the Side of a Hill its Attraction will be the greatest.* It occupies pages 1...14 of the volume: it was read Nov. 11, 1779.

The memoir proposes to find at what point on the surface of a hill the *horizontal* component of the attraction of the hill is greatest. The problem was naturally suggested by the operations on the mountain Schehallien.

Hutton supposes that the vertical section of his mountain is a triangle, and that the mountain extends to infinity in the horizontal direction on both sides of the point considered. Thus his problem may be stated in these words: find a point on a given face of a triangular prism of infinite length where the attraction of the prism resolved parallel to another given face is greatest.

Hutton's solution is wrong. I will briefly indicate the correct method.

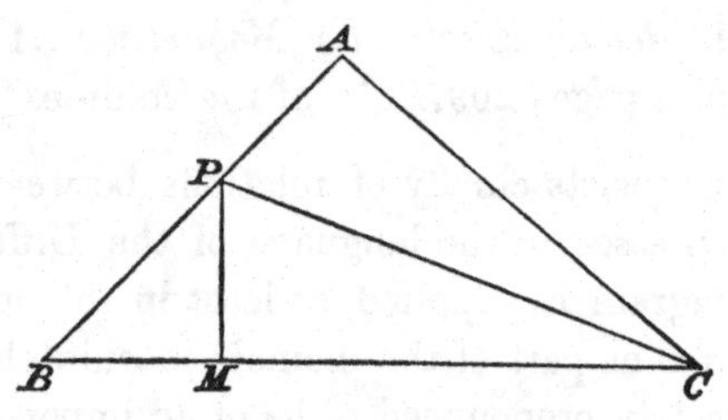

Let ABC be a section of the prism at right angles to its edges. Let P be any point on the side AB.

Then we may divide the prism into two prisms, one corresponding to PBC, and the other to PAC, and estimate the attraction of each separately.

Suppose P the origin of polar coordinates, in the plane of the triangle; and take the initial line parallel to BC.

The attraction of the infinite rod parallel to the edges of the prism which corresponds to the polar element $rdr\,d\theta$ is $\frac{2}{r}\,rdr\,d\theta$, where the density is taken to be unity.

Consider first the prism corresponding to PBC.

The attraction of a rod will be $2\cos\theta dr d\theta$ parallel to BC, and $2\sin\theta dr d\theta$ perpendicular to BC. Let the perpendicular PM be denoted by h. Then we must integrate for r from 0 to $\frac{h}{\sin\theta}$; and we must integrate for θ from α to β, where α is equal to PCB, and β exceeds α by BPC. Thus the component attractions are $2h\log\frac{\sin\beta}{\sin\alpha}$ parallel to BC, and $2h(\beta-\alpha)$ perpendicular to BC.

Similarly for the prism corresponding to PAC we find that the component attractions are $2h_1\log\frac{\sin\beta_1}{\sin\alpha_1}$ parallel to AC, and $2h_1(\beta_1-\alpha_1)$ perpendicular to AC; where h_1 denotes the perpendicular from P on AC, and α_1 is equal to PCA, and $\beta_1-\alpha_1$ to APC.

Hence the attraction of the whole prism corresponding to ABC parallel to BC is

$$2h \log \frac{\sin \beta}{\sin \alpha} + 2h_1 \cos C \log \frac{\sin \beta_1}{\sin \alpha_1} + 2h_1 \sin C \,(\beta_1 - \alpha_1).$$

This expression may be made to involve only one variable α; and then we can seek the maximum value by putting the differential coefficient with respect to α zero. The equations which serve to express the other variables in terms of α are

$$h\,(\cot B + \cot \alpha) = BC,$$
$$h\,BC + h_1\,AC = \text{twice the area of } ABC,$$
$$\beta = \pi - B,$$
$$\alpha_1 + \alpha = C,$$
$$\beta_1 - \alpha_1 = B + \alpha.$$

Hutton instead of taking the whole prism supposes another plane to pass through PM, and to make an infinitesimal angle with the plane of the paper. Thus he obtains a slice in the form of a double wedge; he estimates the resolved attraction of the slice, and assumes that this will represent the attraction of the whole prism. He says:

And then from the foregoing suppositions it is evident that in whatever point of AB the attraction of ABC is greatest, there also will the attraction of the whole hill be the greatest.

This assertion is unjustifiable. After I wrote this I found that to this sentence Hutton adds the words "very nearly," in the abridgement of the memoir which is given in the *Philosophical Transactions abridged* by Hutton, Shaw and Pearson; and also in the republication of the memoir in Hutton's *Tracts*, Vol. II.

Hutton considers especially the case in which the triangle ABC is equilateral.

Suppose, for example, that P is at B. Put c for AB. Then $h = 0$, $h_1 = \frac{c\sqrt{3}}{2}$, $\alpha_1 = \frac{\pi}{3}$, $\beta_1 = \frac{2\pi}{3}$. The attraction parallel to BC

$$= 2 \cdot \frac{c\sqrt{3}}{2} \cdot \frac{\sqrt{3}}{2} \cdot \frac{\pi}{3} = \frac{c\pi}{2}.$$

Next suppose that P is at the middle point of AB. Then

$$h_1 = h = \frac{c\sqrt{3}}{4}, \quad \alpha = \frac{\pi}{6}, \quad \beta = \frac{2\pi}{3}, \quad \alpha_1 = \frac{\pi}{6}, \quad \beta_1 = \frac{2\pi}{3}.$$

The attraction parallel to BC

$$= \frac{c\sqrt{3}}{2}\log\sqrt{3} + \frac{c\sqrt{3}}{4}\log\sqrt{3} + \frac{c\sqrt{3}}{2}\,\frac{\sqrt{3}}{2}\cdot\frac{\pi}{2} = c\,\frac{3\sqrt{3}}{4}\log\sqrt{3} + \frac{3c\pi}{8}$$

The ratio of the latter to the former

$$= \frac{3}{4} + \frac{3\sqrt{3}}{2\pi}\log\sqrt{3} = \frac{3}{4}\left(1 + \frac{\sqrt{3}}{\pi}\log 3\right)$$

This is obviously the ratio of the whole attraction at B to the whole attraction midway between A and B; for each whole attraction is inclined at 30° to the horizon.

This will be found 1 204 approximately.

Hutton's theoretical expression for the ratio is quite different; but numerically it is nearly the same: he gives almost 1·206. The coincidence is curious.

737. Here we may be said to finish the first part of our work; the history has been carried through a period of nearly a century, from the publication of the *Principia*, in which Newton laid the solid foundations of the Theories of Attraction and of the Figure of the Earth. Maclaurin and Clairaut continued the work with great success; the former mainly devoting himself to the Theory of Attraction, and the latter to that of the Figure of the Earth. The incessant labours of D'Alembert effected more indirectly than directly; they kept up an interest in the subjects, and probably suggested the more fortunate efforts of Legendre and Laplace. In the second volume we shall trace the progress of our Theories under the influence of the powerful analysis of these two great mathematicians.

738. Some publications are recorded in La Lande's *Bibliographie Astronomique* bearing on our subject, which I have not been able to consult; although I believe they are of small importance, I will quote the titles here, adding the pages of La Lande's work where they are given.

1735. *Paris, in*-12. Proposition d'une mesure de la terre, dont il résulte une diminution considérable dans la circonférence de l'équateur, par M. D'Anville, géographe ordinaire du roi...... Page 400.

1738. ... *in*-12. Anecdotes physiques et morales. Page 407.

A notice of this in La Lande's work follows immediately after that of the *Examen desintéressé;* he says it is on the same subject: see Art. 143.

1740. *Bologna, in*-8°. Lettera contenente l'aviso delle operazioni fatte nell' America meridionale dai matematici spagnuoli e francesi, per cui venne a conchiudersi la gran controversia sopra la figura della terra. Page 412.

1743. *Viennæ, in*-8°. De figurâ telluris dialogus à scholasticis universitatis Viennensis. Page 422.

1744. *Stockholm, in*-8°. ...Klingenstierna, ...donna le problème suivant: *Trouver la figure de la terre par la comparaison de deux degrés.* Page 424.

1748. *London, in*-8°. A new theory of the figure of the earth, wherein are demonstrated the mechanical causes of its figure as it is determined by the observations of Rowland Jackson. Page 433.

1748. *Tournay, in*-8°. Discours sur la figure de la terre, par M. le baron de Grante. Page 435.

1763. *Manhemii, in*-4°. Basis Palatina, anno 1762 bis dimensa, hoc anno 1763 novis mensuris aucta et confirmata, à Christiano Mayer.... Page 483; see also page 503.

1766. *Pisa, in*-8°. Ragionamento filosofico-historico sopra la figura della terra, dal S^r. Ant. Mattuni. Page 496.

1769. *Vlissengen, in*-8°. ...On trouve dans le troisième volume un mémoire de M. Hennert sur la figure de la terre, ... Page 509.

1778. *Utrecht, in*-8°. Dissertations physiques et mathématiques, sur la figure de la terre, les comètes, l'attraction, &c., par M. Hennert. Page 562.

1778. *Varsaviæ, in*-8°. Michaelis Hube, De telluris formâ. Page 563.

739. I hope that few of the memoirs which are contained in the collections of the various Academies and Scientific Societies have been overlooked. In seeking for these the well-known *Repertorium Commentationum* of J. D. Reuss affords most valuable assistance. Here I find memoirs by four authors recorded which I have not been able to consult, namely Celsius, Klingenstierna, Mallet, and Fester: the titles occur in the fifth volume of the work on the pages 80, 82, and 83.

740. There are three matters considered in the present volume to which it will be convenient to allude here.

The difficulty which is mentioned in Art. 124 is discussed in Art. 725.

Since Art. 228 was printed I have been informed by M. O. Struve, that Delisle's original manuscripts were found some years since at St Petersburg: those at Paris were copies.

Since Art. 542 was printed I have published some remarks on the modern South African arc in the *Monthly Notices of the Royal Astronomical Society*, Vol. XXXIII. pages 27...34.

END OF VOLUME I.

CAMBRIDGE: PRINTED BY C. J. CLAY, M.A., AT THE UNIVERSITY PRESS.

Printed in the United States
By Bookmasters